MULTIMEDIA SIGNALS AND SYSTEMS

**THE KLUWER INTERNATIONAL SERIES
IN ENGINEERING AND COMPUTER SCIENCE**

MULTIMEDIA SIGNALS AND SYSTEMS

Mrinal Kr. Mandal
University of Alberta, Canada

SPRINGER SCIENCE+BUSINESS MEDIA, LLC

Additional material to this book can be downloaded from http://extra.springer.com.

Library of Congress Cataloging-in-Publication Data

Mandal, Mrinal Kr.
    Multimedia Signals and Systems / Mrinal Kr. Mandal.
        p.cm.—(The Kluwer international series in engineering and computer science; SECS 716)
    Includes bibliographical references and index.
    ISBN 978-1-4020-7270-3   ISBN 978-1-4615-0265-4 (eBook)
    DOI 10.1007/978-1-4615-0265-4
        1. Multimedia systems. 2. Signal processing—Digitial techniques. I. Title. II. Series.

    QA76.575 .M3155 2002
    006.7—dc21

                                                        2002034047

MATLAB® is a registered trademark of the MathWorks, Inc.

Printed on acid-free paper.

Table of Contents

PREFACE

Multimedia computing and communications have emerged as a major research and development area. Multimedia computers in particular open a wide range of possibilities by combining different types of digital media such as text, graphics, audio and video. The emergence of the World Wide Web, unthinkable even two decades ago, also has fuelled the growth of multimedia computing.

There are several books on multimedia systems that can be divided into two major categories. In the first category, the books are purely technical, providing detailed theories of multimedia engineering, with an emphasis on signal processing. These books are more suitable for graduate students and researchers in the multimedia area. In the second category, there are several books on multimedia, which are primarily about content creation and management.

Because the number of multimedia users is increasing daily, there is a strong need for books somewhere between these two extremes. People with engineering or even non-engineering background are now familiar with buzzwords such as JPEG, GIF, WAV, MP3, and MPEG files. These files can be edited or manipulated with a wide variety of software tools. However, the curious-minded may wonder how these files work that ultimately provide us with impressive images or audio.

This book intends to fill this gap by explaining the multimedia signal processing at a less technical level. However, in order to understand the digital signal processing techniques, readers must still be familiar with discrete time signals and systems, especially sampling theory, analog-to-digital conversion, digital filter theory, and Fourier transform.

The book has 15 Chapters, with Chapter 1 being the introductory chapter. The remaining 14 chapters can be divided into three parts. The first part consists of Chapters 2-4. These chapters focus on the multimedia signals, namely audio and image, their acquisition techniques, and properties of human auditory and visual systems. The second part consists of Chapters 5-11. These chapters focus on the signal processing aspects, and are strongly linked in order to introduce the signal processing techniques step-by-step. The third part consists of Chapters 12-15, which presents a few select multimedia systems. These chapters can be read independently. The objective of including this section is to introduce readers to the intricacies of a few select frequently used multimedia systems.

including this section is to introduce readers to the intricacies of a few select frequently used multimedia systems.

The chapters in the first and second parts of the book have been organized to enable a hierarchical study. In addition to the Introductory Chapter, the following reading sequence may be considered.

i) Text Representation: Chapter 6
ii) Audio Compression: Chapters 2, 4, 5, 6, 7
iii) Audio Processing: Chapters 2, 4, 5, 10
iv) Image Compression: Chapters 3, 4, 5, 6, 7, 8
v) Video Compression: Chapters 3, 4, 5, 6, 7, 8, 9
vi) Image & Video Processing: Chapters 3, 4, 5, 11
vii) Television Fundamentals: Chapters 3, 4, 5, 6, 7, 8, 9, 12

Chapters 13-15 can be read in any order.

A major focus of this book is to illustrate with examples the basic signal processing concepts. We have used MATLAB to illustrate the examples since MATLAB codes are very compact and easy to follow. The MATLAB codes of most examples, wherever appropriate, in the book are provided in the accompanying CD so that readers can experiment on their own.

Any suggestion and concern regarding the book can be emailed to the author at the email address: mandal@ee.ualberta.ca. There would be a follow-up website (http://www.ee.ualberta.ca/~mandal/book-multimedia/) where future updates will be posted.

I would like to extend my deepest gratitude to all my coworkers and students who have helped in the preparation of this book. Special thanks are due to Sunil Bandaru, Alesya Bajoria, Mahesh Nagarajan, Shahid Khan, Hongyu Liao, Qinghong Guo, and Sasan Haghani for their help in the overall preparation. I would also like to thank Drs. Philip Mingay, Bruce Cockburn, Behrouz Nowrouzian, and Sethuraman Panchanathan (from Arizona State University) for their helpful suggestions to improve the course content. Jennifer Evans and Anne Murray from Kluwer Academic Publishers have always lent a helping hand. Last but not least, I would like to thank Rupa and Geeta, without whose encouragement and support this book would not be completed.

August 2002   Mrinal Kr. Mandal

Chapter 1

Introduction

Communication technology has always had a great impact on modern society. In the pre-computer age, newspaper, radio, television, and cinema were the primary means of mass communication. When personal computers were introduced in the early 1980s, very few people imagined their tremendous influence on our daily lives. But, with the technological support from network engineers, global information sharing suddenly became feasible through the now ubiquitous World Wide Web. Today, for people to exploit efficiently the computer's potential, they must present their information in a medium that maximizes their work. In addition, their information presentation should be efficiently structured for storage, transmission, and retrieval applications. In order to achieve these goals, the field of multimedia research is now crucial.

Multimedia is one of the most exciting developments in the field of personal computing. Literally speaking, a medium is a substance, such as water and air, through which something is transmitted. Here, *media* means the representation and storage of information, such as text, image, video, newspaper, magazine, radio, and television. Since the term "multi" means multiple, *multimedia* refers to a means of communication with more than one medium. The prefix "multi," however, is unnecessary since *media* is already plural and refers to a combination of different mediums. Interestingly, the term is now so popular (a search on the *Google* web search engine with the keyword "multimedia" produced more than 13 million hits in July 2002, compared to an established but traditional subject "physics" which produced only 9 million hits), it is now unlikely to change.

The main reason for the multimedia system's popularity is its long list of potential applications that were not possible even two decades ago. A few examples are shown in Fig. 1.1. Limitless potential of applications such as the World Wide Web, High Definition and Interactive Television, Video-on-demand, Video conferencing, Electronic Newspapers/Magazines, Games and E-Commerce are capturing people's imaginations. Significantly, multimedia technology can be considered the key driving force for these applications.

1.1 DEVELOPMENT OF MULTIMEDIA SYSTEMS

A brief history of the development of multimedia systems is provided in Table 1.1. The newspaper is probably the first mass communication medium, which uses mostly text, graphics and images. In late 1890s, Guglielmo Marconi demonstrated the first wireless radio transmission. Since then, radio has become the major medium for broadcasting. Movies and televisions were introduced around 1930s, which brought video to the viewers, and again changed the nature of mass communications. The concept of the World Wide Web was introduced around the 1950s, but supporting technology was not available at that time and did not resurface until the early 1980s. Current Multimedia system technologies became popular in the early 1990s due to the availability of low-cost computer hardware, broadband networks, and hypertext protocols.

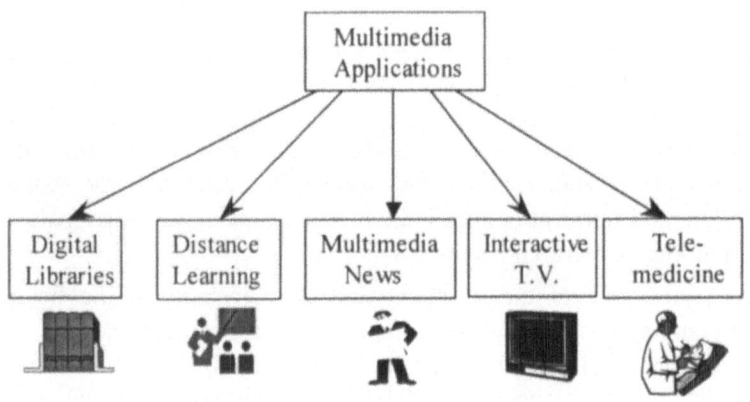

Figure 1.1. Multimedia applications.

Today's multimedia technology is possible because of technological advances in several diverse areas, including telecommunications, consumer electronics, audio and movie recoding studios, and publishing houses.

Furthermore, in the last few decades, telephone networks have changed gradually from analog to digital networks. Correspondingly, separate broadband data networks have been established for high-speed computer communication.

Consumer electronics industries continue to make important advances in areas such as high fidelity audio systems, high quality video and television systems, and storage devices (*e.g.*, hard disks, CDs). Recording studios in particular have noticeably improved consumer electronics, especially high quality audio and video equipment.

Table 1.1. Brief history of multimedia systems.

Year	Events
Pre-Computer Age	Newspaper, radio, television, and cinema were the primary means of mass communications.
Late 1890s	Radio was introduced.
Early 1900s	Movie was introduced.
1940s	Television was introduced.
1960s	Concept of hypertext systems was developed.
Early 1980s	Personal computer was introduced.
1983	Internet is born, TCP/IP protocol was established. Audio-CD was introduced.
1990	Tim Berners-Lee proposed the World Wide Web. HTML (Hyper Text Markup Language) is developed.
1980-present	Several digital audio, image and video coding standards have been developed.
Mid 1990s	High Definition Television standard established in North America.
1993-present	Several web-browsers, hypertext languages have been developed.

As well, publication houses assisted the development of efficient formats for data representation. Note that the hypertext markup language (HTML) of the World Wide Web was preceded by the development of generalized markup languages for creating machine independent document structures.

1.2 CLASSIFICATION OF MEDIA

We have noted that multimedia represents a variety of *media*. These *media* can be classified according to different criteria.

Perception: In a typical multimedia environment, the information is ultimately presented to people (e.g., in a cinema). This information representation should exploit our five senses: hearing, seeing, smell, touch and taste (see Fig. 2). However, most current multimedia systems only employ the audio and visual senses. The technology for involving the three other (minor) senses has not yet matured. Some work has been carried out to include smell and taste in multimedia systems [11], but it needs more research to become convenient and cost effective. Hence, in the current multimedia framework, text, image, and video can be considered visual media, whereas music and speech can be considered auditory media.

Representation: Here, the media is characterized by internal computer representation, as various formats represent media information in a computer. For example, text characters may be represented by ASCII code; audio signals may be represented by PCM samples; image data may be represented by PCM or JPEG format; and video data may be represented in PCM or MPEG format.

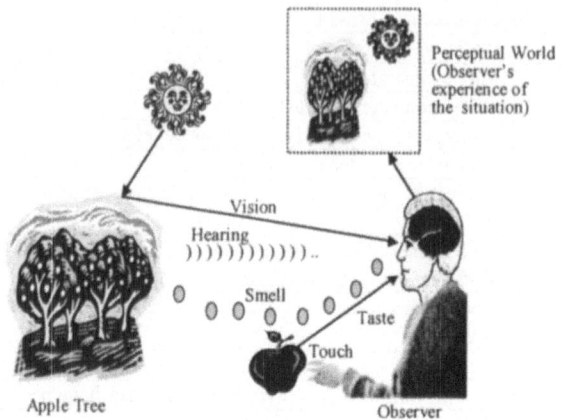

Figure 1.2: Sensory Perception.

Presentation: This refers to the tools and devices for the input and output of information. The paper, screen, and speakers are the output media, while the keyboard, mouse, microphone, and camera are the input media.

Storage: This refers to the data carrier that enables the storage of information. Paper, microfilm, floppy disk, hard disk, CD, and DVD are examples of storage media.

Transmission: This characterizes different information carriers that enable continuous data transmission. Optical fibers, coaxial cable, and free air space (for wireless transmission) are examples of transmission media.

Discrete/Continuous: Media can be divided into two types: time-independent or discrete media, and time-dependent or continuous media. For time-independent media (such as text and graphics), data processing is not time critical. In time-dependent media, data representation and processing *is* time critical. Figure 1.3 shows a few popular examples of discrete and continuous media data, and their typical applications. Note that the multimedia signals are not limited to these traditional examples. Other signals can also be considered as multimedia data. For example, the output of different sensors such as smoke detectors, air pressure, and temperature can be considered continuous media data.

1.4 PROPERTIES OF MULTIMEDIA SYSTEMS

Literally speaking, any system that supports two or more media should be called a multimedia system. Using this definition, a newspaper is a multimedia presentation because it includes text and images for illustration. However, in practice, a different interpretation often appears. Nevertheless, a multimedia system should have the following properties:

Combination of Media: It is well-known that a multimedia system should include two or more media. Unfortunately, there is no exclusive way to specify the media types. On one hand, some authors [1] suggest that there should be at least one continuous (time-dependent) and one discrete (time independent) media. With this requirement, a text processing system that can incorporate images may not be called a multimedia system (since both media are discrete). On the other hand, some authors [3] prefer to relax this interpretation, and accept a more general definition of multimedia.

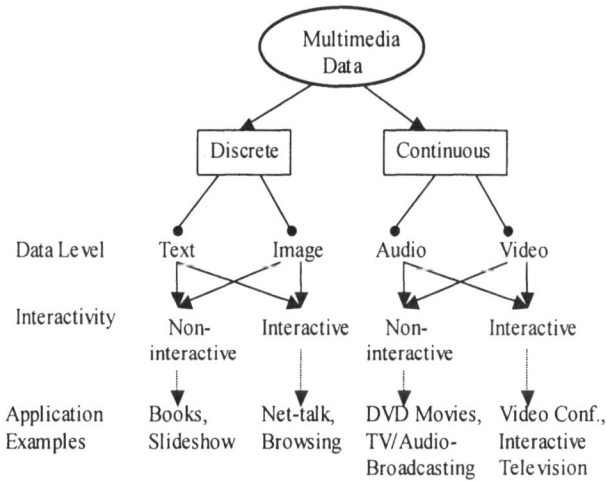

Figure 1.3. Different types of multimedia and their typical applications.

Independence: Different media in a multimedia system should have a high degree of independence. This is an important criterion for a multimedia system, as it enables independent processing of different media types, and provides the flexibility of combining media in arbitrary forms. Most conventional information sources that include two or media will fail this test. For example, the text and images in a newspaper are tightly coupled; so are the audio and video signals in a VHS cassette. Therefore, these systems do not satisfy the independence criteria, and are not multimedia systems.

Computer Supported Integration: In order to achieve media independence, computer-based processing is almost a necessity. The computers provide another important feature of a multimedia system: *integration*. The different media in a multimedia system should be integrated. A high level of integration ensures that changing the content of one media causes corresponding changes in other media.

Communication Systems: In today's highly-networked world, a multimedia system should be capable of communicating with other multimedia systems. The multimedia data transmitted through a network may be discrete (*e.g.*, a text document, or email) or continuous (*e.g.*, streamed audio or video) data.

1.5 MULTIMEDIA COMPUTING

Multimedia computing is the core module of a typical multimedia system. In order to perform data processing efficiently, high-speed processors and peripherals are required to handle a variety of media such as text, graphics, audio and video. Appropriate software tools are also required in order to process the data.

In the early 1990s, the "multimedia PC" was a very popular term used by personal computer (PC) vendors. To ensure the software and hardware compatibility of different multimedia applications, the *Multimedia PC Marketing Council* developed specifications for Multimedia PC, or MPC for short [4]. The first set of specifications (known as MPC Level 1) was published in 1990. The second set of specifications (MPC Level 2) was specified in 1994, and included the CD-ROM drive and sound card. Finally, the MPC Level 3 (MPC3) specifications were published in 1996, with the following specifications:

- CPU speed: 75 MHz (or higher) Pentium
- RAM: 8 MB or more
- Magnetic Storage: 540 MB hard drive or larger
- CD-ROM drive: 4x speed or higher
- Video: Super VGA (640x480 pixels, 16 bits (*i.e.*, 65,536) colors)
- Sound card: 16-bit, 44.1 kHz stereo sound
- Digital video: Should support delivery of digital video with 352×240 pixels resolution at 30 frames/sec (or 352×288 at 25 frames/sec). It should also have MPEG1 support (hardware or software).
- Modem: 28.8 Kbps or faster to communicate with the external world.

Note that most PCs available on the market today far exceed the above specifications. A typical multimedia workstation is shown in Fig. 1.4.

Today's workstations contain rich system configurations for multimedia data processing. Hence, most PCs can technically be called MPCs. However, from the technological point of view, there are still many issues that require the full attention of researchers and developers. Some of the more critical aspects of a multimedia computing system include [2]:

Processing Speed: The central processor should have a high processing speed in order to perform software-based real-time multimedia signal processing. Note that among the multimedia data, video processing requires the most computational power, especially at rates above 30 frames/sec. A distributed processing architecture may provide an expensive high-speed multimedia workstation [5].

Architecture: In addition to the CPU speed, efficient architectures are required to provide high-speed communication between the CPU and the RAM, and between the CPU and the peripherals. Note that the CPU speed is constantly increasing over the years. Several novel architectures, such as Intelligent RAM (IRAM), and Computational RAM have been proposed to address this issue [6]. In these architectures, the RAM has its own processing elements, and hence the memory bandwidth is very high.

Operating System: High performance real-time multimedia operating systems are required to support real-time scheduling, efficient interrupt handling, and synchronization among different data types [7].

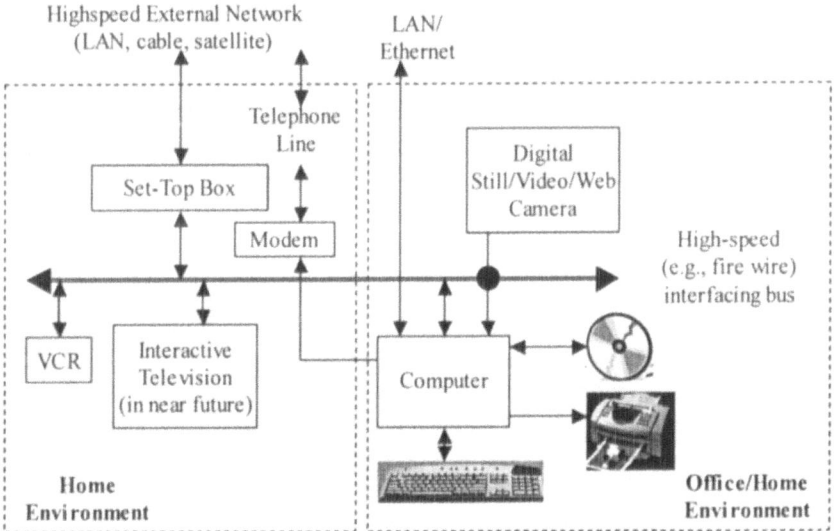

Figure 1.4. Multimedia Workstations at home and office environment.

Storage: High capacity storage devices are required to store voluminous multimedia data. The access time should be fast for interactive applications. Although magnetic devices (such as hard disks) are still generally used for storing multimedia data, other technologies such as CD/DVD, and smart memories are becoming popular for their higher portability [8].

Database: The volume of multimedia data is growing exponentially. Novel techniques are essential for designing multimedia databases so that content representation and management can be performed efficiently [9].

Networking: Efficient network architecture and protocols are required for multimedia data transmission [10]. The network should have high bandwidth, low latency, and reduced jitter.

Software Applications: From a consumer's viewpoint, this is the most important aspect of a multimedia system. A normal user is likely to be working with the software tools without paying much attention to what is inside the computer. Efficient software tools, with easy to use graphical interfaces, are desirable for multimedia applications.

Different Aspects of Multimedia

Multimedia is a broad subject that can be divided into four domains [1]: *device*, *system*, *application*, and *cross* domains. The device domain includes *storage media*, and *networks*, and basic concepts such as *audio*, *video*, *graphics*, and *images*. Conversely, the system domain includes the database systems, operating systems and communication systems. The application domain includes the *user interface* through which various *tools, applications, and documents* are made accessible to the multimedia users. Finally, the cross domain includes the integration of various media. In a multimedia system, the continuous media have to be synchronized. *Synchronization* is the temporal relationship among various media, and it relates to all three domains mentioned above.

The main focus of this book is the device domain aspect of the multimedia system. There are fourteen chapters (Chapters 2-15) in the book, which can be divided into three parts. Chapters 2-4 present the characteristics of audio signals, the properties of our ears and eyes, and the digitization of continuous-time signals. The data and signal processing concepts for various media types, namely text, audio, images and video, are presented in Chapters 5-11. The details of a few select systems – namely television, storage media, and display devices – are presented in Chapters 12, 14, and 15. Lastly, a brief overview of multimedia content creation and management, which lies in the application domain, is presented in Chapter 13.

REFERENCES

1. R. Steinmatz and K. Nahrstedt, *Multimedia: Computing, Communications and Applications*, Prentice Hall, 1996.
2. B. Furht, S. W. Smoliar, and H. Zhang, *Video and Image Processing in Multimedia Systems*, Kluwer Academic Publishers, 1995.
3. N. Chapman and J. Chapman, *Digital Multimedia*, John Wiley & Sons, 2000.
4. W. L. Rosch, *Multimedia Bible*, SAMS Publishing, Indianapolis, 1995.
5. K. Dowd, C. R. Severance, M. Loukides, *High Performance Computing*, O'Reilly & Associates, 2nd edition, August 1998.
6. C. E. Kozyrakis and D. A. Patterson, "A new direction for computer architecture research," *IEEE Computer*, pp. 24-32, Nov 1998.
7. A. Silberschatz, P. B. Galvin, and G. Gagne, *Operating System Concepts*, John Wiley & Sons, 6th Edition, 2001.
8. B. Prince, *Emerging memories: technologies and trends*, Kluwer Academic Publishers, Boston, 2002.
9. V. Castelli and L. D. Bergman, *Image Databases: Search and Retrieval of Digital Imagery*, John Wiley & Sons, 2002.
10. F. Halsall, *Multimedia Communications: Applications, Networks, Protocols, and Standards*, Addison-Wesley Publishing, 2000.
11. T. N. Ryman, "Computers learn to smell and taste," *Expert Systems*, Vol. 12, No. 2, pp. 157-161, May 1995.

QUESTIONS

1. What is a multimedia system? List a few potential multimedia applications that are likely to be introduced in the near future, and twenty years from now.
2. Do you think that integrating the minor senses with the existing multimedia system will enhance its capability? What are the possible technical difficulties in the integration process?
3. Classify the media with respect to the following criteria – i) perception, ii) representation, and iii) presentation.
4. What are the properties of a multimedia system?
5. What is continuous media? What are the difficulties of incorporating continuous media in a multimedia system?
6. List some typical applications that require high computational power.
7. Why is real-time operating system important for designing an efficient multimedia system?
8. Explain the impact of high-speed networks on multimedia applications.
9. Explain with a schematic the four main domains of a multimedia system.

Chapter 2

Audio Fundamentals

Sound is a physical phenomenon produced by the vibration of matter, such as a violin string, a hand clapping, or a vocal tract. As the matter vibrates, the neighboring molecules in the air vibrate in a spring-like motion creating pressure variations in the air surrounding the matter. This alteration of high pressure (compression) and low pressure (rarefaction) is propagated through the air as a wave. When such a wave reaches a human ear and is processed by the brain, a sound is heard.

2.1 CHARACTERISTICS OF SOUND

Sound has normal wave properties, such as reflection, refraction, and diffraction. A sound wave has several different properties [1]: pitch (or frequency), loudness (or amplitude/intensity), and envelope (or waveform).

Frequency

The frequency is an important characteristic of sound. It is the number of high-to-low pressure cycles that occurs per second. In music, frequency is known as *pitch*, which is a musical note created by an instrument. The frequency range of sounds can be divided into the following four broad categories:

Infra sound	0 Hz - 20 Hz
Audible sound	20 Hz – 20 KHz
Ultrasound	20 KHz – 1 GHz
Hypersound	1 GHz – 10 GHz

Different living organisms have different abilities to hear high frequency sounds. Dogs, cats, bats, and dolphins can hear up to 50 KHz, 60 KHZ, 120 KHZ, and 160 KHZ, respectively. However, the human ear can hear sound waves only in the range of 20 Hz-20 kHz. This frequency range is called the *audible band*. The exact audible band differs from person to person. In addition, the ear's response to high frequency sound deteriorates with age. Middle-aged people are fortunate if they are able to hear sound frequencies above 15 KHz. Sound waves propagate at a speed of approximately 344 m/s in humid air at room temperature (20^0C). Hence, audio wavelengths typically vary from 17 m (corresponding to 20 Hz) to 1.7 cm (corresponding to 20 KHz).

There are different compositions of sounds such as natural sound, speech, or music. Sound can also be divided into two categories: *periodic* and *nonperiodic*. Periodic sounds are repetitive in nature, and include whistling wind, bird songs, and sound generated from musical instruments. Nonperiodic sound includes speech, sneezes, and rushing water. Most sounds are complex combinations of sound waves of different frequencies and waveshapes. Hence, the spectrum of a typical audio signal contains one or more fundamental frequencies, their harmonics, and possibly a few cross-modulation products. Most of the fundamental frequencies of sound waves are below 5 KHz. Hence, sound waves in the range 5 KHz-15 KHz mainly consist of harmonics. These harmonics are typically smaller in amplitude compared to fundamental frequencies. Hence, the energy density of an audio spectrum generally falls off at high frequencies. This is a characteristic that is exploited in audio compression or noise reduction systems such as Dolby.

The harmonics and their amplitude determine the *tone quality* or *timbre* (in music, *timbre* refers to the quality of the sound, *e.g.* a flute sound, or a cello sound) of a sound. These characteristics help to distinguish sounds coming from different sources such as voice, piano, or guitar.

Sound Intensity

The sound intensity or amplitude of a sound corresponds to the loudness with which it is heard by the human ear. For sound or audio recording and reproduction, the sound intensity is expressed in two ways. First, it can be expressed at the acoustic level, which is the intensity perceived by the ear. Second, it can be expressed at an electrical level after the sound is converted to an electrical signal. Both types of intensities are expressed in decibels (dB), which is a relative measure.

The acoustic intensity of sound is generally measured in terms of the sound pressure level.

$$\text{Sound intensity (in dB)} = 20 * \log_{10}(P/P_{Ref}) \tag{2.1}$$

where P is the acoustic power of the sound measured in *dynes/cm^2*, and P_{Ref} is the intensity of sound at the threshold of hearing. It has been found that for a typical people, $P_{Ref} = 0.0002\, d/cm^2$. Hence, this value is used in Eq. (2.1) to measure the sound intensity. Note that the human ear is essentially insensitive to sound pressure levels of less than P_{Ref}. Table 2.1 shows intensities of several naturally occurring sounds.

The intensity of an audio signal is also measured in terms of the electrical power level.

$$\text{Sound intensity (in dBm)} = 10\log_{10}(P/P_0) \qquad (2.2)$$

where P is the power of the audio signal, and $P_0 = 1\,mW$. Note that the suffix m in dBm is because the intensity is measured with respect to $1\,mW$.

Table 2.1. Pressure levels of various sounds. 0 dB corresponds to SPL of 0.0002 dynes/cm^2 (or microbar)

Intensity	Typical Examples
0 dB	Threshold of hearing
25 dB	Recoding studio (ambient level)
40 dB	Resident (ambient level)
50 dB	Office (ambient level)
70 dB	Typical conversation
90 dB	Home audio listening level
120 dB	Threshold of Pain
140 dB	Rock singer screaming into microphone

Envelope

An important characteristic of a sound is its envelope. When a sound is generated, it does not last forever. The rise and fall of the intensity of the sound (or a musical note) is known as the envelope. A typical envelope has four sections: *attack*, *decay*, *sustain* and *release*, as shown in Fig. 2.1. During the attack, the intensity of a note increases from silence to a high level. The intensity then decays to a middle level where it is sustained for a short period of time. The intensity drops from the *sustain* level to zero during the release period.

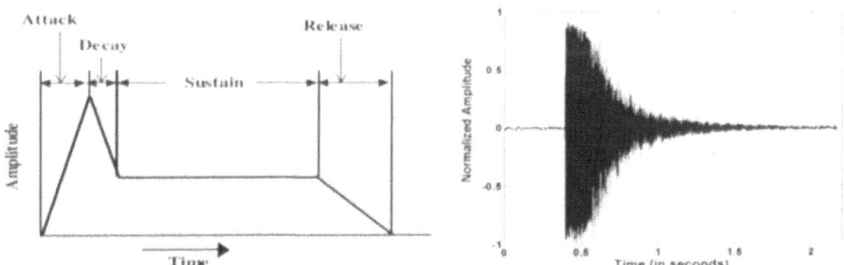

Figure 2.1. Musical note. a) a typical envelope, b) the waveform of a signal produced by a bell. Note that the bell was hit suddenly, resulting in a sharp attack, and a gradual decay. Typical notes, however, will have a more gradual attack.

Each musical instrument has a different envelope. Violin notes have slower attacks but a longer *sustain* period, whereas guitar notes have quick attacks and a slower release. Drum hits have rapid attacks and decays.

Human speech is certainly one of the most important categories of multimedia sound. For efficient speech analysis, it is important to understand the principles of the human vocal system, which is beyond the scope of this book. Here, we are more interested in effective and efficient speech representation and to do this it is helpful to understand the properties of the human auditory system. In the next section, the properties of the human auditory system are briefly presented.

2.2 THE HUMAN AUDITORY SYSTEM

The ear and its associated nervous system is a complex, interactive system. Over the years, the human auditory system has evolved incredible powers of perception. A simplified anatomy of the human ear is shown in Fig. 2.2. The ear is divided into three parts: *outer*, *middle* and *inner* ear. The outer ear comprises the external ear, the ear canal and the eardrum. The external ear and the ear canal collect sound, and the eardrum converts the sound (acoustic energy) to vibrations (mechanical energy) like a microphone diaphragm. The ear canal resonates at about 3 KHz, providing extra sensitivity in the frequency range critical for speech intelligibility. There are three bones in the middle ear – *hammer*, *anvil* and *stirrup*. These bones provide impedance matching to efficiently convey sounds from the eardrum to the fluid-filled inner ear. The coiled basilar membrane detects the amplitude and frequency of sound. These vibrations are converted to electrical impulses, and sent to the brain as neural information through a bundle of nerve fibers. To determine frequency, the brain decodes the period of the stimulus and point of maximum stimulation along the basilar membrane. Examination of the basilar membrane shows that the ear contains roughly 30,000 hair cells arranged in multiple rows along the basilar membrane, which is roughly 32 mm long.

Although the human ear is a highly sophisticated system, it has its idiosyncrasies. On one hand, the ear is highly sensitive to small defects in desirable signals; on the other hand, it ignores large defects in signals it assumes are irrelevant. These properties can be exploited to achieve a high compression ratio for the efficient storage of the audio signals.

It has been found that sensitivity of the ear is not identical throughout the entire audio spectrum (20Hz-20KHz). Fig. 2.3 shows the experimental results with the human auditory system using sine tones [2]. The subjects were people mostly 20 years of age. First, a sine tone was generated at 20 dB intensity (relative to 0.0002 dyne/cm^2 pressure level) at 1 KHz frequency, and the loudness level was recorded. Then the sine tones were generated at other frequencies, and the amplitude of the tones were changed such that the intensity of the tones were identical. The amplitude at other frequencies resulted in the second bottom-most curve (represented by 20 dB at 1 KHz). The experiment was repeated for 40, 60, 80, 100, and 120 dB. The equal loudness contours show that the ear is nonlinear with respect to frequency and loudness. The bottommost curve represents the minimum audible field (MAF) of the human ear. It is observed from these contours that the ear is most sensitive within the frequency range of 1KHz - 5KHz.

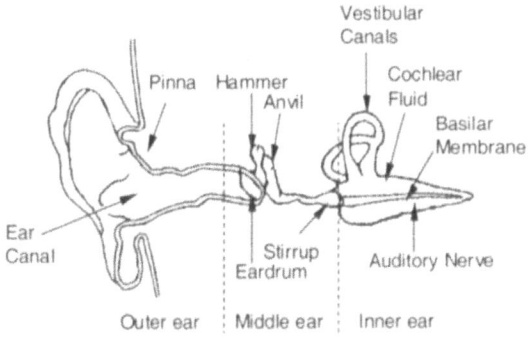

Figure 2.2. Anatomy of human ear. The coiled cochlea and basilar membrane are straightened for clarity of illustration.

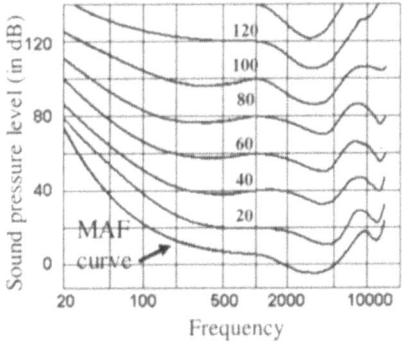

Figure 2.3. Equal loudness contours (for age 20 years). The curves show the relative sound pressure levels at different frequencies that will be heard by the ear with similar loudness (adapted from [3]).

The resonant behavior of the basilar membrane (in Fig. 2.2) is similar to the behavior of a transform analyzer. According to the uncertainty principle of transforms, there is a tradeoff between frequency resolution and time resolution. The human auditory system has evolved a compromise that balances frequency resolution and time resolution. The imperfect time resolution arises due to the resonant response of the ear. It has been found that a sound must be sustained for at least 1 *ms* before it becomes audible. In addition, even if a given sound ceases to exist, its resonance affects the sensitivity of another sound for about 1 *ms*.

Due to the imperfect frequency resolution, the ear cannot discriminate closely-spaced frequencies. In other words, the sensitivity of sound is reduced in the presence of another sound with similar frequency content. This phenomenon is known as *auditory masking*, which is illustrated in Fig. 2.4. Here, a strong tone at a given frequency can mask weaker signals corresponding to the neighboring frequencies.

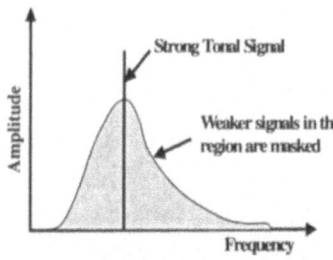

Figure 2.4. Audio noise masking. A strong tone
masks other closely-spaced weaker tones.

The frequency resolution of the ear is not uniform throughout the entire audio spectrum (20Hz-20KHz). The sensitivity is highest at the low frequencies, and decreases at higher frequencies. At low frequencies, the ear can distinguish tones a few Hz apart. However, at high frequencies, the tones must differ by hundreds of Hz to be distinguished. Hence, the human ear can be considered as spectrum-analyzer with logarithmic bands. Experimental results show that the audio spectrum can be divided into several critical bands (as shown in Table 2.2). Here, a critical band refers to a band of frequencies that are likely to be masked by a strong tone at the center frequency of the band. The width of the critical bands is smaller at lower frequencies. It is observed in Table 2.2 that the critical band for a 1 KHz sine tone is about 160 Hz in width. Thus, a noise or error signal that is 160 Hz wide and centered at 1 KHz is audible only if it is greater than the same level of a 1 KHz sine tone.

When the frequency sensitivity and the noise masking properties are combined, we obtain the threshold of hearing as shown in Fig. 2.5. Any audio signal whose amplitude is below the masking threshold is inaudible to the human ear. For example, if a 1 KHz, 60 dB tone and a 1.1 KHz, 25 dB tone are simultaneously present, we will not be able to hear the 1.1 KHz tone; it will be masked by the 1 KHz tone.

Table 2.2. An example of critical bands in the human hearing range showing increase in the bandwidth with absolute frequency. A critical band will arise at an audible sound at any frequency. For example, 220 Hz strong tone is likely to mask the frequencies in the band 170-270 Hz.

Critical Band Number (Bark)	Lower Cut-off Frequency (Hz)	Upper Cut-off Frequency (Hz)	Critical Band (Hz)	Center Frequency (Hz)
1	---	100	---	50
2	100	200	100	150
3	200	300	100	250
4	300	400	100	350
5	400	510	110	450
6	510	630	120	570
7	630	770	140	700
8	770	920	150	840
9	920	1080	160	1000
10	1080	1270	190	1170
11	1270	1480	210	1370
12	1480	1720	240	1600
13	1720	2000	280	1850
14	2000	2320	320	2150
15	2320	2700	380	2500
16	2700	3150	450	2900
17	3150	3700	550	3400
18	3700	4400	700	4000
19	4400	5300	900	4800
20	5300	6400	1100	5800
21	6400	7700	1300	7000
22	7700	9500	1800	8500
23	9500	12000	2500	10500
24.	12000	15500	3500	13500
25	15500	22050	6550	18775

■ Example 2.1

In this example, the psychoacoustics characteristics are demonstrated using a few test audio signals. The following MATLAB code generates three test audio files (in wav format) each with duration of two seconds. The first one-second audio consists of a pure sinusoidal tone with a frequency of 2000 Hz. The next one-second of audio signals contain a mixture of 2000 Hz and 2150 Hz tones. The two tones have similar energy in the *test1* audio file. However, the 2000 Hz tone has 20 dB higher energy than the 2150 Hz tone in the *test2* audio file. In the *test3* audio file, the 2000 Hz tone has 40 dB higher energy than the 2150 Hz tone. The power spectral density of the *test3* (for duration 1-2 seconds) is shown in Fig. 2.6.

```
fs = 44100 ;                          % sampling frequency
nb =16 ;                              % 16-bit/sample
sig1 = 0.5*sin(2*pi*(2000/44100)*[1:1*44100]) ; % 2000 Hz, 1 sec audio
sig2 = 0.5*sin(2*pi*(2150/44100)*[1:1*44100]) ; % 2150 Hz, 1 sec audio
sig3 = [sig1  sig1+sig2] ;            % 2000 Hz and 2150 hz tones are equally strong
sig4 = [sig1  sig1+0.1*sig2] ;        % 2000 Hz is 20 dB stronger than 2150 hz
sig5 = [sig1  sig1+0.01*sig2] ;       % 2000 Hz is 40 dB stronger than 2150 hz
wavwrite(sig3,fs,nb,'f:\test1.wav');
wavwrite(sig4,fs,nb,'f:\test2.wav');
wavwrite(sig5,fs,nb,'f:\test3.wav');
```

It can be easily verified by playing the files that the transition from the pure tone (first one second) to the mixture (the next one second) is very sharp in the test1 audio file. In the second file (*i.e.*, test2.wav), the transition is barely identifiable. In the third file, the 2150 Hz signal is completely inaudible. ∎

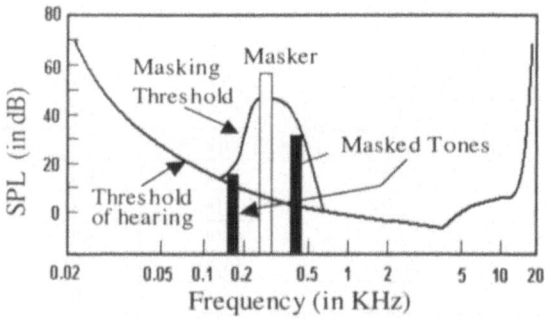

Figure 2.5. Audio masking threshold. The threshold of hearing determines the weakest sound audible by the human ear in the 20-20KHz range. A masker tone (at 300 Hz) raises this threshold in the neighboring frequency range. It is observed that two tones at 180 and 450 Hz are masked by the masker, i.e., these tones will not be audible. SPL: sound pressure level

Similarly, it can be demonstrated that when a low frequency and high frequency tones are generated with equal amplitude, the high frequency tones do not seem to be as loud as the low frequency tones.

2.3 AUDIO RECORDING

In our daily lives, sound is generated and processed in various ways. During speech, the sound wave is generated by the person, and is heard by the listeners. In this case, no automatic processing of sound is required. However, processing and storage of sound is necessary in many applications such as radio broadcasting and music industry. In these applications, the audio signals produced are stored for future retrieval and playback.

Acoustics

Sound typically involves the sound source, a listener, and the environment. The sound is generally reflected from the surrounding objects. The listener hears the reflected sound as well as the sound coming directly from the source. These other sound components contribute to what is known as the *ambience* of the sound.

The ambience is caused by the reflections in the closed spaces, such as a concert hall. In a smaller place there may be multiple reflections, none of which is delayed enough to be called an *echo* (which is discrete repetition of a portion of a sound), but the sound continues to bounce around the room until it eventually dies out because of the partial absorption that occurs at each reflection. For example, when you shout "hello" in an empty auditorium, most likely you will hear "hello-o-o-o-o-o." This phenomenon is known as *reverberation*.

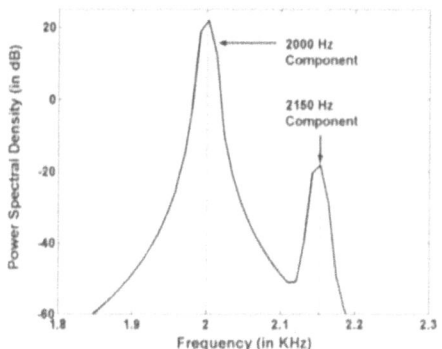

Figure 2.6. Power spectral density of test3 audio signal. The 2000 Hz component has 40 dB higher energy than the 2150 hz component.

Reverberation contributes to the feeling of space, and is important in sound reproduction. For example, if the sound is picked up directly from a musical instrument with no reverberation, the sound will appear dead. This can be corrected by adding artificial reverberation, which is usually done by digital processing.

Multichannel Audio

A brief introduction to the human auditory system was provided in section 2. When sound is received by the ears, the brain decodes the two resulting signals, and determines the directivity of the sound. Historically, sound recording and reproduction started with a single audio channel, popularly known as mono audio. However, it was soon discovered that the directivity of the sound could be improved significantly using two audio channels. The two-channel audio is generally called *stereo* and is widely used in recording and broadcasting industries. The channels are called *left* (L) and *right* (R), corresponding to the speaker locations for reproduction.

The concept of using two channels was natural, given that we have two ears. For a long time, there was the common belief (and many people still believe it today) that with two ears all we need are two channels. However, researchers have found that more audio channels (see Table 2.3) can enhance the spatial sound experience further. Four or more channels audio has been in use in cinema applications since the 1940s. However, four-channel (quadraphonic) audio was introduced for home listeners only in the 1970s. It did not become popular because of the difficulty of storing four channels in audio-cassettes using the available analog technology.

Table 2.3. History of Multichannel audio for home and cinema applications

Decades	Major Milestones
1930s	Experiment with three channel audio at Bell Laboratories
1950s	4-7 channels (cinema)
	2 channel stereo audio (home)
1970s	Stereo surround (cinema)
	Four channel stereo audio (home)
	Mono and stereo video cassettes (home)
1980s	3-4 channel surround audio (home)
	2 channel digital CD audio (home)
1990s	5.1 channel surround audio (home)

It has been found that more realistic sound reproduction can be obtained by having one or more reproduction channels that emit sound behind the listener [4]. This is the principle of *surround sound*, which has been widely used in movie theater presentations, and has recently become popular for home theater systems. There is a variety of configurations for arranging the speakers around the listener (see Table 2.4). The most popular configuration for today's high-end home listening environment is the standard surround system that employs 5 channels with 3 speakers in the front and two speakers at the rear (see Fig. 2.7). For cinema application, however, more rear speakers may be necessary depending on the size of the theater hall.

Table 2.4 shows that standard surround sound can be generated with 5 full audio channels (with up to 20 KHz bandwidth). However, it has been observed that adding a low bandwidth (equivalent of 0.1) subwoofer channel (termed as LFE in Fig. 2.7) enhances the quality of the reproduction. These systems are typically known as 5.1 channels – *i.e.*, five full bandwidth channels and one low bandwidth channel, and have become very popular for high-end home audio systems.

Table 2.4: Configuration of speakers in surround sound system. The code (p/q) refers to the speaker configuration in which p speakers are in the front, and q speakers are at the rear. "x" indicates the presence of a speaker in a given configuration. F-L: front left, F-C: front center, F-R: front right, M-L: mid left, M-R: mid right, R-L: rear left, R-C: rear center, R-R: rear right

Name	Codes	Speaker positions							
		F-L	F-C	F-R	M-L	M-R	R-L	R-C	R-R
Mono	1/0								
Stereo	2/0	x		x					
3-ch. Stereo	3/0	x	x	x					
3-ch surround	2/1	x		x				x	
Quadrphonic surround	2/2	x		x			x		x
Standard surround	3/2	x	x	x			x		x
Enhanced surround	5/2	x	x	x	x	x	x		x

Multi-track Audio

Figure 2.8(a) shows a schematic of a simple stereo audio recording appropriate for a musical performance. The sound waves are picked by the two microphones and converted to electrical signals. The resulting sound quality is affected by the choice of microphones and their placement. The two-track recorder records the two channels (left and right) separately into two different tracks. For playback, the recorded signal is fed to a power amplifier, and then driven in electrical form to two speakers.

Figure 2.8(b) shows the audio recording of a more complex musical performance with several microphones [5]. Here, each microphone is placed close to each singer or instrument. To obtain a balanced musical recording, all the microphones are plugged into a mixer that can control individually the volume of signals coming from each microphone. The output of the mixer can be recorded on a multi-track tape for future editing, but the sound editing might require playback of the music several times for fine adjustment of individual components. On completion of the editing process, the audio signal can be recorded on a two-track stereo tape or one-track mono tape (see Fig. 2.8(c)).

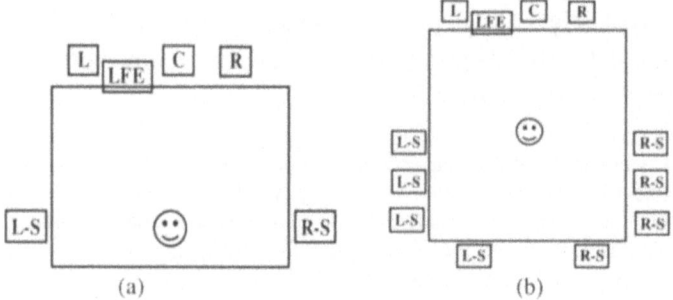

Figure 2.7. Multichannel audio playback. a) at home, b) at cinema. L: left channel, C: center, R: right, L-S: left surround, R-S: right surround, LFE: low frequency effects.

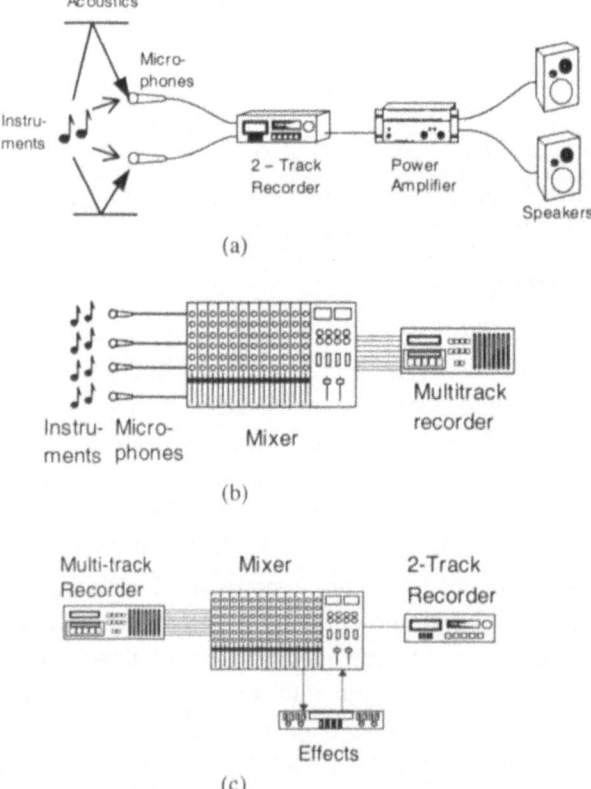

Figure 2.8. Live audio recording. a) Two-track recorder, (b) Four-track recorder, (c) conversion of four-track recorded signal to two-track.

The advantage of multi-track audio is its flexibility. A track can be made ON or OFF during recording and/or playback. Consider a scenario where, after a performance had been recorded in a studio, it was found that the piano signal was not blending well with the other components. With only one or two recorded tracks, one might have to repeat the entire musical performance. However, in multi-track audio, the track corresponding to the piano component can be substituted by a new recording of just the piano component.

2.4 AUDIO SIGNAL REPRESENTATION

There are primarily two methods of representing an audio signal – waveform and parametric methods. The *waveform representation method* focuses on the exact representation of the audio signal produced, whereas the *parametric representation method* focuses on the modeling of the signal generation process. The choice of the digital representation of audio signals is governed by three major considerations: processing complexity, information rate (*e.g.* bit-rate) and flexibility. There are primarily two types of parametric methods: i) speech synthesis by modeling human vocal system, and ii) music synthesis using the octave chart. The former method mostly has been applied to achieve very low bit-rate speech compression, and is currently not used in general purpose high quality audio coding. However, the second method is widely used in the framework of the MIDI standard. The next two sections present a brief introduction of the waveform method and the MIDI standard.

2.4.1 Waveform method

A typical audio generation and playback schematic is shown in Fig. 2.9. In this method, one or more microphones are used to convert the acoustic energy (sound pressure levels) to electrical energy (watts). The voltage produced at the output of the microphone is then sampled and quantized. The digital audio, thus produced, is then stored as an audio file or transmitted to the receiver for immediate playback. While being played back, the digital audio is converted to a time-varying analog voltage that drives one or more loud speakers. The sound is thus reproduced for listening.

In order to obtain a desirable quality of reproduced audio signal, the different components of Fig. 2.9 have to be designed properly. In this book, we primarily concentrate on the principles involved in sampling, digitization, and storage. The detailed procedures for sampling and digitization are presented in Chapter 4, while the compression techniques for storage and transmission techniques are presented in Chapter 7.

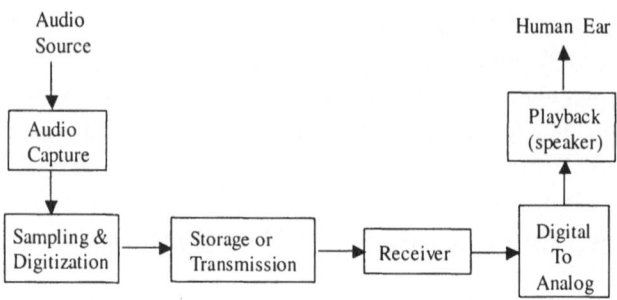

Figure 2.9. Audio generation and playback.

2.4.2 Musical Instrument Digital Interface

Musical sound differs from other sounds in the way that it can be generated. Once a musical sound has been created, it can be played by different musicians by following the corresponding octave chart. This has led to the development of a standard known as the *musical instrument digital interface* (MIDI) standard [6, 7]. In this standard, a given piece of music is represented by a sequence of numbers that specify how the musical instruments are to be played at different time instances. A MIDI studio typically has the following subsystems:

Controller: A musical performance device that generates a MIDI signal (*e.g.*, keyboards, drum pads) when played. A MIDI signal is simply a sequence of numbers that encodes a series of notes.

Synthesizer: A piano-style keyboard musical instrument that simulates the sound of real musical instruments. It generally creates sounds electronically with oscillators.

Sequencer: A device or a computer program that records a MIDI signal corresponding to a musical performance.

Sound module: A device that produces a pre-recorded samples when triggered by a MIDI controller or sequencer.

Fig. 2.10 shows a MIDI system [5] where music is being played by a musician on a MIDI controller (*e.g.*, a keyboard). As the musician plays the keyboard, the controller sends out the corresponding computer code detailing the sequence of events for creating the music. This code is received by a *sound module* that has several tone generators. These tone generators can create sounds corresponding to different musical instruments, such as piano, guitar and drums. When the tone generators synthesize sound according to the MIDI signal and the corresponding electrical signal is driven to the speakers, we hear the sound of the music. The sound module can also be connected to a sequencer that records the MIDI signal, which can be saved on a floppy disk, a CD or a hard disk.

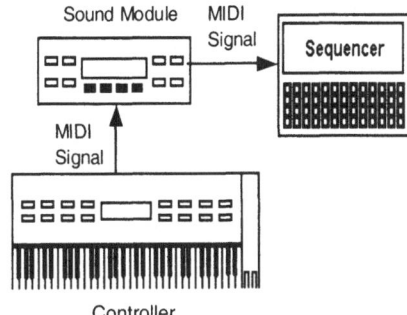

Figure 2.10. A simple MIDI System.

Figure 2.11 shows the bit-stream organization of a MIDI file. The file starts with the header chunk, which is followed by different tracks. Each track contains a track header and a track chunk. The format of the header and track chunks is shown in Table 2.5. The header chunk contains four-bytes of chunk ID which is always "MThd". This is followed by chunk size, format type, number of tracks, and time division. There are three types of standard MIDI files:

- Type 0 - which combines all the tracks or staves into a single track.
- Type 1 - saves the files as separate tracks or staves for a complete score with the tempo and time signature information only included in the first track.
- Type 2 - saves the files as separate tracks or staves and also includes the tempo and time signatures for each track.

The header chunk also contains the time-division, which defines the default unit of delta-time for this MIDI file. The time-division is a 16-bit binary value, which may be in either of two formats, depending on the value of the most significant bit (MSB). If the MSB is 0, then bits 0-14 represent the number of delta-time units in each quarter-note. However, if the MSB is 1, then bits 0-7 represent the number of delta-time units per SMTPE frame, and bits 8-14 form a negative number, representing the number of SMTPE frames per second (see Table 2.6).

The track chunks contains a chunk ID (which is always "MTrk"), chunk size, and track event data. The *track event data* contains a stream of MIDI events that defines information about the sequence and how it is played. This is the actual music data that we hear. Musical control information such as playing a note and adjusting a MIDI channel's modulation value are

defined by *MIDI channel events*. There are three types of events: MIDI
Control Events, System Exclusive Events and Meta Events.

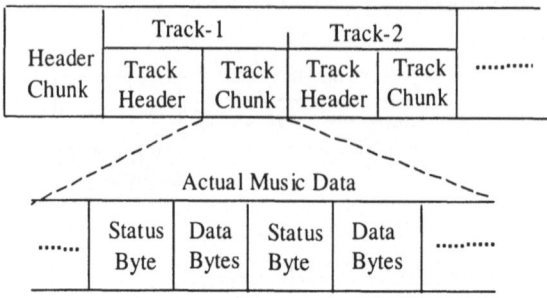

Figure 2.11. MIDI file organization.

Table 2.5. Format of Header and Track Chunks

	Offset	Length	Type	Description	Value
	0x00	4	char[4]	chunk ID	"MThd" (0x4D546864)
Header	0x04	4	dword	chunk size	6 (0x00000006)
Chunk	0x08	2	word	format type	0 - 2
	0x10	2	word	number of tracks	1 - 65,535
	0x12	2	word	time division	in ticks/frame
Track	0x00	4	char[4]	chunk ID	"MTrk" (0x4D54726B)
Chunk	0x04	4	dword	chunk size	size of track data
	0x08		track event data (see following text)		

Table 2.6. Time division information format

Bit:	15	14 ... 8	7 ... 0
<division>	0	ticks per quarter note	
	1	-frames/second	ticks / frame

The MIDI channel event format is shown in Table 2.7. It is observed that
each MIDI channel event consists of a variable-length delta time and 2-3
bytes description that determines the MIDI channel it corresponds to, the
type of event it is, and one or two event type specific values. A few selected
MIDI Channel Events, and their numeric value and parameters are shown in
Table 2.8.

Table 2.7. MIDI channel event format

Delta Time	Event Type Value	MIDI Channel	Parameter 1	Parameter 2
Variable-length	4 bits	4 bits	1 byte	1 byte

In MIDI, a new event is recorded by storing a Note On message. The
velocity in Table 2.8 indicates the force with which a key is struck, which in

turn relates to the volume at which the note is played. However, specifying a velocity of 0 for a Note On event is the same as using the Note Off event. Most MIDI files use this method as it maximizes running mode, where a command can be omitted and the previous command is then assumed.

Table 2.8. MIDI Channel Events

Event Type	Value	Parameter 1	Parameter 2
Note Off	0x8	note number	velocity
Note On	0x9	note number	velocity
Note Aftertouch	0xA	note number	aftertouch value
Controller	0xB	controller number	controller value
Program Change	0xC	program number	not used
Channel Aftertouch	0xD	aftertouch value	not used
Pitch Bend	0xE	pitch value (LSB)	pitch value (MSB)

Note that when a device has received a Note Off message, the note may not cease abruptly. Some sounds, such as organ and trumpet sounds will do so. Others, such as piano and guitar sounds, will decay (fade-out) instead, albeit more quickly after the note-off message is received.

A large number of devices are employed in a professional recording environment. Hence, MIDI protocol has been designed to enable computers, synthesizers, keyboards, and other musical devices to communicate with each other. In the protocol, each musical device is given a number. Table 2.9 lists the MIDI instruments and their corresponding numbers.

Table 2.9 shows the names of the instruments whose sound would be heard when the corresponding number is selected on MIDI synthesizers. These sounds are the same for all MIDI channels except channel 10, which has only percussion sounds and some sound "effects."

On MIDI channel 10, each MIDI Note number (*e.g.*, "Key#") corresponds to a different drum sound, as shown in Table 2.10. While many current instruments also have additional sounds above or below the range shown here, and may even have additional "kits" with variations of these sounds, only these sounds are supported by General MIDI Level 1 devices.

■ **Example 2.2**

In this example, we demonstrate the MIDI format using a small one-track MIDI file. Consider a MIDI file (consisting of only 42 bytes) as follows.

```
4D 54 68 64 00 00 00 06 00 01 00 01 00 78 4D 54 72 6B 00 00 00 14
01 C3 02 01 93 43 64 78 4A 64 00 43 00 00 4A 00 00 FF 2F 00
```

Table 2.9. General MIDI Instrument Sounds

ID	Sound	ID	Sound	ID	Sound
0	Acoustic grand piano	43	Contrabass	86	Lead 7 (Fifths)
1	Bright acoustic piano	44	Tremolo strings	87	Lead 8 (bass+lead)
2	Electric grand piano	45	Pizzicato Strings	88	Pad 1 (New age)
3	Honky-tonk piano	46	Orchestral harp	89	Pad 2 (Warm)
4	Rhodes piano	47	Timpani	90	Pad 3 (Polysynth)
5	Chorused piano	48	String Ensemble 1	91	Pad 4 (Choir)
6	Harpsichord	49	String ensemble 2	92	Pad 5 (Bowed)
7	Clarinet	50	Synth strings 1	93	Pad 6 (Metallic)
8	Celesta	51	Synth strings 2	94	Pad 7 (Halo)
9	Glockenspiel	52	Choir aahs	95	Pad 8 (sweep)
10	Music box	53	Voice oohs	96	FX 1 (Rain)
11	Vibraphone	54	Synth voice	97	FX 2 (Soundtrack)
12	Marimba	55	Orchestra hit	98	FX 3 (Crystal)
13	Xylophone	56	Trumpet	99	FX 4 (Atmosphere)
14	Tubular bell	57	Trombone	100	FX 5 (Brightness)
15	Dulcimer	58	Tuba	101	FX 6 (Goblins)
16	Hammond organ	59	Muted trumpet	102	FX 7 (Echoes)
17	Percussive organ	60	French horn	103	FX 8 (Sci Fi)
18	Rock organ	61	Brass section	104	Sitar
19	Church organ	62	Synth brass 1	105	Banjo
20	Reed organ	63	Synth brass 2	106	Shamisen
21	Accordion	64	Soprano saxophone	107	Koto
22	Harmonica	65	Alto Saxophone	108	Kalimba
23	Tango accordion	66	Tenor saxophone	109	Bagpipe
24	Acoustic guitar (nylon)	67	Baritone saxophone	110	Fiddle
25	Aoustic guitar (steel)	68	Oboe	111	Shanai
26	Electric Guitar (jazz)	69	English horn	112	Tinkle bell
27	Electric guitar (clean)	70	Bassoon	113	Agogo
28	Electric guitar (muted)	71	Clarinet	114	Steel drums
29	Overdriven Guitar	72	Piccolo	115	Wood block
30	Distortion guitar	73	Flute	116	Taiko drum
31	Guitar harmonics	74	Recorder	117	Melodic tom
32	Acoustic bass	75	Pan flute	118	Synth drum
33	Electric bass (finger)	76	Bottle blow	119	Reverse cymbal
34	Electric bass (pick)	77	Shakuhachi	120	Guitar fret noise
35	Fretless Bass	78	Whistle	121	Breath noise
36	Slap Bass 1	79	Ocarina	122	Seashore
37	Slap Bass 2	80	Lead 1 (Square)	123	Bird tweet
38	Synth Bass 1	81	Lead 2 (Saw tooth)	124	Telephone ring
39	Synth Bass 2	82	Lead 3 (Calliope lead)	125	Helicopter
40	Violin	83	Lead 4 (Chiff lead)	126	Applause
41	Viola	84	Lead 5 (Charang)	127	Gunshot
42	Cello	85	Lead 6 (Voice)		

Table 2.9. General MIDI Percussion Key Map

ID	Sound	ID	Sound	ID	Sound
35	Acoustic bass drum	51	Ride cymbal 1	67	High agogo
36	Bass drum 1	52	Chuinese cymbal	68	Low agogo
37	Side stick	53	Ride bell	69	Cabasa
38	Acoustic snare	54	Tambourine	70	Maracas
39	Hand clap	55	Splash cymbal	71	Short whistle
40	Electric snare	56	Cowbell	72	Long whistle
41	Low floor tom	57	Crash cymbal 2	73	Short guiro
42	Closed high-hat	58	Vibraslap	74	Long guiro
43	High-floor tom	59	Ride cymbal 2	75	Claves
44	Pedal high-hat	60	High bongo	76	High wood block
45	Low tom	61	Low bonga	77	Low wood block
46	Open high-hat	62	Mute high conga	78	Mute cuica
47	Low-mid tom	63	Open high conga	79	Open cuica
48	High-mid tom	64	Low conga	80	Mute triangle
49	Crash cymbal 1	65	High timbale	81	Open triangle
50	High tom	66	Low timbale		

The explanation of different bytes is shown in Fig. 2.12. The first 22 bytes contain the header chunk and the track header, whereas the next 20 bytes contain the music information such as the delta-time, the instrument and the note's volume.

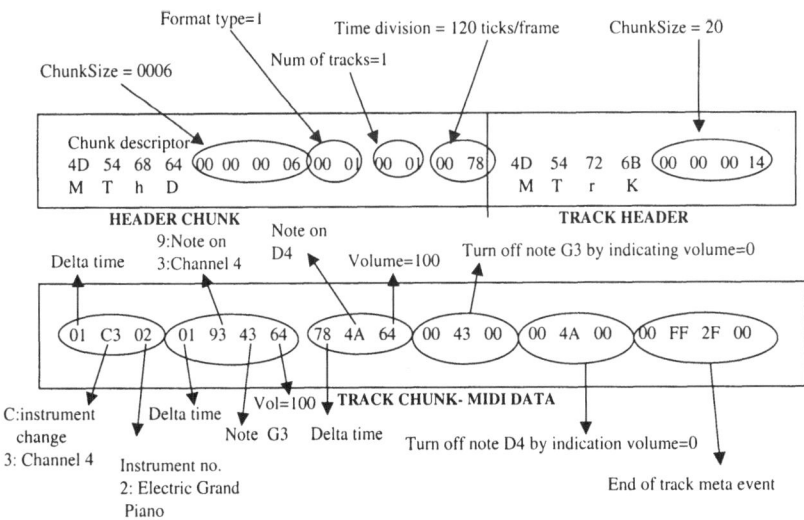

Figure 2.12. Explanation of the MIDI file in Example 2.2.

After the track header, the first byte represents the delta-time for the following note. The next byte is 0xC3 (0x denotes the *hexadecimal* format), which indicates change of instrument (denoted by 0xC) on channel 4. The next byte is 0x02, which corresponds to the Electric Grand Piano. Different instruments can also be set to different channels (there are 16 channels: 0 to F) and can be played simultaneously.

The MIDI file can be created using the following MATLAB code:

```
data=hex2dbytes('4D546864000000060001000100784D54726B0000001401C30201934
364784A64004300004A0000FF2F00');
fid=fopen('F:\ex2_2.mid','wb');
fwrite(fid,data);
fclose('all');
```

Note that "hex2dbytes" is a MATLAB function (included in the CD) that creates a stream of bytes from the hex data, which is then written as a MIDI file. The *ex2_2.mid* file is included in the CD. The readers can verify that the file creates a musical tone from the piano. ∎

Most conventional music can be represented very efficiently using the MIDI standard. However, a MIDI file may produce different qualities of sound depending on the playback hardware. For example, a part of the MIDI sequence may correspond to a particular note on piano. If the quality of the piano is poor, or the piano is improperly tuned, the sound produced from the piano may not be of high quality. Another problem with the MIDI standard is that spoken dialogs or songs are difficult to represent parametrically. Hence, the human voice may need to be represented by sampled data in a MIDI file. However, this will increase the size of the MIDI file.

REFERENCES

1. M. T. Smith, *Audio Engineer's Reference Book*, 2nd Edition, Focal Press, Oxford, 1999.

2. K. C. Pohlmann, *Principles of Digital Audio*, McGraw-Hill, New York 2000.

3. D. W. Robinson and R. S. Dadson, "A redetermination of the equal-loudness relations to for pure tones," *British Journal of Applied Physics*, Vol. 7, pp. 166-181, 1956.

4. D. R. Begault, *3D Sound for Virtual Reality and Multimedia*, Academic Press, 1994.

5. B. Bartlett and J. Bartlett, *Practical Recording Techniques*, 2nd Edition, Focal Press, 1998.

6. J. Rothstein, *MIDI: A Comprehensive Introduction*, 2nd Edition, Madison, Wis., 1995.

7. M. Joe, *MIDI Specification 1.0*, http://www.ibiblio.org/emusic-l/info-docs-FAQs/MIDI-doc/.

QUESTIONS

1. What are the characteristics of sound?
2. What is audible frequency range? Does the audible frequency range of a person remain same throughout his life?
3. How is the loudness of sound measured?
4. What is envelope of a sound? Explain the differences between the envelopes of a few musical instruments.
5. Explain the anatomy of the human ear. How does it convert the acoustic energy to electrical energy?
6. Draw the block schematic of a typical audio (stereo) recording system. Explain briefly the function of each block.
7. Explain the functionality of the human ear.
8. What are the critical bands? Does the width of a critical band depend on its center frequency?
9. What is the difference between echo and reverberation?
10. What is multi-channel audio? How does it improve sound quality?
11. Explain how does surround sound improve the sound perception compared to the two-channel audio.
12. What is multi-track recording? How does it help in sound recording and editing?
13. What is MIDI? What are the advantages of the MIDI representation over digital audio?
14. How many different types of instruments are included in the General MIDI standard?
15. Create a single-track MIDI file that plays one note of an electric guitar. Change the volume from a moderate to loud level.
16. Why do we need a percussion key map?

Chapter 3

The Human Visual System and Perception

It was demonstrated in Chapter 2 that when a distant object vibrates, it creates contraction and expansion of the surrounding medium. This creates a sound wave that is detected by the human ear to sense the object. However, if the object does not vibrate, or there is no medium to convey the vibration, the ear will not be able to sense the object's presence. Nature has gifted us another sensor, namely the eye, which can detect electromagnetic waves coming from the object. It is widely believed that more than 70% of our information is collected by vision. Hence, vision is the most important sensor (compared to the other sensors for hearing, smell, touch or taste) of a normal human being. Since, the output of a multimedia system is generally viewed through our visual system, it is important to know the characteristics of the human visual system (HVS). This will enable us to make effective use of the multimedia technology. It will be demonstrated in later chapters that these properties are used extensively for image and video compression, and for designing display systems. In this chapter, we present the properties of the human visual system and visual perceptions.

3.1 INTRODUCTION

Electromagnetic (EM) waves can have widely different wavelength starting from long wavelengths (corresponding to the dc/power band) to extremely short wavelengths (gamma rays). However, the eye can detect the waves of a narrow spectrum, extending from approximately 400 to 700 nanometers. Sunlight has a spectrum in this wavelength range, and is the major source of the Earth's EM waves. Indeed, it is believed that the evolution of the HVS was influenced by the presence of the sunlight.

Fig. 3.1 shows a typical object detection process. If $L(\lambda)$ is the incident energy distribution coming from a light source, the light received by an observer from an object can be expressed as

$$I(\lambda) = \rho(\lambda)L(\lambda) \tag{3.1}$$

where $\rho(\lambda)$ is the reflectivity or transmissivity of the object. If the light source, the object, and the observer are stationary, the spectrum of $I(\lambda)$ will have the same frequency range as $L(\lambda)$, but with a different spectrum that is determined by $\rho(\lambda)$. This is true for most practical applications, except possibly for astronomical applications where the Doppler shift is an important issue.

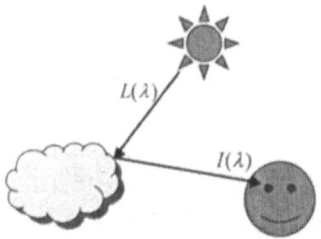

Figure 3.1. Incoming light energy from an object.

When the light energy distribution $I(\lambda)$ is received by a viewer, a complex process is started in the visual system to form an image from light energy. The image formed depends on the different visual systems for various living organisms. Here, we are mostly interested in the mechanism of the human visual system. It is important to learn the characteristics of the HVS to efficiently design systems for the transmission or display of pictorial information. It is a well-established principle that the system need not transmit or display what the eye does not see. This is done in order to reduce the cost of transmission and make the display devices inexpensive. In this chapter, we present a brief introduction of the HVS, and describe how objects are perceived by human eyes.

3.2 THE HUMAN VISUAL SYSTEM

The human visual system can be considered an optical system (see Fig. 3.2). When light from an object falls on the eye, the pupil of the eye acts as an aperture, and an image is created on the *retina*, and the viewer see the object. The perceived size of an object depends on the angle it creates on the retina. The retina can resolve detail better when the retinal image is larger and spread over more of its photoreceptors. The pupil controls the amount of light entering the eye. For typical illumination, the pupil is about 2 mm in diameter. For low illumination, the size increases to allow more light, whereas for high illumination the size decreases to limit the amount of light.

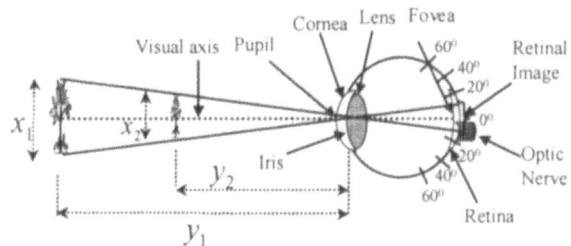

Figure 3.2. Image formation at human eye. The angle $\theta = 2\arctan(x_1 / 2y_1) = 2\arctan(x_2 / 2y_2)$ determines the retinal image size.

The retina contains two types of photoreceptors called *rods* and *cones*. There are approximately 100 millions rods in a normal eye, and these cells are relatively long and thin. These cells provide *scotopic* vision, which is the visual response of the HVS at low illumination. The cone cells are shorter, thicker, and less sensitive than the rods, and are fewer in numbers (approximately 6-7 millions). These cells provide *photopic* vision, the visual response at higher illumination. In the *mesopic* vision (i.e., the intermediate illumination), both rods and cones are active. Since the electronic displays are well lighted, the photopic vision is of primary importance in display engineering. The color vision is provided by the *cones*. Because these receptors are not active at low illumination, the color of an object cannot be detected properly at low illumination.

The distribution of rod and cone cells (per sq. mm) is shown in Fig. 3.3. It is observed that the retina is not a homogeneous structure. The cone cells are densely packed in the fovea (a circular area of about 1.5 mm diameter in the center), and fall off rapidly outside a circle of 1^0 radius. There are no rods or cones in the vicinity of the optic nerve, and thus the eye has a blind spot in this region. When a light stimulates a rod or cone, a photochemical transition occurs producing a nerve impulse. There are about 0.8 millions nerve fibers. In many regions of the retina, the rods/cones are interconnected to the nerve fiber on a many-to-one basis.

The central area, known as the foveal area, is capable of providing high-resolution viewing, but performance falls off in the extra-foveal region. The foveal area subtends only an angle of approximately $1-2^0$ degrees, while the home television generally subtends $5-15^0$ (when a viewer sits at a distance of about 6 times the height of the TV receiver, a vertical viewing angle of 9.5^0 is created) depending on viewing distance. Hence, home televisions are generally watched with extrafoveal vision.

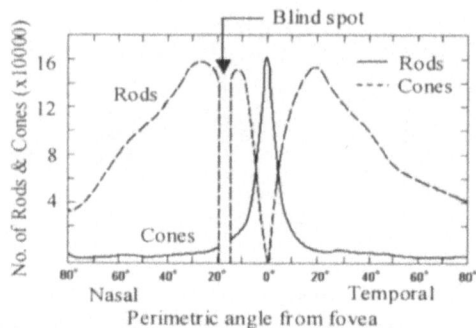

Figure 3.3. Distribution of rod and cone receptors in the retina.

3.2.1 Relative Luminous Efficiency

When a radiant energy distribution $I(\lambda)$ is received, the different wavelengths do not have identical effects on the HVS. Figure 3.4 shows the *relative luminous efficiency* functions for the foveal (within 2 degrees of the fovea) vision of the HVS. It is observed that the HVS acts as a bandpass filter to radiant energy, the peak response being obtained around 555 nm.

The *luminance* or *intensity* of a spatially-distributed object with incoming light energy distribution $I(\lambda)$ can be calculated using the following equation:

$$L = \int_0^\infty I(\lambda)V(\lambda)d\lambda \qquad (3.2)$$

where $V(\lambda)$ is the relative luminous efficiency function.

The luminance is more important than the radiant energy distribution in the image formation process. A high $I(\lambda)$ might appear dark (if the energy lies at the two extreme of the bell-shaped curve), but light with a high L always appears bright to a standard human observer. The luminance is expressed in cd/m^2 where cd (candela) is the unit of luminous intensity. A CRT display has a luminance value in the range 0.1-1000 cd/m^2.

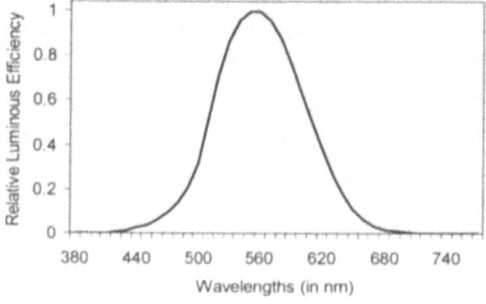

Figure 3.4. Typical relative luminous efficiency function.

The luminance as defined in Eq. (3.2) is an important factor for determining if a picture area will be dark or bright. However, the luminance of the surrounding areas also plays an important role in the appearance of a given picture area. This is demonstrated in Fig. 3.5 where a circle of a given luminance has been placed onto two different backgrounds. It is observed that the circle on the dark background appears to be brighter than the circle on the gray background. The perceived luminance of an object – called the *brightness* – depends on the luminance of the surround [2].

 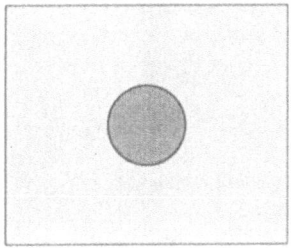

Figure 3.5. Simultaneous contrast: small circles in the middle have equal luminance values, but have different brightness.

3.2.2 Weber's Law

It has been demonstrated that two objects with different surroundings could have identical luminance, but different brightness values. The primary reason is that our perception is sensitive to *luminance contrast* rather than the absolute luminance values themselves. *Contrast* is a term commonly used to describe differences in luminance values of two neighboring picture areas.

Figure 3.6 shows a small circle of luminance $L + \Delta L$ surrounded by a larger circle with a luminance of L. If ΔL is zero, the two regions will be indistinguishable. If we slowly increase ΔL starting from zero, for a certain value of ΔL, the contrast between the two regions will just be noticeable. Let this value be denoted by ΔL_N. As a rough approximation, the relationship between L and ΔL_N can be expressed as

$$\frac{\Delta L_N}{L} = k \tag{3.3}$$

where k is known as the Weber constant, and the above relationship is known as the Weber-Fechner's law, or simply the Weber's law. Typically, the value of the Weber's constant is about 0.01-0.02. The above equation states that if a picture area has a high luminance value, a higher ΔL_N value is required to notice the contrast between the two regions. We can express Eq. (3.3) as

$$c \frac{dL}{L} = dB \tag{3.4}$$

where c is a constant. Integrating both sides, we obtain $B = c \log L + d$ where d is another constant. This relationship states that the brightness is proportional to the logarithm of the luminance values.

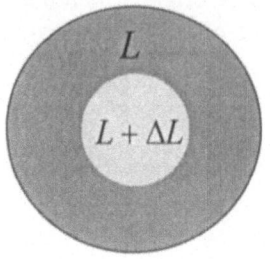

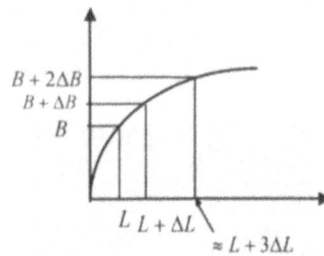

Figure 3.6. Contrast and luminance   Figure 3.7. Brightness (B) and luminance (L)

3.2.3 Modulation Transfer Function

It was shown in Chapter 4 that an audio signal could be expressed as a superposition of sinusoidal waves of different frequencies (in Hz). Similarly, an image can be considered as a superposition of gratings of different spatial frequencies. Figure 3.8 shows two examples of one-dimensional square wave gratings. Note that the spatial frequencies are generally expressed in *cycles/degree*. A periodic grating will have a frequency of M cycles/degree if M wavelengths of the grating make an angle of one degree at the observer's eye, at a specified distance between the eyes and the scene. The same grating will have a different spatial frequency if it is moved closer or further from the eyes.

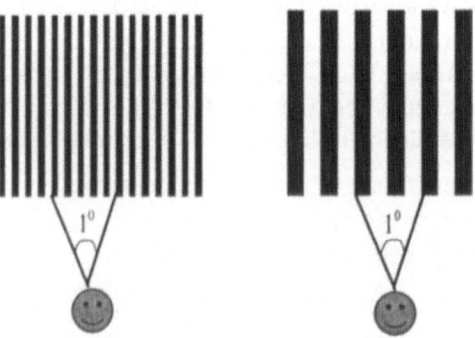

Figure 3.8. Square wave gratings. The left grating (5 cycles/deg) has a higher frequency than the right (2 cycles/deg).

The spatial frequency response of the eye is known as the *modulation transfer function* (MTF). A direct measure of the MTF is possible by considering a sinusoidal grating of varying contrast and spatial frequency. In Fig. 3.9, an oscillating high frequency pattern is shown in two-dimensional space. The pattern has been generated such that any line drawn horizontally across the picture is a sinusoidal wave with constant amplitude but whose frequency increases exponentially from left to right. If a line is drawn vertically in the picture, the amplitude of the wave decreases exponentially from bottom to top. Let us hold the picture at arm's length, then go up from the bottom of each vertical stripe and mark a point where the pattern just becomes invisible. If we plot the marked points on the vertical stripes, we will obtain a curve similar to that shown in Fig. 3.10(a). The plot shows that the middle stripes have the highest sensitivity and the lower (left stripes) and higher (right stripes) frequencies will be visible only when their amplitudes are relatively high. The stripe where the sensitivity will be highest depends on the distance between the picture and the eyes. If we increase the distance, the peak shifts to the left, because the spatial frequency associated with the stripes will change (increase if the distance is increased). Scientists have performed many experiments to determine the spatial frequency response. The spatial frequencies at which the eyes have the highest sensitivity have been reported to vary from 2-10 cycles/degree depending on the experimental set-up (e.g., luminance of the image, choice of subjects).

Note that Fig. 3.10(a) shows only the one-dimensional spatial frequency response. Assuming that the HVS is isotropic in spatial domain (*i.e.*, eyes do not have any preferential direction towards horizontal, vertical or diagonal), the 2-D frequency response can be easily calculated from Fig. 3.10(a), which is shown in Fig. 3.10(b).

Figure 3.9. Contrast versus spatial frequency sinusoidal grating.

It is observed in Fig. 3.10(a) that HVS response is relatively poor at higher frequencies. It is important to know the maximum frequency that can be perceived by eyes. This can be calculated from the *visual acuity* of the

the eye, which is defined as the smallest angular separation at which the individual lines (if we assume that the image consists of horizontal and vertical lines of pixels) can be distinguished. Visual acuity varies widely, as much as $0.5'$ to $5'$ (minutes of arc) depending on the contrast ratio and the keenness of the individual's vision. An acuity of $1.7'$ is often assumed for practical purposes, and this corresponds to approximately 18 $(=60/(2\times1.7'))$ cycles/degree.

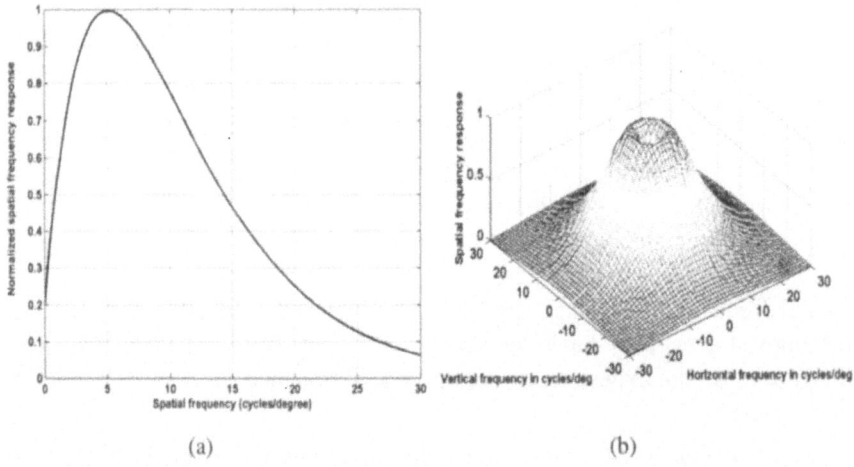

(a)   (b)

Figure 3.10: Frequency response of the eye. a) 1-D and b) 2-D frequency response.

The impulse response of the system can be obtained by calculating the inverse Fourier transform of the modulation transfer function (shown in Fig. 3.10). The one-dimensional impulse response function is known as the *line-spread function*. It provides the response of the system to an infinitely thin light, represented by an impulse sheet $\delta(x)$ or $\delta(y)$. A typical line-spread function of the HVS is shown in Fig. 3.11 [4]. It is observed that the (infinitely) thin light is seen by the HVS as having a finite width. This might be explained as the limitation of a physical system. However, another strange property is observed in Fig. 3.11. At about 0.08^0, we observe negative brightness (which apparently should have been zero). This property creates a special interaction between the luminance from an object and its surrounding area, which is known as the Mach band effect [3]. Figure 3.12 shows vertical strips of increasing luminances. However, the brightness (*i.e.*, the perceived luminance) is not strictly a monotonic function. There is overshoot and undershoot at the edges (see Fig. 3.12(c) and Fig 3.12(d).

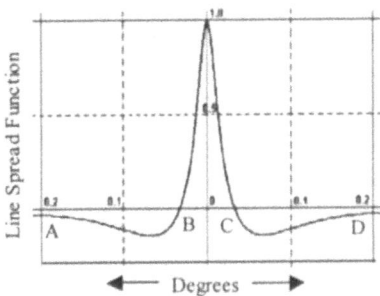

Figure 3.11. Line spread function of the human eye.

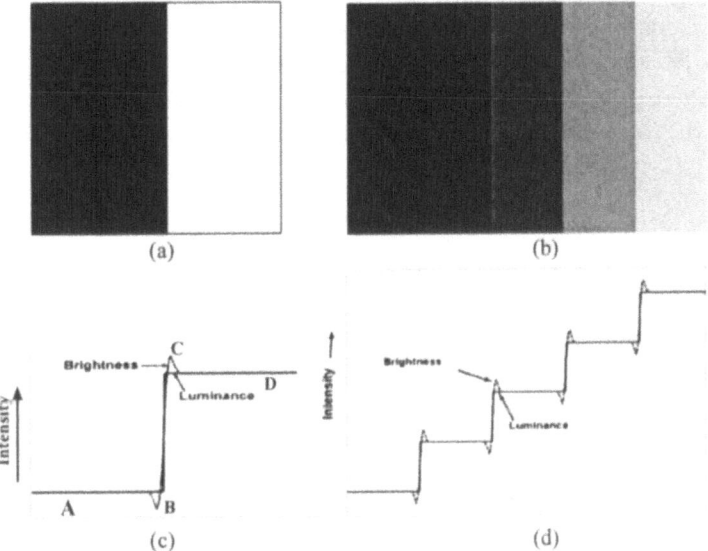

Figure 3.12: Mach band effect. (a) the unit step function, (b) the stair-case function, (c) brightness corresponding to image (a), (d) brightness corresponding to image (b).

■ Example 5.1

In this example, the Mach band effect at Fig. 3.12 (c) is justified from the line-spread function of Fig. 3.11. The line-spread function (*i.e.*, the impulse response in one spatial dimension) of the HVS is shown in Fig. 3.11. The input shown in Fig. 3.12(a) can be considered as the step input. Assuming that the HVS is linear and time invariant, the step response of the system can be obtained by integrating the impulse response of the system. If we integrate the line spread function in Fig. 3.11, a function similar to Fig. 3.12(c) is obtained. Note that points A, B, C, and D in the two figures roughly correspond to each other. ■

3.2.4 HVS Model

So far we have discussed several properties of the HVS. In order to design an efficient system for image processing or analysis, the development of a model for the HVS is very important. A few assumptions are generally made to simplify the model. First, it is assumed that HVS is a linear system, which is generally true for low contrast images. Second, it may be assumed that the HVS is isotropic in spatial domain, *i.e.,* eyes do not have any preferential direction (towards horizontal, vertical or diagonal). Although it has been found that the sensitivity of eyes in the diagonal direction is less than that in the horizontal or vertical direction, this assumption is used in most models.

The HVS can be considered a system with several subsystems. First, when light enters the eye, the pupil acts as an aperture that is equivalent to a lowpass filter. Next, the spectral response of the eye (that is the luminous efficiency function $V(\lambda)$) is applied on the incoming light, and luminance of an image is obtained. The nonlinear response of the rods and cones, and the modulation transfer function, provide the contrast and lateral inhibition.

3.3 COLOR REPRESENTATION

The study of color is important in the design and development of color vision systems. The use of color is not only more pleasing, it also enables us to grasp quickly more information. Although we can differentiate only about a hundred gray levels, we can easily differentiate thousands of colors. There are three main perceptual attributes of colors: brightness, hue, and saturation. As previously mentioned, *brightness* is the perceived luminance. The *hue* is an attribute we commonly describe as blue, red, yellow, *etc.* *Saturation* is our impression of how different the color is from an achromatic (white or gray) color. The pastel colors are of low saturation, and spectral colors are of high saturation. Supersaturated colors are more saturated than the purest spectral colors. A perceptual representation of color space is shown in Fig. 3.13. Brightness varies along the vertical axis, hue varies along the circumference, and saturation varies along the radial axis. A pictorial explanation of these three attributes is provided in the Chapter 3 supplement included in the CD.

3.3.1 Three-Receptor Model

There are several thousands natural colors that can be distinguished by the HVS. It is difficult to design a system that will be able to individually display such a large number of colors. Fortunately, the special properties of the HVS make it possible for engineers to design a simplified color display system. It has been long known that any color can be reproduced by mixing an appropriate set of three primary colors. Subsequently, it has been discovered that there are three different types of cone cells in the human retina. When light falls on the retina, it excites the cone cells. The excitation of different types of cone cells determines the color seen by the observer.

The absorption spectra of three types of cone cells, $S_R(\lambda)$, $S_G(\lambda)$, and $S_B(\lambda)$ with respect to the incident wavelength are shown in Fig. 3.14. If the spectral distribution of the incoming color light is $C(\lambda)$, the excitation of the three types of cone cells can be calculated as:

$$\alpha_p(C) = \int S_p(\lambda)C(\lambda)d\lambda, \qquad p \in \{R, G, B\} \qquad (3.5)$$

Note that if the two different spectral distributions $C_1(\lambda)$ and $C_2(\lambda)$ produce identical $\{\alpha_R, \alpha_G, \alpha_B\}$, the two colors will be perceived as identical. Hence, two colors that look identical can have different spectral distributions.

Not all humans, however, are so lucky – many people are color blind. Monochromatic people have rod cells and one type of cone cells. These people are completely color blind. Dichromatic people have only two types of cones (*i.e.,* one cone type is absent). They can see some colors perfectly, and misjudge other colors (depending on which type of cone cells is absent).

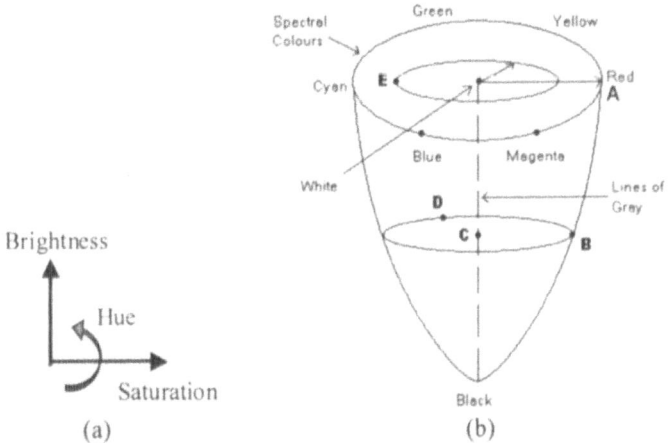

Figure 3.13. Perceptual representation of the color space. a) hue, brightness, and saturation coordinates, b) the corresponding 3-D color space. Various colors are shown in Fig. (b) with uppercase letters. A: highly saturated bright red, B: highly saturated moderate bright red, C: gray, D: highly saturated moderate bright red, E: bright blue with moderate saturation.

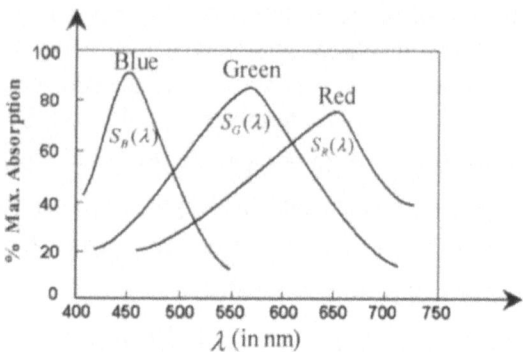

Figure 3.14. Typical absorption spectra of three types of cones in the human retina.

3.3.2 Color Matching

Most color reproduction systems exploit the three-receptors model of color vision. In order to implement such a system, one must know the proportion of the primary colors to be mixed to reproduce a given color. The science of color measurement is known as *colorimetry*. Some of the laws employed for color matching are [5]:

1. Any color can be matched by mixing at most three colored lights.
2. The luminance of a color mixture is equal to the sum of the luminance of its components.
3. Color Addition: If colors A and B match with colors C and D, respectively, then color (A+B) matches color (C+D).
4. Color Subtraction: If color (A+B) matches color (C+D), and color A matches color B, then color B matches color C.

There are different ways to perform color matching experiment. A typical approach is to set-up two screens side by side. The test color is projected on one screen. Controlled R, G, and B lights are projected on the second screen such that the colors corresponding to the two screens are identical. The intensities of the R, G, and B channels corresponding to the second screen are the amount of color primaries needed to reproduce the test color.

Figure 3.15 shows a few composite colors by mixing the three primaries in equal amount. It is observed that mixtures of (blue, red), (blue, green) and (red, green) produce the colors magenta, cyan and yellow, respectively. When all red, green and blue are mixed in equal proportion, white color is obtained.

In the experiment shown in Fig. 3.15 when different primaries are added, they excite the cone receptors differently, and we see a wide range of colors. This is known as additive color mixing. However, when paints of two different colors are mixed, we observe a different composite color. Figure 3.16 shows an experiment mixing yellow and blue paint. It can be observed that the composite color is green instead of white as seen in Fig. 3.15. The reason is explained in Fig. 3.16(c). Yellow paint is seen as yellow because it reflects lights of wavelength in the range 500-600 nm, and absorbs all other wavelengths. On the other hand, blue paint is blue because it reflects lights of wavelength in the range 425-525 nm, and absorbs all other wavelengths. When yellow and blue paints are mixed, only a narrow band of wavelengths (500-530 nm) are reflected and all other wavelengths are absorbed. Hence, we see the green color. Note that most display devices employ additive color mixing to reproduce different colors, while the printing industry employs subtractive color mixing to reproduce colors. Figure 3.16(b) shows a few other colors obtained by mixing different proportions of cyan, magenta and yellow dyes.

3.3.3 Tristimulus Value

The primary sources recommended by CIE (Commission Internationale de l'Eclairage - the international commission on color standards) are three monochromatic colors at wavelength 700 nm (red), 546.1 nm (green), and 435.8 nm (blue). Let the amount of k-*th* primary needed to produce a color C, and reference white color be denoted by β_k, and ω_k, respectively. Then β_k / ω_k is called the *tristimulus* values of color C. The *tristimulus value* of a color is the relative amount of primaries required to match the color.

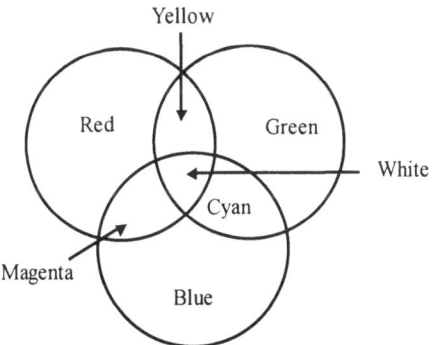

Figure 3.15. Primary colors can be added to obtain different composite colors. The corresponding color figure is included in "Chapter 3 supplement" in the CD.

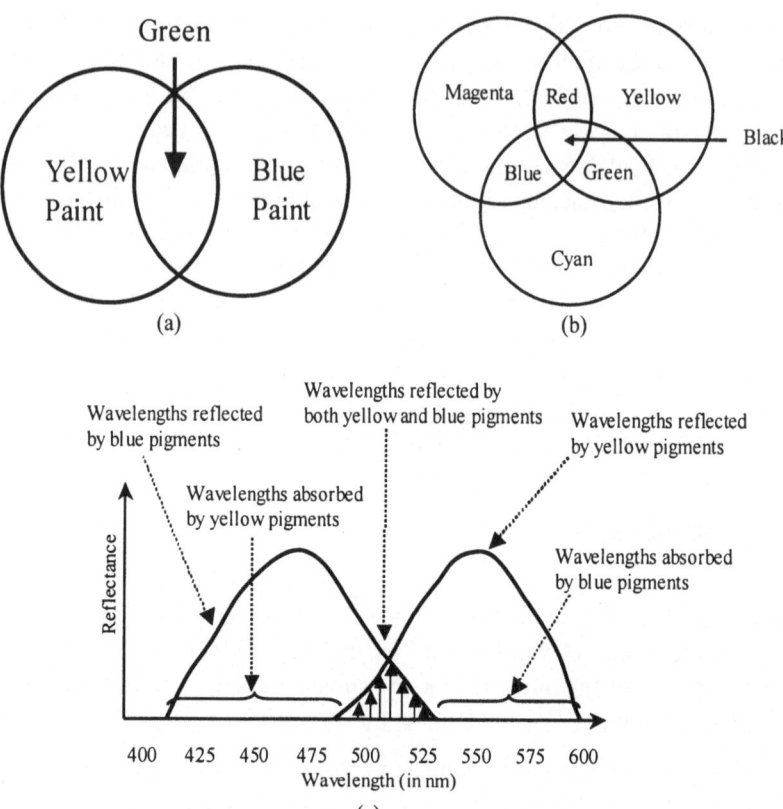

Figure 3.16. Subtractive color mixing. a) mixture of yellow and blue paint produces green color, b) mixture of cyan, magenta, and yellow colors, c) composite color is the difference between two added colors. The corresponding color figure is included in "Chapter 3 supplement" in the CD.

Figure 3.17(a) shows the necessary amounts of three primaries to match all the wavelengths of the visible spectrum. Note that some colors have to be added in negative amounts in order to produce certain colors. In other words, these primary sources do not yield a full gamut of reproducible colors. In fact, no practical set of three primaries has been found that can produce all colors. This has led to the development of many color coordinate systems with their own advantages, and disadvantages. Of particular interest is the CIE $\{X, Y, Z\}$ system with hypothetical primary sources such that all the spectral tristimulus values are positive. Here, X, Y, and Z roughly correspond to supersaturated red, green, and blue, respectively. Figure 3.17(b) shows the corresponding amount of three primaries needed to match all the wavelengths of the visible spectrum [2].

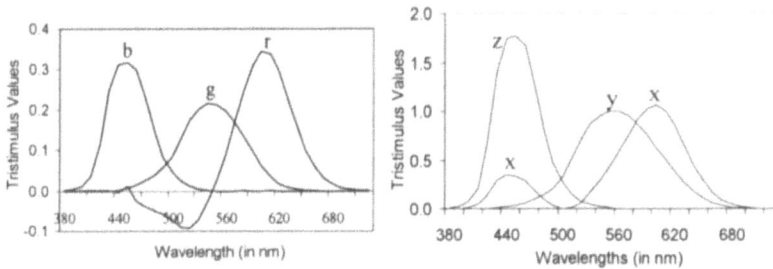

Figure 3.17. Tristimulus curves for a) CIE spectral primary system, and b) CIE {X,Y,Z} system.

3.3.4 Chromaticity Diagram

As a further convenience, colors are often specified by their chromaticity. Chromaticity is expressed in terms of chromaticity coordinates. For {X,Y,Z} system, the chromaticity coordinates can be represented as:

$$x = \frac{X}{X+Y+Z}$$
$$y = \frac{Y}{X+Y+Z} \tag{3.6}$$
$$z = \frac{Z}{X+Y+Z} = 1-x-y$$

Thus, x, y, z represent the proportions of the X primary, Y primary, and Z primary respectively in a given color mixture. If x is large (i.e., close to 1), the color mixture contains a significant portion of (supersaturated) red light, and may appear as orange, red, or reddish-purple. If both x and y are small, the primary Z component is dominant, and the mixture will appear as blue, violet, or purple. The chromaticity diagram of CIE {X, Y, Z} system is shown in Fig. 3.18. The outer bounded curve shows the chromaticity of monochromatic colors at different wavelengths. These are the deepest and most intense colors that are seldom found in natural images. As we move towards the center, lighter colors, such as, pink, light green, and pale blue appear. The central region represents the unsaturated colors including the white color.

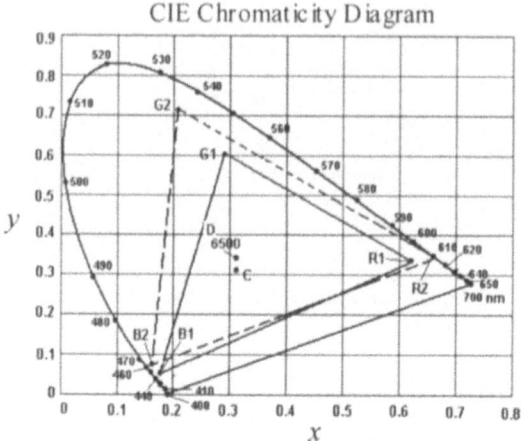

Figure 3.18. The 1931 CIE chromaticity diagram. $\{R_1, G_1, B_1\}$ and $\{R_2, G_2, B_2\}$ are the locations of the red, green and blue color television primaries for PAL system I and NTSC system, respectively [2].

3.3.5 Color Models and Transformation of Primaries

We have so far considered the $\{R, G, B\}$, and $\{X, Y, Z\}$ color systems. Although these systems are capable of representing a large number of colors, several other color systems are also used in practice. Different color systems have their own advantages and disadvantages, and a particular system can be chosen depending on the application. This is somewhat comparable to signal transforms where a signal can be represented in several transform coordinates. However, one transform may provide a superior representation over other transforms for a given application.

It is important to devise a procedure for transforming one color system to another because a single application may use two different color representations of the same visual signal. Figure 3.19 shows how this can be done for transforming the $\{R, G, B\}$ system to the $\{X, Y, Z\}$ system using a geometrical or vector representation. Note that the R, G, and B (or, X, Y, and Z) coordinates need not be orthogonal in the three-dimensional space. However, they must specify three unique directions in the space. It can be shown that the $\{X, Y, Z\}$ can be written in a matrix form as follows:

$$\begin{bmatrix} X \\ Y \\ Z \end{bmatrix} = \begin{bmatrix} X_R & X_G & X_B \\ Y_R & Y_G & Y_B \\ Z_R & Z_G & Z_B \end{bmatrix} \begin{bmatrix} R \\ G \\ B \end{bmatrix} \qquad (3.7)$$

The matrix

$$M_{RGB\to XYZ} = \begin{bmatrix} X_R & X_G & X_B \\ Y_R & Y_G & Y_B \\ Z_R & Z_G & Z_B \end{bmatrix} \tag{3.8}$$

is termed as transformation matrix. Note that $M^T{}_{RGB\to XYZ} = M_{XYZ\to RGB}$.

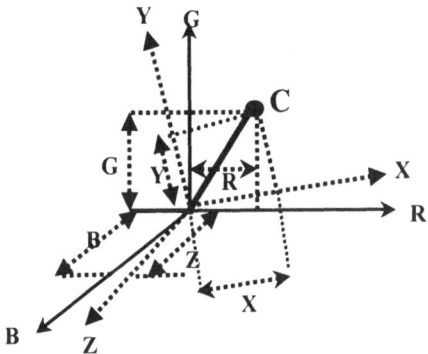

Figure 3.19. Geometrical representation of a color C in $\{R, G, B\}$ and $\{X, Y, Z\}$ space.

The coefficients of the matrix $M_{XYZ\to RGB}$ can be calculated as follows. We note that $X = X_R.R$ when $G = B = 0$. So, project one unit of R-vector on the $\{X, Y, Z\}$ coordinates. The projections on the $\{X, Y, Z\}$ coordinates will provide the values of X_R, Y_R, Z_R. Similarly, the projection of a unit G-vector on the $\{X, Y, Z\}$ coordinates will provide the values of X_G, Y_G, Z_G. The coefficients X_B, Y_B, Z_B can be obtained by projecting a unit B-vector on the $\{X, Y, Z\}$ coordinates. Table 3.1 shows the transformation matrices corresponding to several color coordinates.

A brief introduction to the color systems that are widely used in practice will now be presented.

3.3.5.1 NTSC Receiver Primary

The National Television Systems Committee's (NTSC) receiver primary system (R_N, G_N, B_N) is the standard for television receivers in North America. It has adopted three phosphor primaries that glow in the red, green and blue regions of the visible spectrum. Since the apparent color of an object is a function of illumination, the true color of an object is revealed only under ideal white light, which is difficult to produce. Three reference white sources are generally used in laboratory experiments. Illuminant A corresponds to a tungsten filament lamp, illuminant B corresponds to midday sunlight, and illuminant C corresponds to typical daylight. For

NTSC reference white (which was chosen as the illuminant C), $R_N = G_N = B_N = 1$. The relationship between this color system and the $\{X, Y, Z\}$ system is shown in Table 3.1.

Table 3.1. A few selected color coordinate systems.

Color coordinate systems	Definition/Transformation Matrix	Comments
CIE spectral primary system: $\{R, G, B\}$	Monochromatic primary sources, red=700 nm, green =546.1 nm, blue=435.8 nm.	Reference white has flat spectrum and R=G=B=1.
CIE $\{X, Y, Z\}$ system. Y=luminance	$\begin{bmatrix} X \\ Y \\ Z \end{bmatrix} = \begin{bmatrix} 0.490 & 0.310 & 0.200 \\ 0.177 & 0.813 & 0.011 \\ 0.000 & 0.010 & 0.990 \end{bmatrix} \begin{bmatrix} R \\ G \\ B \end{bmatrix}$	All tristimulus values are positive.
CIE uniform chromaticity scale (UCS): U, V, W	$\begin{bmatrix} U \\ V \\ W \end{bmatrix} = \begin{bmatrix} 0.67 & 0 & 0 \\ 0 & 1 & 0 \\ -0.5 & 1.5 & 0.5 \end{bmatrix} \begin{bmatrix} X \\ Y \\ Z \end{bmatrix}$	Mac Adam ellipses are almost circle.
NTSC receiver primary system R_N, G_N, B_N	$\begin{bmatrix} R_N \\ G_N \\ B_N \end{bmatrix} = \begin{bmatrix} 1.910 & -0.533 & -0.288 \\ -0.985 & 2.000 & -0.028 \\ 0.058 & -0.118 & 0.896 \end{bmatrix} \begin{bmatrix} X \\ Y \\ Z \end{bmatrix}$	Linear transformation of $\{X, Y, Z\}$ is based on TV phosphor primaries.
NTSC trans. system: Y= luminance I, Q= chrominance	$\begin{bmatrix} Y \\ I \\ Q \end{bmatrix} = \begin{bmatrix} 0.299 & 0.587 & 0.114 \\ 0.596 & -0.274 & -0.322 \\ 0.211 & -0.523 & 0.312 \end{bmatrix} \begin{bmatrix} R_N \\ G_N \\ B_N \end{bmatrix}$	Used for TV transmission in North America.

3.3.5.2 NTSC Transmission System

Although NTSC employs the $\{R_N, G_N, B_N\}$ color system for television receivers, the transmission system employs the $\{Y, I, Q\}$ system to facilitate the use of monochrome television channels without increasing the bandwidth. The Y coordinate is the luminance of the color, and simultaneously acts as the monochrome channel in a monochrome receiver. The I and Q channels jointly represent the hue and saturation of the color. Note that the bandwidth required for I and Q is much smaller than that of the Y channel [6]. Hence, the $\{Y, I, Q\}$ color channels provide a better bandwidth utilization compared to $\{R_N, G_N, B_N\}$ color channels. The transformation matrices between $\{R_N, G_N, B_N\}$ and other color coordinates are shown in Table 3.2.

The chromaticity of PAL and NTSC color primaries is shown in Fig. 3.18. On one hand, the PAL system employs the primaries $\{R_1, G_1, B_1\}$ with the $\{x, y\}$ chromaticity coordinates $\{(0.64, 0.33), (0.29, 0.60), (0.15, 0.06)\}$. On the other hand, the NTSC system employs the primaries $\{R_2, G_2, B_2\}$ with the chromaticity coordinates $\{(0.67,0.33),(0.21,0.71),(0.14,0.08)\}$. The camera output signal of the PAL system is normalized with respect to the reference white D_{6500} whose chromaticity coordinate is $(0.31,0.33)$, while the output of an NTSC camera is normalized with respect to the reference white C whose chromaticity coordinate is $(0.31,0.32)$.

Table 3.2. Transformations from NTSC receiver primaries $\{R_N, G_N, B_N\}$ to different coordinate systems.

Color System	Output Matrix	Transformation Matrix
CIE Spectral primary	$\begin{bmatrix} R \\ G \\ B \end{bmatrix}$	$\begin{bmatrix} 1.167 & -0.146 & -0.151 \\ 0.114 & 0.753 & 0.159 \\ -0.001 & 0.059 & 1.128 \end{bmatrix}$
NTSC Transmission System	$\begin{bmatrix} Y \\ I \\ Q \end{bmatrix}$	$\begin{bmatrix} 0.299 & 0.587 & 0.114 \\ 0.596 & -0.274 & -0.322 \\ 0.211 & -0.523 & 0.312 \end{bmatrix}$
CIE UCS tristimulus system	$\begin{bmatrix} U \\ V \\ W \end{bmatrix}$	$\begin{bmatrix} 0.405 & 0.116 & 0.133 \\ 0.299 & 0.587 & 0.144 \\ 0.145 & 0.827 & 0.627 \end{bmatrix}$
CIE X,Y,Z system	$\begin{bmatrix} X \\ Y \\ Z \end{bmatrix}$	$\begin{bmatrix} 0.607 & 0.174 & 0.201 \\ 0.299 & 0.587 & 0.114 \\ 0.000 & 0.066 & 1.117 \end{bmatrix}$

The chromaticity diagram is very useful for color mixing experiments. Note that the chromaticity of any mixture of two primaries lies on a straight line joining the primaries on a chromaticity diagram. Consider a straight-line connecting G and B. The colors corresponding to various points on the straight-line will represent various shades of G and B. Next, consider a straight-line connecting C (the reference white for NTSC system) and any monochrome color X. The colors corresponding to various points on the straight-line will represent the hue X with different saturation values. If the point is closer to C, the corresponding color will be less saturated.

We have noted that the TV receiver produces the color by mixing the primaries $\{R_1, G_1, B_1\}$ and $\{R_2, G_2, B_2\}$ for PAL and NTSC systems, respectively. Hence, the chromaticity of a mixture of three primaries is bounded by the triangle whose vertices are the primaries. It is observed in Fig. 3.18 that the PAL or, NTSC system provides only about half of all possible reproducible colors (compare the triangle areas with the area bound by the outer curve). However, this does not degrade the quality of the TV video as "real world" scenes contain mostly unsaturated colors represented by the central region.

■ Example 5.2

The NTSC receiver primary *magenta* corresponds to the tristimulus values $R_N = B_N = 1, G_N = 0$. The tristimulus and chromaticity values of magenta are determined in i) CIE spectral primary and ii) $\{X,Y,Z\}$ coordinate systems.

CIE Spectral primaries

The tristimulus values of magenta can be obtained using the transformation matrix (corresponding to CIE primary system) shown in Table 3.2.

$$R = 1.167R_N - 0.146G_N - 0.151B_N = 1.167 - 0.151 = 1.016$$

$$G = 0.114R_N + 0.753G_N + 0.159B_N = 0.114 + 0.159 = 0.273$$

$$B = -0.001R_N + 0.059G_N + 1.128B_N = -0.001 + 1.128 = 1.127$$

The chromaticity values can be calculated from the tristimulus values as follows:

$$r = \frac{R}{R+G+B} = \frac{1.016}{1.016 + 0.283 + 1.127} = 0.42 , \; g = \frac{0.273}{2.416} = 0.11, \text{ and}$$
$$b = 1 - r - g = 0.47$$

X,Y,Z coordinates

The tristimulus values of magenta can be obtained using the transformation matrix (corresponding to $\{X,Y,Z\}$ system) shown in Table 3.2.

$$X = 0.607R_N + 0.174G_N + 0.201B_N = 0.607 + 0.201 = 0.808$$

$$Y = 0.299R_N + 0.587G_N + 0.114B_N = 0.299 + 0.114 = 0.413$$

$$Z = 0.000R_N + 0.066G_N + 1.117B_N = 1.117$$

The chromaticity values corresponding to the $\{X,Y,Z\}$ coordinates can be calculated as follows:

$$x = \frac{X}{X+Y+Z} = \frac{0.808}{0.808 + 0.413 + 1.117} = \frac{0.808}{2.338} = 0.34 , \; y = \frac{0.413}{2.338} = 0.18,$$

$$z = 1 - x - y = 0.48 \quad ■$$

3.3.5.3 1960 CIE-UCS Color coordinates

Color spaces such as $\{R, G, B\}$ and $\{Y, I, Q\}$ represent colors efficiently, and are widely used in practice. However, these color spaces have a major disadvantage. The Euclidian distance between two color points in these spaces may not correspond to the perceptual distance between the two colors. Figure 3.20 demonstrates this with a simple MATLAB example. Figure 3.20(a) displays two concentric green circles (the color figures are included in the CD). The color corresponding to the inner circle is represented by r=0.2, g=0.6, and b=0.2 (normalized values) in {R,G,B} color space. The corresponding pixel values, with 8 bit representation, are r=51, g=153, and b=51. The color corresponding to the outer circle is represented by r=0.2, g=0.62, and b=0.2 with pixel values r=51, g=158, and b=51. Note that the two colors have a Euclidian distance of 0.2 (=0.62-0.6). The distance is very small, and hence the colors are perceptually very similar, but can still be distinguished. Figure 3.20(b) displays two concentric blue circles. The color corresponding to the inner circle is represented by r=0.2, g=0.2, and b=0.6 in {R,G,B} color space. The corresponding pixel values, with 8 bit representation, are r=51, g=51, and b=153. The color corresponding to the outer circle is represented by r=0.2, g=0.2, and b=0.62 with pixel values r=51, g=51, and b=158. Note that the two colors also have a Euclidian distance of 0.2 (=0.62-0.6). However, the colors are perceptually indistinguishable (the two circles seem to have the same color).

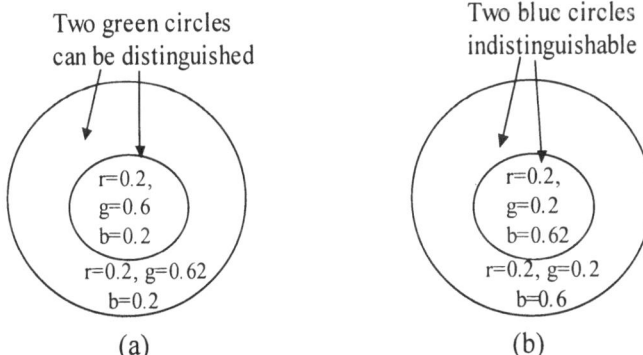

(a)   (b)

Figure 3.20. Perceptual distance experiment using two concentric circles in R,G,B color space. a) Inner circle: r=0.2, g=0.6, b=0.2; outer circle: r==0.2, g=0.62, b=0.2, b) Inner circle: r=0.2, g=0.2, b=0.6; outer circle: r==0.2, g=0.2, b=0.62. The values of r, g, and b are normalized. The Euclidian distance in both cases is 0.02. The corresponding color figure is included in "Chapter 3 supplement" in the CD.

In many applications, such as color-based feature extraction, a uniform color space may provide a superior performance. The *1960 CIE Uniform Chromaticity Scale* (UCS) diagram was derived from the 1931 CIE chromaticity diagram by stretching it unevenly such that the chromaticity distance corresponded more closely to the perceptible difference. The three components of a color in UCS color space is typically denoted by {U,V,W}. The corresponding transformation matrix is provided in row-4 of Table 3.2.

3.3.5.4 CMY Model

Most display devices, such as television receivers and computer monitors, employ additive color mixing. Hence, the three color primaries $\{R, G, B\}$ are almost exclusively used for display devices since it is easier to reproduce these colors on phosphor screens. However, the printing application employs the subtractive color mixing, and hence the $\{R, G, B\}$ color model is not suitable. Instead, the three primaries used for color printing are cyan (=white-red), magenta (white-red), and yellow (white-cyan), which was shown in Fig. 3.16(b). The CMY model can be expressed as ("1" represents the white color):

$$\begin{bmatrix} C \\ M \\ Y \end{bmatrix} = \begin{bmatrix} 1-R \\ 1-G \\ 1-B \end{bmatrix}$$

It can be shown that a large number of colors can be reproduced by mixing these primary dyes, and the color model is known as CMY. In addition, an additional color channel K has been added to reproduce black color since most printing is in black.

3.4 TEMPORAL PROPERTIES OF VISION

It has been observed in section 2 that the sensitivity of the HVS is low for high spatial frequencies. This property has been exploited in various image processing applications. When considering video, a different problem is encountered. A video camera generally shoots a video by shooting images (or video frames) in quick succession. The display device (e.g., television, computer monitor) displays the individual frames at the same rate. If the rate of displaying the individual frames is above a certain threshold (i.e., critical frame rate), the HVS cannot distinguish the individual frames, and the video appears continuous. This critical frame rate is important in the design of video capturing and display devices. We note that this critical frame rate is dependent on several factors such as the viewing conditions, and the video content.

A large number of studies have been made to determine the critical frame rate and its dependence on the test stimulus. A simple experimental setup is explained here. In this experiment, a point source of light, spatially constant, is made to fluctuate about a mean value of luminance L according to the following equation:

$$L + \Delta L \cos(2\pi f t)$$

where ΔL is the peak amplitude of the fluctuation , and f is its frequency (in Hz). If ΔL is large enough, and f is not too high, such a source appears to flicker. For any given value of f, the value of ΔL that produces a threshold sensation of flicker is determined experimentally. Let this value be denoted by ΔL_{T_t}. The ratio $L/\Delta L_T$ is known as the temporal contrast sensitivity of the HVS. Figure 3.21 shows the temporal contrast sensitivity for three illumination levels as a function of f. In this experiment, the illumination has been expressed in *trolands*, which is the unit of retinal illuminance. As an approximation, 100 trolands is equivalent to 10 cd/m^2 on the display [2].

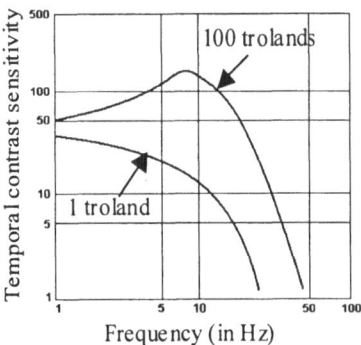

Figure 3.21. Flicker sensitivity of the eye for a 2^0 field at
1 troland and 100 trolands illuminance levels.

It is observed in Fig. 3.21 that the eye is more sensitive to flicker at high luminance. Moreover, the eye acts as a lowpass filter for temporal frequencies. The flicker sensitivity is negligible above 50/60 Hz. This property is the basis of designing television systems, and computer displays that employs at least 50 Hz field or frame rate.

REFERENCES

1. R. C. Gonzalez and Richard E. Woods, *Digital Image Processing*, Addison Wesley, 1993.
2. A. K. Jain, *Fundamentals of Digital Image Processing*, Prentice Hall, 1989.
3. T. N. Cornsweet, *Visual Perception*, Academic Press, New York, 1970.
4. D. E. Pearson, *Transmission and Display of Pictorial Information*, Pentech Press, 1975.
5. H. G. Grassman, "Theory of compound colors," *Philosophic Magazine*, Vol. 4, No. 7, pp. 254-264, 1954.
6. A. N. Netravali and B. G. Haskell, *Digital Pictures: Representation, Compression and Standards*, Plenum Press, New York, 1994.

QUESTIONS

1. Explain the importance of studying the HVS model for designing a multimedia system.
2. Explain the image formation process on the retina. How many different types of receptors are present in retina?
3. What is relative luminous efficiency of the HVS?
4. Define luminance and brightness. Even if two regions of an image have identical luminance values, they may not have same brightness. Why?
5. Define spatial frequencies and modulation transfer function. What is the sensitivity of the HVS to different spatial frequencies?
6. Consider a sinusoidal grating being viewed by an observer on a computer monitor. The distance between the observer and the monitor is X, and the grating has a frequency of 2 cycles/degree. If the observer moves forward (towards the monitor) by X/4, what will be the new grating frequency? Assume a narrow angle vision.
7. Explain the frequency sensitivity of the HVS, and the Mach band effect.
8. Why do we need vision models? Explain a simple monochrome vision model.
9. What are the three perceptual attributes of color?
10. What are the three primary colors that are used for additive color reproduction? What is the basis of this model from the HVS perspective?
11. What are tristimulus values of a color?
12. Why is there a large number of color spaces? Compare the {R,G,B}, {X,Y,Z}, and {Y,I,Q} color spaces.
13. What is the principle of subtractive color mixing? What are the three primaries typically used? What will be the reproduced color for i) yellow-green mixture, and ii) yellow-red?
14. Which method (subtractive or additive) of color reproduction is used in i) printing, ii) painting, iii) computer monitors, and iv) television monitors?
15. What is a chromaticity diagram? How does it help in color analysis? Which colors are found near the edges of the diagram? Which colors are found near the center?
16. Given a spectral color on the chromaticity diagram, how would you locate the coordinates of the color for different saturation value?
17. Determine the chromaticity values of i) green, ii) cyan, and iii) blue in CIE Spectral Primary coordinate systems.
18. The reference yellow in the NTSC receiver primary system corresponds to $R_N = G_N = 1$, $B_N = 0$. Determine the chromaticity coordinates of yellow in the CIE spectral primary system, and the {X,Y,Z} coordinates.
19. Repeat the previous problem for the reference white that corresponds to $R_N = G_N = B_N = 1$.
20. A 2×2 color image is represented in CIE spectral primary system with the following R, G, and B values (in 8-bit representation).
$$R = \begin{bmatrix} 20 & 220 \\ 150 & 80 \end{bmatrix}, \ G = \begin{bmatrix} 20 & 220 \\ 150 & 30 \end{bmatrix}, \ B = \begin{bmatrix} 220 & 20 \\ 120 & 160 \end{bmatrix}$$
    What are the colors of the four pixels? Convert the color image to a gray scale image. What would be the relative intensity of the gray pixels?
21. What is the temporal contrast (or flicker) sensitivity of the HVS? Does the sensitivity depend on the illumination level?

Chapter 4

Multimedia Data Acquisition

The fundamentals of the audio and image data were presented in Chapters 2 and 3. For multimedia data analysis, we need to store the multimedia data and process it. The audio sources (*e.g.*, microphones) or the image sources (*e.g.*, camera) generally produce continuous-time analog signals (*e.g.*, output voltage). In order to store audio or video data into a computer, we must sample and digitize the data. This will convert the continuous data into a stream of numbers. In this chapter, we primarily focus on the principles of sampling and digitization of multimedia data.

4.1 SAMPLING OF AUDIO SIGNALS

Consider the audio signal "test44k" shown in Fig. 4.1. The signal appears to be changing rapidly. For storing the signal, it is physically impossible to determine the value of the signal at each time instance and store it. Hence, we need to sample the signals at regular time intervals. When sampling is done, the value of the signal at times other than the sampling instances are apparently lost forever. The amount of information lost due to sampling will increase with the sampling interval. A critical question is whether it is possible to sample in such a way that the information present in a signal remains intact or, if the information loss is at least tolerable.

Figure 4.2 shows the amplitude spectrum of the audio signal shown in Fig. 4.1. It is observed that the spectrum is negligible above 12 KHz. Since the high frequency components are very small, it can be said that the signal is not changing at a high rate. In this case, it might be possible to sample the signal and recover it completely by interpolation if the sampling frequency is high enough. The required sampling frequency can be determined using Nyquist criterion.

Consider an arbitrary audio signal $g(t)$. The Fourier transform of the signal is defined as

$$G(\omega) = \int_{-\infty}^{\infty} g(t)e^{-j\omega t} dt \qquad (4.1)$$

If $G(\omega) = 0$ for $|\omega| > 2\pi B$, $g(t)$ can be said to be band limited to B Hz. The signal $g(t)$ is sampled uniformly at a rate of f_s samples per second (*i.e.*, sampling interval $T = 1/f_s$ sec). The sampled signal $g_s(t)$ can be represented as

$$g_s(t) = g(t)s(t) = \sum_{k=-\infty}^{\infty} g(kT)\delta(t - kT) \tag{4.2}$$

where $s(t)$ is the sampling function (a train of impulse functions) defined as follows.

$$s(t) = \sum_{k=-\infty}^{\infty} \delta(t - kT)$$

It can be shown that [] the Fourier transform of $g_s(t)$ is

$$G_s(\omega) = \frac{1}{T} \sum_{k=-\infty}^{\infty} G(\omega - 2\pi k f_s) \tag{4.3}$$

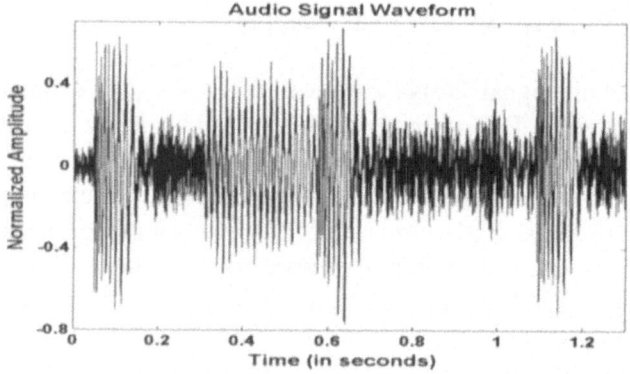

Figure 4.1. The waveform of the audio signal "test44k." The full signal is of 30 second duration of which the first 1.3 seconds is shown here.

Note that $G_s(\omega)$ is a periodic extension $G(\omega)$ (as shown in Fig 4.3). The two consecutive periods of $G_s(\omega)$ will not overlap if $f_s > 2B$. Hence, $G(\omega)$ can be obtained exactly from $G_s(\omega)$ by lowpass filtering if the following condition is satisfied.

$$f_s > 2B \tag{4.4}$$

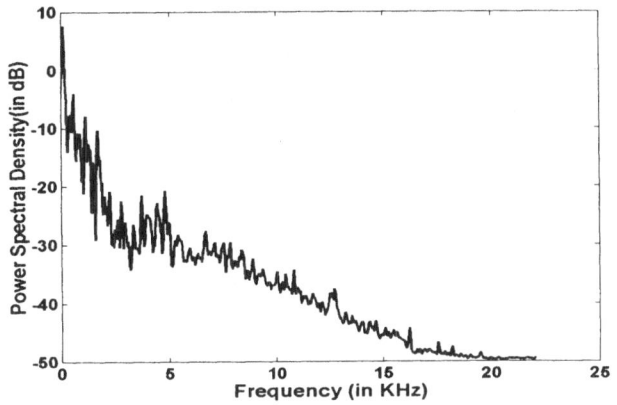

Figure 4.2. The amplitude spectrum of the audio signal shown in Fig. 4.1.

The above condition is known as the Nyquist criterion [3]. In summary, the Nyquist criterion says that a signal $g(t)$ with a bandwidth B Hz can be reconstructed exactly from $g(kT)$ if the sampling frequency $f_s (=1/T) > 2B$. The minimum sampling frequency for the perfect reconstruction (in this case it is $2B$ samples/sec) is known as the Nyquist frequency. A graphical explanation of the Nyquist criterion is shown in Fig. 4.3.

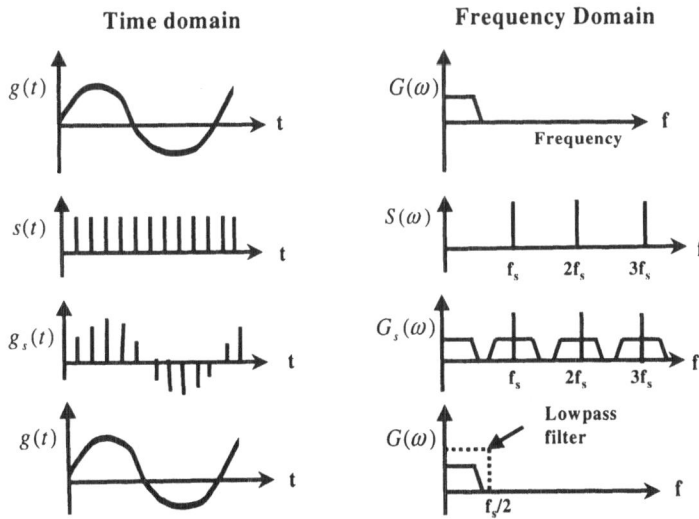

Figure 4.3 Time domain and frequency domain signals illustrate the process of bandlimited sampling. $g(t)$: input signal, f_s: sampling frequency, $s(t)$: sampling waveform, $g_s(t)$: sampled signal. $G(\omega)$, $S(\omega)$, and $G_s(\omega)$ are the Fourier transforms of $g(t)$, $s(t)$, and $g_s(t)$, respectively.

In order to reconstruct the original signal, the sampled signal must be passed through an ideal lowpass with a cut-off frequency of $f_s/2$. This will remove the high frequency imaging components from the $G_s(\omega)$, and thus the original signal with Fourier spectrum $G(\omega)$ is obtained.

Alternatively, the audio signal can be reconstructed directly from the samples using the following time-domain interpolation formula [3]:

$$g(t) = \sum_{k=-\infty}^{\infty} g(kT) \frac{\sin(f_s t - k)\pi}{(f_s t - k)\pi} \qquad (4.5)$$

Eq. (4.5) states that the signal values between the non-sampling instances can be calculated exactly using a weighted sum of all sampled values. Note that both frequency and time-domain approaches, although apparently different, are equivalent since the *sinc* function at the right side of Eq. (4.5) is the impulse response of the ideal lowpass filter.

■ **Example 4.1**

Consider the following audio signal with a single sinusoidal tone at 4.5 KHz.

$$g(t) = 5\cos(2\pi * 4500t)$$

Sample the signal at a rate of i) 8000 samples/sec, and ii) 10000 samples/sec. Reconstruct the signal by passing it through an ideal lowpass filter with cut-off frequencies of one-half of the sampling frequency. Assume that the passband gains of the two filters are i) 1/8000, and ii) 1/10000. Determine the reconstructed audio signal in both cases.

The Fourier transform of the input signal can be expressed as

$$G(\omega) = 5\pi \left[\delta(\omega - 9000\pi) + \delta(\omega + 9000\pi) \right]$$

Case 1:

Using Eq. (4.3), the Fourier transform of the sampled signal can be expressed as

$$G_s(\omega) = \frac{1}{T_1} \sum_{n=-\infty}^{\infty} G(\omega - 2\pi n f_s) = 8000 \sum_{n=-\infty}^{\infty} G(\omega - 16000\pi * n) \quad [\because 1/T_1 = 8000]$$

$$= 8000 * \left[\ldots\ldots + G(\omega - 16000\pi) + G(\omega) + G(\omega + 16000\pi) + \ldots\ldots \right]$$

$$= 40000\pi * \left[\ldots + \delta(\omega - 9000\pi) + \delta(\omega - 7000\pi) + \delta(\omega + 7000\pi) + \delta(\omega + 9000\pi) + \ldots \right]$$

The cutoff frequency of the lowpass filter is half of the sampling frequency, i.e., 4000 Hz. Hence, the transfer function of the filter is

$$H_1(\omega) = \begin{cases} 1/8000 & |\omega| \leq 8000\pi \\ 0 & \text{otherwise} \end{cases}$$

When the sampled signal is passed through the lowpass filter, the Fourier transform of the output signal will be:

$$G_1(\omega) = G_s(\omega)H_1(\omega)$$
$$= 5\pi\left[\delta(\omega - 7000\pi) + \delta(\omega + 7000\pi)\right]$$

Hence, the output signal $g_1(t) = \Im^{-1}\{G_1(\omega)\} = 5\cos(2\pi * 3500t)$

Case 2:

The Fourier transform of the sampled signal can be expressed as

$$G_s(\omega) = \frac{1}{T_2}\sum_{n=-\infty}^{\infty}G(\omega - 2\pi nf_s) = 10000\sum_{n=-\infty}^{\infty}G(\omega - 20000\pi * n) \quad [\because 1/T_2 = 10000]$$
$$= 10000 * [........ + G(\omega - 20000\pi) + G(\omega) + G(\omega + 20000\pi) + ...]$$
$$= 50000\pi * [.... + \delta(\omega - 11000\pi) + \delta(\omega - 9000\pi) + \delta(\omega + 9000\pi) + \delta(\omega + 11000\pi) + ...]$$

The cutoff frequency of the lowpass filter is 5000 Hz. The transfer function

$$H_2(\omega) = \begin{cases} 1/10000 & |\omega| \leq 10000\pi \\ 0 & \text{otherwise} \end{cases}$$

When the sampled signal is passed through the lowpass filter, the Fourier transform of the output signal will be:

$$G_2(\omega) = G_s(\omega)H_2(\omega)$$
$$= 5\pi\left[\delta(\omega - 9000\pi) + \delta(\omega + 9000\pi)\right]$$

Hence, the output signal $g_2(t) = \Im^{-1}\{G_2(\omega)\} = 5\cos(2\pi * 4500t)$. ∎

Note that the reconstructed signal is identical to the original signal in Case 2. However, in Case 1, the reconstructed signal is different from the original signal. The original and reconstructed signals are plotted in Fig 4.4. It is observed that the two signals are identical only at the sampling instances. At other time instances, the two signals differ because of the aliasing.

Aliasing

When an arbitrary audio signal is sampled at a rate of f_s samples/sec, the frequency range that can be reconstructed exactly is

$$0 \leq f \leq f_s/2 \tag{4.6}$$

Frequencies greater than $f_s / 2$ will be aliased, and appear as frequencies below $f_s / 2$, which is called the *folding* frequency. A sinusoidal tone with a frequency f_t, with $f_t > f_s / 2$, will appear as a tone with a lower frequency f_a that can be calculated from the following relationship [3]:

$$f_a = f_t - mf_s, \text{ where } m \text{ is an integer such that } |f_a| \le f_s / 2 \qquad (4.7)$$

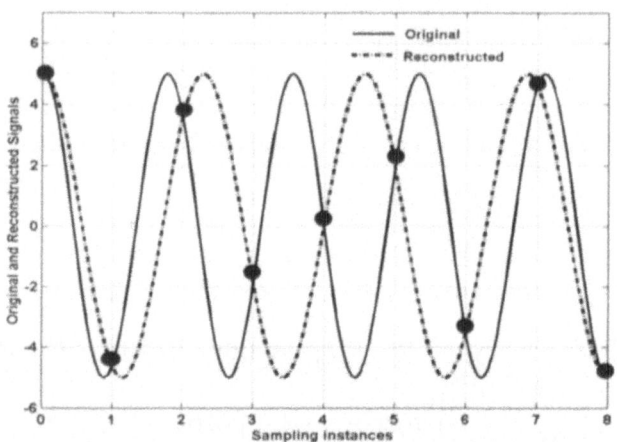

Figure 4.4. Original Signal and reconstructed audio signals (for case 1) in Example 4.1.

If an audio signal is sampled at a rate of 6000/samples/sec, the frequency of a few arbitrary signals and the corresponding reconstructed signals are shown in Table 4.1.

Table 4.1. Frequency of a few selected signals and the corresponding frequency of the reconstructed signals.

Original frequency (in Hz)	$\|f_t - mf_s\|$	Frequency of Reconstructed signals	Comments
500		500	No aliasing
2500		2500	No aliasing
2900		2900	No aliasing
3001	$\|3001 - 1*6000\|$	2900	Aliasing
3500	$\|3500 - 1*6000\|$	2500	Aliasing
10000	$\|10000 - 2*6000\|$	2000	Aliasing
20000	$\|20000 - 3*6000\|$	2000	Aliasing
1000000	$\|1000000 - 167*6000\|$	2000	Aliasing

In practice, the sampling frequency of an audio signal generally depends on the application at hand. Since the audible bandwidth is typically 20 KHz, a sampling frequency of 40 KHz is generally sufficient for high quality audio applications. For audio CD application, the sampling frequency is 44.1 KHz. In this case, a margin of 4.1 KHz makes the design of the anti-aliasing filter easier. For low-end applications, a lower sampling frequency is employed. For example, a sampling frequency of 8 KHz is typically used for telephony systems.

4.2 SAMPLING OF TWO-DIMENSIONAL IMAGES

In the previous section, the sampling of an audio signal was considered. Nyquist sampling theory can also be extended for discretizing the images. Let an image be represented by $i(x, y)$ where x and y are two spatial dimensions, expressed in degrees. The 2-D Fourier transform of the image can be written as

$$I(\omega_h, \omega_v) = \int\limits_{-\infty}^{\infty} \int\limits_{-\infty}^{\infty} i(x, y) e^{-j(\omega_h x + \omega_v y)} dx dy \qquad (4.8)$$

where ω_h and ω_v are the horizontal and vertical spatial frequencies, respectively, and are expressed in *radian/degree*. Note that $\omega_h = 2\pi f_h$ and $\omega_v = 2\pi f_v$ where f_h and f_v are horizontal and vertical frequencies in *cycles/degree* (see section 3.2.3 for a definition). Let the image be considered bandlimited, and the maximum horizontal and vertical frequencies be denoted by f_H and f_V. Assuming a rectangular sampling grid, if the image is sampled at spatial interval of ($\Delta x, \Delta y$), the sampled image can be expressed as

$$i_s(x, y) = i(x, y)s(x, y) = \sum_{n=-\infty}^{\infty} \sum_{n=-\infty}^{\infty} i(m\Delta x, n\Delta y)\delta(x - m\Delta x, y - n\Delta y) \qquad (4.9)$$

where $s(x, y)$ is the 2-D sampling function (shown in Fig. 4.5) and defined as

$$s(x, y) \equiv \sum_{m=-\infty}^{\infty} \sum_{n=-\infty}^{\infty} \delta(x - m\Delta x, y - n\Delta y)$$

Note that the sampling frequencies in the horizontal and vertical directions are $1/\Delta x$, and $1/\Delta y$ (in samples/degree), respectively. Extending the Nyquist criterion to two-dimensions, it can be shown that the image $i(x, y)$ can be obtained exactly if the following conditions are true:

$$\frac{1}{\Delta x} \geq 2f_H \quad \text{and} \quad \frac{1}{\Delta y} \geq 2f_v \tag{4.10}$$

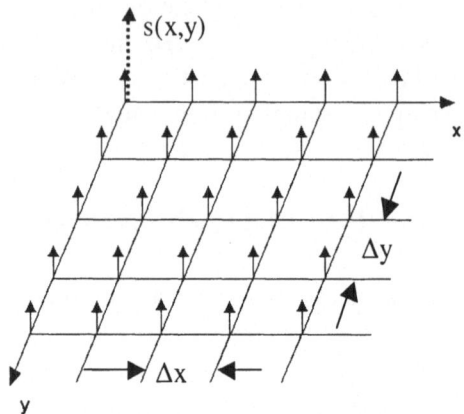

Figure 4.5. Two-dimensional sampling function.

■ Example 4.2

Consider the following image grating with a horizontal and vertical frequency of 4 and 6 cycles/degree, respectively:

$$i(x, y) = 255\cos[2\pi(4x + 6y)] \tag{4.11}$$

Sample the image at a rate of 10 samples/degree in both horizontal and vertical directions. Reconstruct the grating by passing it through an ideal 2-D lowpass filter with the following characteristics:

$$H_{2D}(f_h, f_v) = \begin{cases} 0.01 & -5 \leq f_h, f_v \leq 5 \\ 0 & \text{otherwise} \end{cases} \tag{4.12}$$

Determine the reconstructed grating.

The Nyquist frequencies corresponding to the horizontal and vertical directions are 8 and 12 samples/degree. The actual sampling frequencies at the horizontal and vertical directions are 10 samples/degree. Hence, there will be aliasing in the vertical direction, and no aliasing in the horizontal direction.

The Fourier transform of the input signal is given by

$$I(\omega_h, \omega_v) = 255\pi\left[\delta\left(\omega_h - 4*2\pi, \omega_v - 6*2\pi\right) + \delta\left(\omega_h + 4*2\pi, \omega_v + 6*2\pi\right)\right]$$

Case 1:

Using Eq. (4.3), the Fourier transform of the sampled signal can be expressed as

$$I_s(\omega_h,\omega_v) = \frac{255\pi}{T_1 T_2} \sum_{m=-\infty}^{\infty} \sum_{n=-\infty}^{\infty} I(\omega_h - 20\pi n, \omega - 20\pi m)$$

$$= \frac{255\pi}{0.01} \sum_{m=-\infty}^{\infty} \sum_{n=-\infty}^{\infty} I(\omega_h - 20\pi n, \omega - 20\pi m)$$

The spectrum of the original and sampled image is shown in Fig. 4.6. The circular black dots represent two-dimensional impulse functions. It is observed that the spectrum of the sampled image is periodic. The 2-D filter as specified in Eq. (4.12) is shown as a square in Fig. 4.6(b). The filter will pass through two impulse functions located at (4,-4) and (-4,4) cycles/degree. The filtered spectrum can then be expressed as

$$\hat{I}(\omega_h,\omega_v) = H_{2D}(\omega_h,\omega_v)I_s(\omega_h,\omega_v)$$
$$= 255\pi\left[\delta(\omega_h - 4*2\pi,\omega_v + 4*2\pi) + \delta(\omega_h + 4*2\pi,\omega_v - 4*2\pi)\right]$$

Calculating the inverse Fourier transform of the above equation, we obtain the filtered image as

$$\hat{i}(x,y) = 255\cos[2\pi(4x - 4y)]$$

The filtered image is shown in Fig. 4.7. ∎

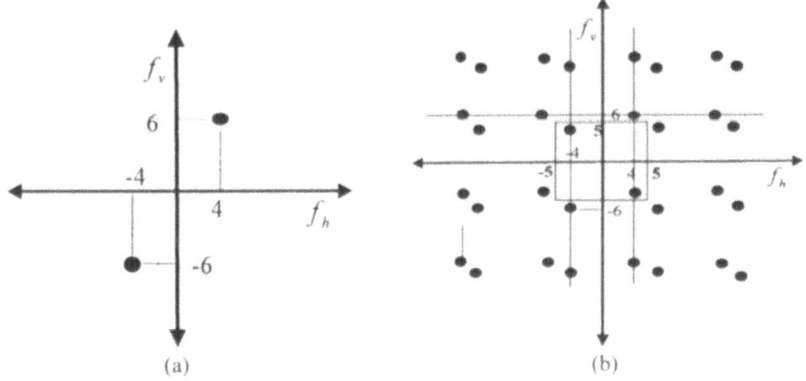

(a)   (b)

Figure 4.6. Fourier spectrum of the a) original image, b) sampled image at 8 samples/degree.

Optimum sampling rate

An important parameter in creating a digital image is the resolution of the image. For example, when a photo is scanned, we have to mention the resolution of the scanned image in *dpi* (dots per inch). In case of web-camera applications (which have become very popular recently), one often has to specify the resolution, such as 640×480, 352×288, and 176×144. A suitable resolution can be determined from Eq. (4.10) and the properties of human visual system discussed in Chapter 3. It was observed in Chapter 3 that the details of an image visible to our eyes depend on the angle the image makes on the retina. It was found that the sensitivity of human eyes is low for frequency components above 20 cycles/degree. Therefore, as a rough approximation, we can assume that the maximum horizontal and vertical frequencies present in an image are 20 cycles/degree, i.e., $f_H = f_V = 20$. The corresponding Nyquist sampling rate will be 40 samples/degree in both directions.

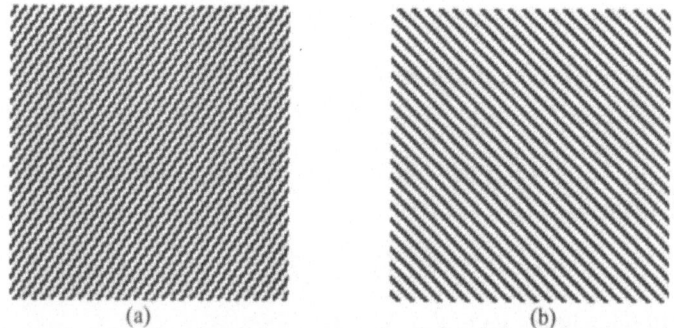

(a)   (b)

Figure 4.7. Example of aliasing in images. a) original image, and b) undersampled image reconstructed by lowpass filtering.

■ **Example 4.3**

A 4"×6" photo is to be scanned. Determine the minimum scanning resolution.

The scanning resolution will depend on the distance between the displayed image and the observer (more specifically the eyes of the observer). Let us assume that the observer will be at a distance of 6 times that of the width of the image (6" side). The digital image will make an angle of approximately $2 * \tan^{-1}((W/2)/6W)$ or 9.5° in the horizontal direction (see Fig. 4.8), and $2 * \tan^{-1}(((4/6) * W/2)/6W)$ or 6.4° in the vertical direction (assuming a landscape photo) at the eyes. Assuming a sampling rate of 40 samples/degree, the digital image should have 380 and 256 pixels in the horizontal and vertical directions, respectively. Since the photo size is 4"×6", a minimum sampling rate of 64 *dpi* is required. ■

4.3 ANTI-ALIASING FILTERS

In practice, a signal is not truly band limited, although the higher frequency components are generally very small. These components introduce alias components, which can be considered as distortion in the sampled signal. To ensure that the audio or image signal is sufficiently band limited, the signal is passed through an anti-aliasing filter (which is basically a low pass filter) to reduce the high frequency components significantly so that the alias components have very small energy.

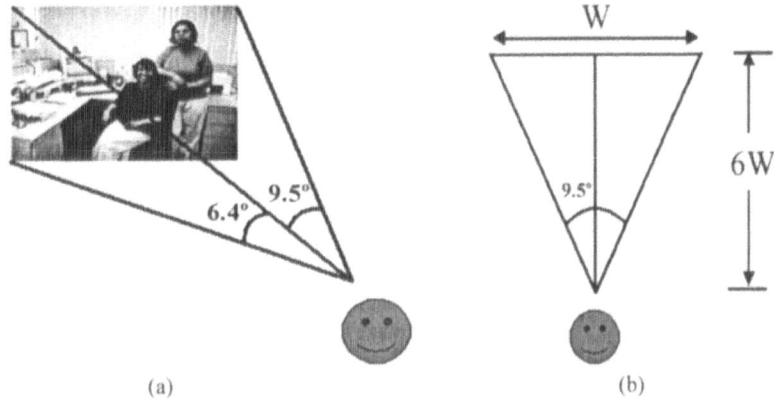

Figure 4.8. Photo scanning. (a) The viewing angles of a 4"×6" image,
(b) The angle formation at eyes.

A typical 1-D anti-aliasing filter characteristic is shown in Fig. 4.9. An ideal lowpass filter with a sharp cut-off frequency of $f_s/2$, where f_s is the sampling frequency, is shown in Fig. 4.9(a). However, this ideal filter is impossible to implement in practice. Fig. 4.9(b) shows the frequency reponse of a typical realizable filter. Here the gain in the passband is close to unity, and the gain in the stopband is close to zero. The transition band allows a gradual decrease in gain from unity to zero.

■ Example 4.4

Consider an audio signal with a spectrum 0-20KHz. The audio signal has to be sampled at 8 KHz. Design an anti-aliasing filter suitable for this application.

The sampling frequency is 8 KHz. An ideal lowpass filter with a sharp cut-off frequency of 4KHz will suppress all the components wth a frequency higher than 4 KHz. However, it is physically impractical to design an ideal filter. In this example, we would design a lowpass filter with the following characteristics:

   i)   The passband is 0-3200 Hz. The gain in the passband, Gp > -2 dB
   ii) The transition band is 3200-4000 Hz
   iii) The stopband is > 4000 Hz. The gain in the stopband, Gs < -20 dB

There is a rich body of literature [3] available for filter design, which is beyond the scope of this book. Consequently, the filter design procedure is briefly mentioned using a MATLAB example. There are several choices to design a filter with the above design constraints. The following MATLAB code (shown in Table 4.2) calculates the transfer function corresponding to the Butterworth, Chebyshev-1, and Chebyshev-2 filters.

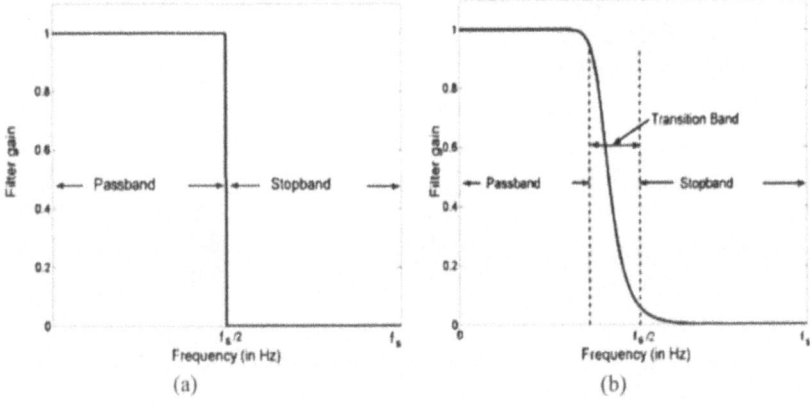

Figure 4.9. Lowpass/anti-alias filter characteristics. (a) An ideal lowpass filter has flat passband gain, instantaneous cutoff and zero stopband gain, and (b) a realizable lowpass filter with a gradual transition from passband to stopband. Note that f_s is the sampling frequency.

The analog filter design is easier to perform in the Laplace transform domain. The Laplace transform $H(s)$ of an impulse response $h(t)$ (of a filter) is defined as:

$$H(s) = \int_{-\infty}^{\infty} h(t)e^{-st} dt \qquad (4.13)$$

Note that $H(s)$ is also known as the transfer function of the filter. Using the MATLAB design procedure, it can be shown that the transfer functions corresponding to the three filters are as follows:

Butterworth

$$H(s) = \frac{4.909 \times 10^{45}}{D(s)}$$

where $D(s) = s^{13} + 27140s^{12} + 3.682 \times 10^8 s^{11} + 3.298 \times 10^{12} s^{10} + 2.171 \times 10^{16} s^9 +$
$1.107 \times 10^{20} s^8 + 1.107 \times 10^{20} s^8 + 4.493 \times 10^{23} s^7 + 1.47 \times 10^{27} s^6 +$
$3.874 \times 10^{30} s^5 + 8.129 \times 10^{33} s^4 + 1.322 \times 10^{37} s^3 + 1.322 \times 10^{37} s^3 +$
$1.322 \times 10^{37} s^3 + 1.579 \times 10^{40} s^2 + 1.245 \times 10^{43} s + 4.909 \times 10^{45}$ (4.14)

Chebyshev-1

$$H(s) = \frac{2.742 \times 10^{16}}{s^5 + 2261 s^4 + 1.536 \times 10^7 s^3 + 2.272 \times 10^{10} s^2 + 4.817 \times 10^{13} s + 2.742 \times 10^{16}}$$ (4.15)

Chebyshev-2

$$H(s) = \frac{1962 s^4 - 1.36 \times 10^{-8} s^3 + 1.196 \times 10^{11} s^2 - 0.2839 s + 1.457 \times 10^{18}}{s^5 + 12670 s^4 + 7.84 \times 10^7 s^3 + 3.215 \times 10^{11} s^2 + 7.672 \times 10^{14} s + 1.457 \times 10^{18}}$$ (4.16)

Table 4.2. MATLAB Code for designing anti-aliasing filter.

```
%MATLAB code for designing           h2 = tf(num,den) ;
lowpass filter                       w=[0:1:8000]; w = w';
Wp=3200; Ws=4000; Gp=-2; Gs=-        [mag2,phase2,w] = bode(num,den,w);
20 ;
%Ideal Filter                        %Chebyshev-2
mag0  =  [ones(1,4001)  zeros(1,     [n, Ws] = cheb2ord(Wp,Ws,-Gp,-Gs,'s') ;
4000)] ;                             [num,den] = cheby2(n,-Gs,Ws,'s') ;
%                                     h3 = tf(num,den) ;
%Butterworth Filter                  w=[0:1:8000]; w = w';
[n,  Wc]  =  buttord(Wp,Ws,-Gp,-     [mag3,phase3,w] = bode(num,den,w);
Gs,'s') ;                            %
[num,den] = butter(n,Wc,'s') ;       plot(w,mag0,w,mag1,w,mag2,w,mag3)
h1 = tf(num,den) ;                   xlabel('Frequency in Hz')
w=[0:1:8000]; w = w';                 ylabel('Filter gain'),grid
[mag1,phase1,w]  =  bode(num,den,    print -dtiff plot.tiff
w);                                   %
%Chebyshev-1                         % Butterworth    12th order, wc=3303 Hz
[n,  Wp]  =  cheb1ord(Wp,Ws,-Gp,-    %Cheby2  order=5, 3903 Hz
Gs,'s') ;
[num,den] = cheby1(n,-Gp,Wp,'s') ;
```

The characteristics of the filters are shown in Fig. 4.10. It is observed that all three filters satisfy the design constraints. The Butterworth filter (13th order, cut-off frequency=3271 Hz) has monotonic gain response. The Chebyshev-1 filter (5th order, Wp=3200, Ws=3903) has ripples in the passband, and monotonic gain in the stopband, whereas the Chebyshev-2 filter (5th order, Wp=3200, Ws=3903) has a monotonic gain in the passband, and ripples in the stopband. Although the response of the Butterworth filter seems to be superior (since there is no ripple), Chebyshev filters have the advantages of lower implementation cost (5th order) compared to the Butterworth filter (13th order). ■

The design of 1-D anti-aliasing filter can be extended to design 2-D anti-aliasing filter for images. In this case, the filter will have two cut-off frequencies, one each for the horizontal and vertical direction.

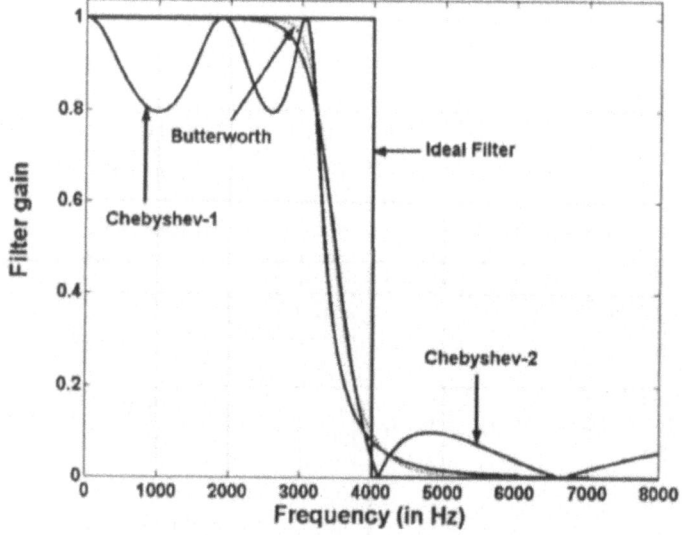

Figure 4.10. Lowpass filter characteristics of the ideal, Butterworth, Chebyshev-1, and Chebyshev-2 filters.

4.4 DIGITIZATION OF AUDIO SIGNALS

Figure 4.11 shows the simplified schematic of an audio recording and storing process. The function of each block is listed in Table 4.3. Typically, the analog voltage produced by a microphone has a low dynamic range. Hence, it is amplified before any processing is done to the signal. When the audio signal is quantized using a small number of bits/sample, the quality of the audio becomes poor. It has been found that if a small amount of noise (known as *dithering noise*), is added before quantization, the subjective quality seems to be better (note that dithering is also applied before image quantization). The anti-aliasing filter, as discussed earlier, is used to reduce the aliasing energy in the sampled audio signal. The sample and hold, and analog-to-digital conversion are applied to obtain the actual digital audio samples.

As discussed in Chapter 2, a typical audio system can have up to 7-8 channels. The above procedure is applied separately to each channel of the audio system. The samples from different audio channels are then multiplexed, compressed, and stored for future playback.

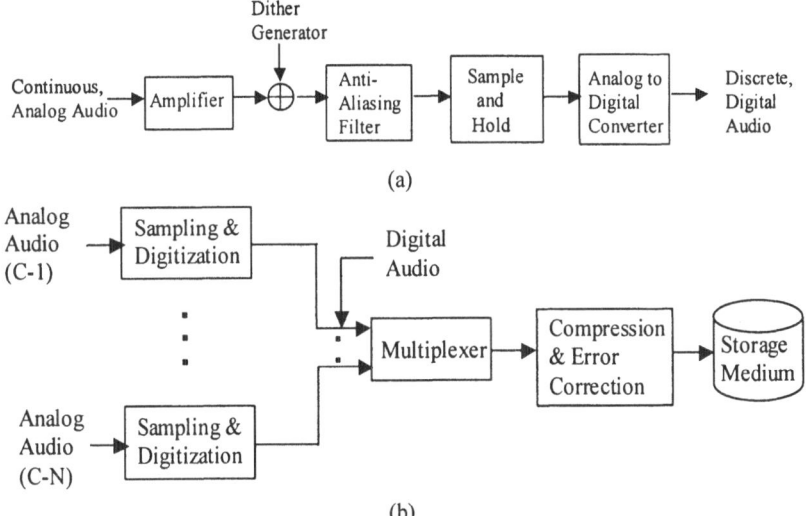

(a)

(b)

Figure 4.11 Digitization of audio signal. a) Sampling and digitization, b) N-channel audio recording and storing process.

4.4.1 Analog to Digital Conversion

The input of the analog-to-digital converter, as shown in Fig. 4.11, is a discrete-time signal whose amplitude is a real number that may require an infinite number of bits/digits for true representation. For digital computer processing, the signal at each time instance has to be converted to a number with finite precision (*e.g.*, 8, 16 or 32 bits). This is done by a quantizer which maps a continuous variable into a discrete variable. The input-output relationship of a typical quantizer is a staircase function as shown in Fig. 4.12.

Suppose a quantizer with N output levels are desired. The decision and the reconstruction (i.e., output) levels are denoted by $\{d_k, 0 \le k \le N\}$ and $\{r_k, 0 \le k \le N-1\}$, respectively where $d_k \le r_k < d_{k+1}$. The output of the quantizer for a given input $g(nT)$ can be calculated by a simple procedure.

$$Q[g(nT)] = r_k \qquad \text{if } d_k \le g(nT) < d_{k+1} \qquad (4.17)$$

If the decision levels are equidistant, i.e., if $(d_{k+1} - d_k)$ is constant for all k, the quantizer is called a uniform quantizer; otherwise it is called a nonuniform quantizer [6].

Table 4.3. Function of different blocks in a typical audio recording and storing system.

Blocks	Functions
Amplifier	Amplifies the audio signal before any noise (*e.g.*, dithering and quantization noise) is introduced.
Dither Generator	Adds a small amount of random noise. Ironically, this increases the perceptual quality when the signal is quantized
Anti-aliasing Filter	It is a lowpass filter to make sure the audio signal is band limited. It eliminates the aliasing component after sampling.
Sample and Hold	Holds the value of the audio signal and then sample the value of the signal at each sampling instances
Analog-to-Digital Converter	Computes the equivalent digital representation of the analog signal
Multiplexer	Multiplexes the bitstream coming from different channels
Compression	Reducing the redundancy and compress the audio file size maintaining an acceptable quality of audio

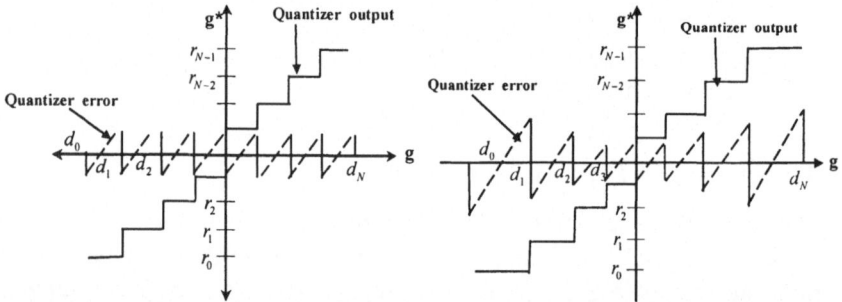

Figure 4.12. Quantization process. a) Uniform quantizer, and b) non-uniform quantizer.

■ Example 4.5

Consider an audio recording system where the microphone generates a continuous voltage in the range [-1,1] volts. Calculate the decision and reconstruction levels for an eight level quantizer.

The decision and reconstruction levels can be calculated from the following equations:

$$d_k = \frac{k-4}{4}, \qquad 0 \le k \le 8$$

$$r_k = d_k + \frac{1}{8} \qquad 0 \le k \le 7$$

The decision and reconstruction levels of the quantizer are shown in Table 4.4.

In the quantization process, an input sample value is represented by the nearest allowable output level, and in this process, some noise is introduced in the quantized signal. The quantization error (also known as quantization noise) is the difference between the actual value of the analog signal and its quantized value. In other words,

$$e(nT) = f(nT) - \hat{g}(nT) = g(nT) - Q[g(nT)] \qquad (4.18)$$

where $g(nT)$ is the input signal, $\hat{g}(nT)$ is the quantizer output, and $e(nT)$ the error signal at sampling instance nT. Figure 4.13 shows the quantization of the audio signal by the quantizer. It is observed that the largest (i.e., worst) quantization error is half of the decision interval, i.e., 0.125. ∎

Table 4.4. Decision and reconstruction levels of the quantizer in Example 4.5.

k	Decision levels	Reconstruction levels
0	-1.00	-0.875
1	-0.75	-0.625
2	-0.50	-0.375
3	-0.25	-0.125
4	0.00	0.125
5	0.25	0.375
6	0.50	0.625
7	0.75	0.875
8	1.00	

Once a signal is quantized, the quantized values can be represented using different methods. Pulse Code Modulation (PCM) is the most widely used representation. Here, each sample is represented using a fixed number of bits, and is encoded independent of others. The total number of possible amplitude levels in PCM is related to the number of bits used to represent each sample as follows:

$$N = 2^B \qquad (4.19)$$

where N=Number of quantizer output levels, and B=No. of bits/sample

From the above relation we see that if a larger number of bits (B) is used to represent a sample, the number of output levels (N) will increase. This will decrease the quantization error, and therefore will provide superior audio quality.

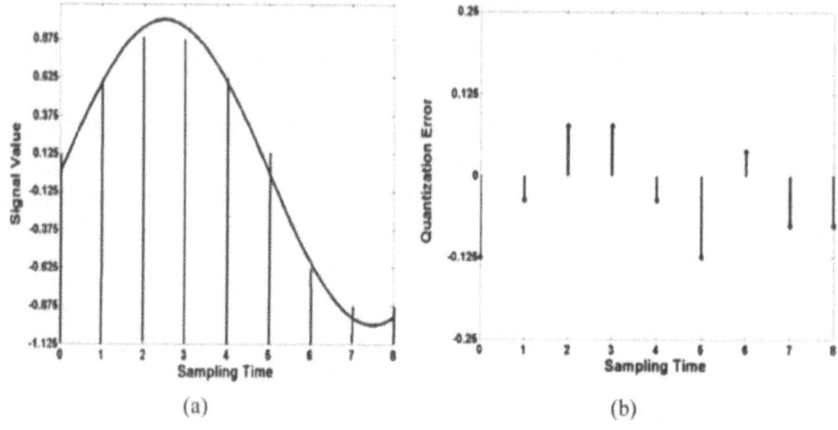

Figure 4.13. Quantization of audio signal. a) Quantizing continuous amplitude samples using the quantizer designed in Example 4.5, and b) The error value at sample times.

Bit-rate

The bit-rate of a quantized audio signal can be calculated as

$$bitrate = F_S * B \ bits / channel / second \qquad (4.20)$$

where F_S is the sampling frequency, and B is the number of bits/sample. For stereo audio, the channels can be considered independent. Therefore, the bit-rate for stereo audio will be twice the bit-rate ($2 * F_S * B \ bits / second$) of the mono audio.

Nonuniform quantization

We have so far considered the uniform quantization process where decision levels are equidistant. Consider the audio signal shown in Fig. 4.1(a). It is observed that the amplitude of the audio signal is likely to be closer to zero statistically than ± 127. Consequently, a nonuniform quantizer would provide a better quality for this signal. In this case, the audio samples with values close to zero will be quantized with a higher precision than the samples values close to ± 127.

4.4.2 Audio Fidelity Criteria

The quality of the quantized audio signal will be superior if the quantization noise is small. This can be achieved if a large number of bits are used to represent each sample (see Eq. (4.19)). However, this will increase the bit-rate of the audio signal. Hence, there is a trade-off between bit-rate and the quality of the digitized audio signal. The choice of the bit-rate and quality depends on the requirement of the audio application under consideration. In order to make an objective judgment, it is important to define the audio fidelity criteria.

Two types of criteria are used for evaluation of audio quality: subjective (*i.e.*, qualitative) and objective (*i.e.*, quantitative). ITU [5] has standardized a five-grade impairment scale for this purpose. The impairment scale measures the quality with respect to the audibility of the distortion in the signal. Given an audio signal, a group of persons (called the subjects) are asked to evaluate the quality (by giving the corresponding point) according to the scale. The opinions are subjective, and the scores from the subject are likely to vary. The average score is known as the mean opinion score (MOS). The MOS requirement depends on the application (see Table 4.5). For telephony, a score of 3 might be sufficient. But for CD quality audio, a goodness MOS value close to 5 (>4.5) is required.

Table 4.5. Impairment scale to represent audio fidelity

Very annoying	1
Annoying	2
Slightly annoying	3
Perceptible but not annoying	4
Imperceptible	5

The subjective quality of an audio signal represents the true quality as judged by the human auditory system. However, it is a cumbersome procedure that it involves experiments with several expert and/or nonexpert subjects. In addition, the subjective measures are influence by several factors such as choice of subjects, and experimental set-up. For this reason, a few quantitative measures have been developed to evaluate the objective quality of an audio signal. The choice of a quantitative distortion measure depends on several factors. First, it should be easily computable. Second, it should be analytically tractable. Finally, it should reflect the true subjective quality of the image. The signal to noise ratio (SNR) and mean square error (MSE) are two popular distortion measures. These are calculated as follows:

$$SNR = \frac{\sum_{n=0}^{N-1} \hat{f}(n)^2}{\sum_{n=0}^{N-1}\left[\hat{f}(n) - f(n)\right]^2} \qquad (4.21)$$

$$SNR \ (in \ dB) = 10\log_{10} SNR \qquad (4.22)$$

$$MSE = \frac{1}{N}\sum_{n=0}^{N-1}\left[\hat{f}(n) - f(n)\right]^2 \qquad (4.23)$$

where $f(n)$ and $\hat{f}(n)$ are the nth sample of the original and quantized audio signal, and N is the number of audio samples.

Traditionally, the signal to noise ratio (SNR) has been the most popular error measure in electrical engineering. It provides useful information in most cases, and is mathematically tractable. For this reason, it is also widely used in audio coding. Unfortunately, the SNR values do not correlate well with the subjective ratings, especially at high compression ratios. Several new distortion measures have been proposed for better adaptation to human auditory system.

The quality of the digital audio sample is determined by the number of bits per sample. If the error $e(nT)$ is assumed to be statistically independent and uniformly distributed in the interval [$-Q/2$ and $Q/2$], mean squared value of the quantization noise for a signal with dynamic range of 1 (*i.e.*, $Q = 2^{-B}$) can be expressed as

$$\dot{E} = \frac{1}{Q}\int_{-Q/2}^{Q/2} e^2 \, de = \frac{Q^2}{12} = \frac{2^{-2B}}{12} \qquad (4.24)$$

where B is the number of bits used to represent each sample. Note that $e(nT)$ is a discrete-time analog signal, and hence the integral has been used in Eq. (4.24). In decibel scale, the mean squared noise can be expressed as

$$E \ (in \ dB) = 10\log\left(Q^2/12\right) = 10\log\left(2^{-2B}/12\right) = -6 * B - 10.8 \qquad (4.25)$$

The above relation shows that each additional bit for analog-to-digital quantization reduces the noise by approximately 6 dB, and thus increases the SNR by the same amount. As a rule of thumb, a signal quantized with B-bits/sample is expected to provide

$$SNR \ (in \ dB) \approx 6 * B \qquad (4.26)$$

As a result, the SNR of a 16-bit quantized audio will be in the order of 96 dB. Typically, an audio signal with more than 90 dB SNR is considered to be of excellent quality for most applications.

■ Example 4.6

Consider a real audio signal "chord" which is a stereo audio signal (*i.e.*, two channel) digitized with 22.050 KHz sampling frequency, with a precision of 16 bits/sample. The normalized sample values (*i.e.*, the dynamic range is normalized to 1) of the left channel are shown in Fig. 7.3 (a). The signal has duration of 1.1 sec, and hence, there are 24231 samples. Estimate the *SNRs* of the signal if the signal is quantized with 5-12 bits/sample.

For this example, we may consider that the 16-bit audio signal is the original signal. A sample with value x ($-0.5 \le x \le 0.5$) can be quantized to m bits with the following operation:

$$y = round(x * 2^b) / 2^b$$

where y is the new sample value with 8-bit precision. All sample values are quantized using the above equation; the noise energy is calculated, and the SNR is calculated in dB. Figure 4.14(b) shows the quantization noise for 8 bit/sample quantization. It is observed that noise has a dynamic range of [-$2^{-(b+1)}, 2^{-(b+1)}$] (in this case, it is [-0.0020,0.0020]). The probability density function of the error values is shown in Fig. 4.14(c). The *pdf* is seen to be close to the uniform density. Finally, the SNR for quantization with different bit-rates is shown in Fig. 4.14(d). It is observed that the SNR increases by 6 dB for each increment of precision by 1 bit. However, the overall SNR is about 15 dB different than what is expected from Eq. (4.26). This is mainly because the SNR also depends on the signal characteristics, which Eq. (4.26) does not take into consideration. ■

Table 4.6 shows the sampling frequency, bits/sample, and the uncompressed data rate for several digital audio applications. From our experience, we can follow the improvement of the audio quality as the bit-rate increases. The telephone and the AM radio employ a lower sampling rate of 8 and 11.025 KHz, respectively, and provide a dark and muffled quality audio. At this bit-rate, it may not be advantageous to use the stereo mode. The FM radio provides a good audio quality because of high bit resolution and stereo mode. However, it produces darker sounding than the CDs because of the lower sampling rate. The CDs and digital audio tape (DAT) generally provide excellent audio quality, and are the recognized standards of audio quality. However, for extremely high quality of video, DVD audio format is used, which has a sampling rate of 192 KHz, and 24 bits/channel/sample resolution. Note that the DVD audio format, mentioned in Table 4.6, is not the same as the Dolby digital audio format used in a DVD movie. It is simply a superb quality audio, without any corresponding video, available in a DVD.

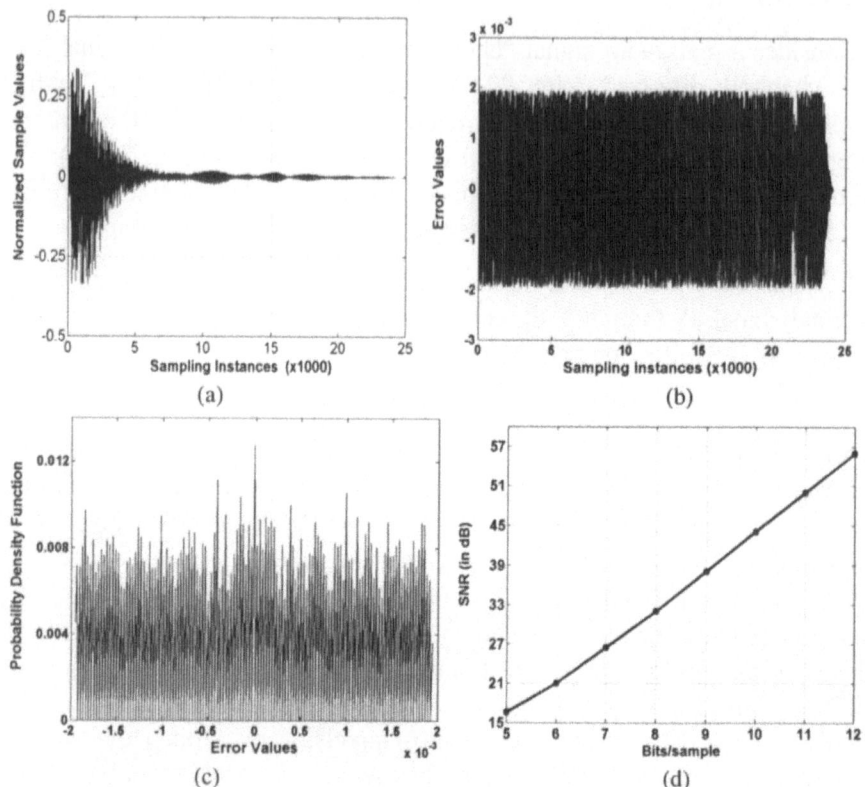

Figure 4.14. The audio signal *chord.wav*. (a) The signal waveform, (b) Quantization error at 8 bits/sample, (c) probability density function of the quantization error at 8 bits/sample, and (d) SNR versus bits/sample.

4.4.3 MIDI versus Digital Audio

It was observed in Chapter 2 that music could be stored as a MIDI signal. In this chapter, we have discussed how an audio signal (musical or otherwise) can be stored as a digitized waveform. Thus, we have two alternatives for representing music, each with their own advantages. The MIDI representation generally requires fewer number of bits compared to the digital audio. However, the MIDI standard assumes standard quality of playback hardwares. If this assumption is violated, the quality of the sound produced might be inferior. In general, the digitized audio is more versatile, and can be used for most applications. However, MIDI may provide a superior performance when i) high quality MIDI sound source is available, ii) complete control over the playback hardware available, and iii) there is no spoken dialog.

4.5 DIGITIZATION OF IMAGES

In the previous section, it was observed that different bit resolution of audio produces different qualities of digital audio, higher resolution producing a better quality. Similar inference can also be made for the digital images. However, the sensitivity of the HVS is poorer than the auditory system. When the auditory system can differentiate thousands of sound pressure levels, the HVS can differentiate only about 200 brightness levels. Hence, a gray level image can be described with an excellent quality with 8-bit resolution that corresponds to 256 gray levels. However, many low resolution sensors produce an image with a lower number of gray levels. Figure 4.15 shows the quality of an image digitized with different bit resolution.

Table 4.6. Digital audio at various sampling rates and resolutions.

Quality	Sample Rate (in KHz)	Bits per Sample	Mono/ Stereo	Data Rate (if uncompressed)	Frequency Band (in Hz)
Telephone	8	8, μ law^2	Mono	8 Kbytes/sec	200-3,400
AM Radio	11.025	8	Mono	11.0 Kbytes/sec	
FM Radio	22.050	16	Stereo	88.2 Kbytes/sec	
CD	44.1	16,linear PCM	Stereo	176.4 Kbytes/sec	20-20,000
DAT^1	48	16	Stereo	192.0 Kbytes/sec	20-20,000
DVD Audio	192	24	Stereo	1152.0 Kbytes/sec	20-20,000

^{1}Digital audio tape, ^{1}more details about μ law in Chapter 7.

The color image typically has three components (as discussed in Chapter 3). For electronic display, the {R, G, B} components are commonly used. Each of these three components is quantized separately. High resolution sensors generally employs 8 bits/channel/pixel. Hence, a total of 24 bits are required to represent a color pixel. Fig. 4.16 shows the Lena image with 24 bits/pixel resolution.

Note that 24 bits/pixel corresponds to 16 million ($= 2^{24}$) colors, which is much more than the human eyes can differentiate. Therefore, many electronic devices use a lower resolution (*e.g.*, 8 bit or 16 bits) display. This is generally obtained using a color map.

4.5.1 Visual Fidelity Measures

The quality of visual data degrades due to quantization of pixel values. In addition, distortion can arise due to the compression of images. As in audio, there are primarily two types of fidelity measures: subjective and objective. An impairment scale similar to that shown in Table 4.5 can be employed to evaluate the quality for images. However, visual testing with real subjects is a cumbersome procedure. Therefore, the objective measures are mostly used in practice.

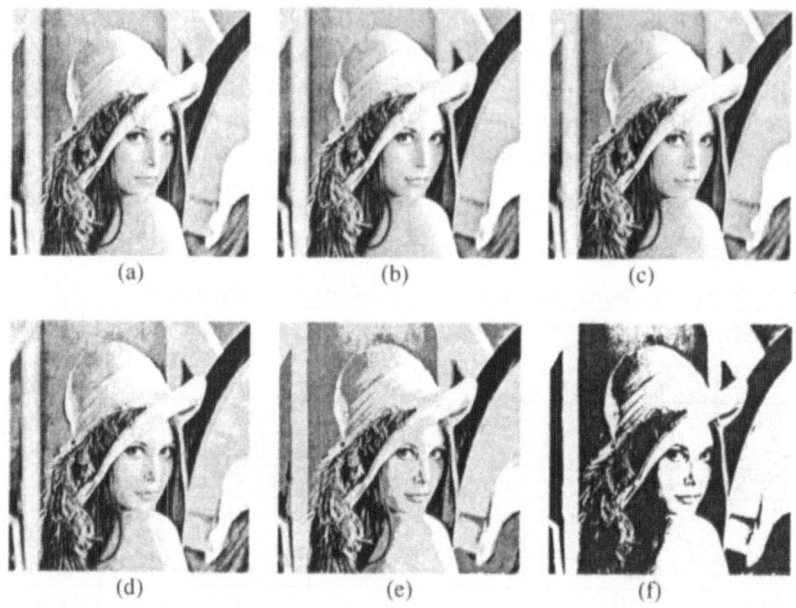

Figure 4.15. Images quantized at different bits/sample. (a) 8 bit, 256 levels, (b) 5 bit, 32 levels, (c) 4 bit, 16 levels, (d) 3 bit, 8 levels, (e) 2 bit, 4 levels, (f) 1 bit, 2 levels.

As in audio, the SNR and MSE are two popular measures used to evaluate the distortion in an image. In addition, peak signal to noise ratio (PSNR) and the mean absolute error (MAE) are two other popular distortion measures. If $f(m,n)$ and $\hat{f}(m,n)$ are the original and distorted images of size $M \times N$, the distortion with respect to various criteria is calculated as follows:

$$SNR \ (in \ dB) = 10\log_{10}\left[\frac{\sum_{m=0}^{M-1}\sum_{n=0}^{N-1}[\hat{f}(m,n)]^2}{\sum_{m=0}^{M-1}\sum_{n=0}^{N-1}\left[\hat{f}(m,n)-f(m,n)\right]^2}\right] \quad (4.27)$$

$$PSNR\,(in\,dB) = 10\log_{10}\left[\frac{\displaystyle\sum_{m=0}^{M-1}\sum_{n=0}^{N-1}[peak\ signal\ value]^2}{\displaystyle\sum_{m=0}^{M-1}\sum_{n=0}^{N-1}[\hat{f}(m,n) - f(m,n)]^2}\right] \qquad (4.28)$$

$$Mean\,Square\,Error\,(MSE) = \frac{1}{M\,N}\sum_{m=0}^{M-1}\sum_{n=0}^{N-1}\left[\hat{f}(m,n)\text{-}f(m,n)\right]^2 \qquad (4.29)$$

$$Mean\,Absolute\;Error\,(MSE) = \frac{1}{M\,N}\sum_{m=0}^{M-1}\sum_{n=0}^{N-1}\left|\hat{f}(m,n)\text{-}f(m,n)\right| \qquad (4.30)$$

Although these objectives measures help to evaluate distortion quickly, it is known that they do not correlate well with the subjective ratings, especially when the distortion is very high. New distortion measures have been proposed for better adaptation to HVS properties.

REFERENCES

1. K. C. Pohlmann, *Principles of Digital Audio*, McGraw-Hill, New York 2000.

2. R. W. Schafer and L. R. Rabiner, "Digital representations of speech signals," *Proc. of the IEEE*, Vol. 63, No. 4, April 1975.

3. B. P. Lathi, *Signal Processing and Linear Systems*, Berkeley Cambridge Press, 1998.

4. G. L. Frendendall and W. L. Behrend, "Picture quality – procedures for evaluating subjective effects of interference," *Proc. of IRE*, Vol. 48, pp. 1030-1034, 1960.

5. ITU-R Rec.BS.1116, "Methods for the subjective assesment of small impairments in audio systems including multichannel sound systems," *International Telecommunication Union*, Geneva, Switzerland, 1994.

6. N. S. Jayant and P. Noll, Digital coding of waveforms: principles and applications to speech and video, Prentice-Hall, New Jersey, 1984.

QUESTIONS

1. Consider an audio signal with a single sinusoidal tone at 6 KHz. Sample the signal at a rate of 10000 samples/sec. Reconstruct the signal by passing it through an ideal lowpass filter with cut-off frequency of 5 KHz. Determine the frequency of the reconstructed audio signal.

2. What is aliasing in digital audio? How can it be avoided?

3. You are recording an audio source that has a bandwidth of 5 KHz. The digitized audio should have an average SNR of at least 60 dB. Assuming that there are two

stereo channels, and no digital compression, what should be the minimum data rate in bits/sec.

4. You are required to design anti-aliasing filters for i) telephony system, ii) FM radio, and iii) CD recording. The audio sampling rates are given in Table 4.6. Determine typical frequency characteristic requirements of the filters for the three applications. Using MATLAB, calculate the transfer functions of the corresponding Butterworth and Chebyshev filters. Which filters will you choose, and why?

5. What should be the typical Nyquist sampling frequencies for full bandwidth audio and image signals?

6. You want to design an image display monitor with an aspect ratio of 4:3 (*i.e.,* width:height). A user is expected to observe the images at a distance of 4 times the monitor diagonal. What should be the minimum monitor resolution? Assume that the HVS sensitivity is circularly symmetric, and is negligible above 25 cycles/degree.

7. Generate a simple color map with 64 colors. To observe the different colors, create images with different combinations of R {=32, 96, 160, 224}, G {=32, 96, 160, 224}, and B {=32, 96, 160, 224} and save them in any standard image format such as tiff, and ppm. Display the colors and verify the additive color mixing property discussed in Chapter 3.

Chapter 5

Transforms and Subband Decomposition

The properties of the audio and video signals, and the digitization process have been discussed in the previous Chapters. When a signal is digitized, further processing of these signals may be needed for various applications, such as compression, and enhancement. The processing of multimedia signal can be done effectively when the limitation of our hearing or visual systems is taken into account. For example, it was shown in Chapter 3 that the human ear is not very sensitive to audio signals with frequencies above 10-12 KHz. Similarly, the eyes also do not respond well above 20 cycles/degree. This dependency of our sensory systems on the frequency spectrum of the audio or visual signals has led to the development of transform and subband-based signal processing techniques. In these techniques, the signals are decomposed into various frequency or scale components. Various components are then suitably modified depending on the application at hand. In this Chapter, we will discuss mainly two types of signal decomposition techniques: transform-based decomposition and subband decomposition.

Although a wide variety of transforms have been proposed in the literature, the unitary transforms are the most popular for representing signals. A unitary transform is a special type of linear transform where certain orthogonal conditions are satisfied. There are several unitary transforms that are used in signal processing. In this Chapter, we present only a few selected transforms, namely Fourier, cosine and wavelet transforms that are widely used for multimedia signal processing. In addition, we also discuss subband decomposition that is very popular for compression and filtering applications. The application of these techniques will be covered in the later Chapters.

Since audio signals are one-dimensional, these signals are represented with one-dimensional (1-D) basis functions. On the other hand, an image can be represented by two-dimensional (2-D) basis functions that are also known basis images. Similarly, a video signal can be represented by three-dimensional basis functions. However, a video signal is generally analyzed using a hybrid approach in order to reduce the computational complexity. In this approach, the spatial information is represented using 2-D transform coefficients, which is then analyzed in the time dimension. Hence, in this Chapter, we will mainly focus on 1-D and 2-D transforms and subband decomposition.

5.1   1-D UNITARY TRANSFORM

For 1-D sequence $f(n)$, $0 \leq n \leq N-1$, a unitary transform (which is also known as orthonormal transform) is defined as [1]:

$$\overline{F} = U\overline{f}, \quad \text{i.e.,} \quad F(k) = \sum_{n=0}^{N-1} U(k,n)f(n) \qquad 0 \leq k \leq N-1 \qquad (5.1)$$

where $\overline{F} = \{F(k)\}$ is the transform coefficient matrix and

$$U = \begin{bmatrix} u(0,0) & \cdots & u(0,N-1) \\ \cdot & \cdots & \cdot \\ \cdot & \cdots & \cdot \\ \cdot & \cdots & \cdot \\ u(N-1,0) & \cdots & u(N-1,N-1) \end{bmatrix} \text{ is a unitary transformation matrix}$$

The inverse transform is given by

$$\overline{f} = U^{*T}\overline{F}, \quad \text{i.e.,} \quad f(n) = \sum_{k=0}^{N-1} F(k)u^{*}(k,n) \qquad 0 \leq n \leq N-1 \qquad (5.2)$$

where the symbols "*" and "T" denote complex conjugation and matrix transpose, respectively. Eq. (5.2) can be considered as a series representation of the sequence $f(n)$. The columns of U^{*T}, i.e., the vectors $\{u^{*}(k,n), 0 \leq n \leq N-1\}$ are called the basis vectors of U.

The matrix $\{u(k,n)\}$ is unitary if the following conditions are true.

$$\sum_{n=0}^{N-1} u(k,n)u^{*}(k_{0},n) = \delta(k - k_{0}) \qquad (5.3)$$

$$\sum_{k=0}^{N-1} u(k,n)u^{*}(k,n_{0}) = \delta(n - n_{0}) \qquad (5.4)$$

where $\delta(k)$ is the discrete-time unit impulse function defined as

$$\delta(k) = \begin{cases} 1 & k = 0 \\ 0 & otherwise \end{cases} \quad (5.5)$$

The above two properties ensure that the basis vectors corresponding to the unitary transform are orthogonal to each other and the set of basis vectors is complete. The completeness property ensures that there exists no non-zero vector (which is not in the set) that is orthogonal to all the basis vectors. The above two conditions will be true if and only if the following condition is true.

$$U^{-1} = U^{*T} \quad (5.6)$$

We now present Fourier and cosine transforms as special cases of unitary transforms.

5.2 1-D DISCRETE FOURIER TRANSFORM

The discrete Fourier transform (DFT) pair corresponding to a sequence $f(n)$ is defined as [2]:

$$F(k) = \frac{1}{\sqrt{N}} \sum_{n=0}^{N-1} f(n) W_N^{kn} \qquad 0 \le k \le N-1 \quad (5.7)$$

$$f(n) = \frac{1}{\sqrt{N}} \sum_{k=0}^{N-1} F(k) W_N^{-kn} \qquad 0 \le n \le N-1 \quad (5.8)$$

where $F(k)$ is the kth DFT coefficient, and $W_N = \exp\left\{\dfrac{-j2\pi}{N}\right\}$. The $N \times N$ forward DFT transformation matrix is as follows (compare Eq. (5.7) with Eq. (5.1)).

$$U = \{u(k,n)\} = \left\{ \frac{1}{\sqrt{N}} W_N^{kn} \right\} = \left\{ \frac{1}{\sqrt{N}} e^{\frac{-j2\pi kn}{N}} \right\} \quad (5.9)$$

Similarly, the inverse matrix is given by:

$$U^{-1} = U^{*T} = \{u^*(n,k)\} = \{u^*(k,n)\} = \left\{ \frac{1}{\sqrt{N}} W_N^{-kn} \right\} = \left\{ \frac{1}{\sqrt{N}} e^{\frac{j2\pi kn}{N}} \right\} \quad (5.10)$$

■ **Example 5.1**

Consider a 4-point sequence f=[2, 5, 7, 6]. Calculate the corresponding DFT coefficients. Reconstruct the input sequence from the DFT coefficients and the DFT basis functions.

The DFT coefficients are

$$
\begin{bmatrix} F(0) \\ F(1) \\ F(2) \\ F(3) \end{bmatrix} = \frac{1}{\sqrt{4}} \begin{bmatrix} 1 & 1 & 1 & 1 \\ 1 & e^{-\frac{j2\pi}{4}} & e^{-\frac{j4\pi}{4}} & e^{-\frac{j6\pi}{4}} \\ 1 & e^{-\frac{j4\pi}{4}} & e^{-\frac{j8\pi}{4}} & e^{-\frac{j12\pi}{4}} \\ 1 & e^{-\frac{j6\pi}{4}} & e^{-\frac{j12\pi}{4}} & e^{-\frac{j18\pi}{4}} \end{bmatrix} \begin{bmatrix} 2 \\ 5 \\ 7 \\ 6 \end{bmatrix} = \begin{bmatrix} 10 \\ -2.5+j0.5 \\ -1 \\ -2.5-j0.5 \end{bmatrix}
\tag{5.11}
$$

The input sequence can be calculated from its DFT coefficients using the inverse DFT transformation matrix.

$$
\begin{bmatrix} f(0) \\ f(1) \\ f(2) \\ f(3) \end{bmatrix} = \frac{1}{\sqrt{4}} \begin{bmatrix} 1 & 1 & 1 & 1 \\ 1 & e^{\frac{j2\pi}{4}} & e^{\frac{j4\pi}{4}} & e^{\frac{j6\pi}{4}} \\ 1 & e^{\frac{j4\pi}{4}} & e^{\frac{j8\pi}{4}} & e^{\frac{j12\pi}{4}} \\ 1 & e^{\frac{j6\pi}{4}} & e^{\frac{j12\pi}{4}} & e^{\frac{j18\pi}{4}} \end{bmatrix} \begin{bmatrix} 10 \\ -2.5+j0.5 \\ -1 \\ -2.5-j0.5 \end{bmatrix} = \begin{bmatrix} 2 \\ 5 \\ 7 \\ 6 \end{bmatrix}
\tag{5.12}
$$

The four basis vectors corresponding to the 4-point transform are the columns of the 4x4 inverse matrix shown in Eq. (5.12). The input sequence can also be reconstructed using the basis vectors as follows:

$$
\begin{bmatrix} f(0) \\ f(1) \\ f(2) \\ f(3) \end{bmatrix} = 10* \begin{bmatrix} 1 \\ 1 \\ 1 \\ 1 \end{bmatrix} + (-2.5+j0.5)* \begin{bmatrix} 1 \\ e^{\frac{j2\pi}{4}} \\ e^{\frac{j4\pi}{4}} \\ e^{\frac{j6\pi}{4}} \end{bmatrix} -1* \begin{bmatrix} 1 \\ e^{\frac{j4\pi}{4}} \\ e^{\frac{j8\pi}{4}} \\ e^{\frac{j12\pi}{4}} \end{bmatrix} + (-2.5-j0.5)* \begin{bmatrix} 1 \\ e^{\frac{j6\pi}{4}} \\ e^{\frac{j12\pi}{4}} \\ e^{\frac{j18\pi}{4}} \end{bmatrix} = \begin{bmatrix} 2 \\ 5 \\ 7 \\ 6 \end{bmatrix} \blacksquare
$$

The DFT of an arbitrary length sequence can be calculated in a similar manner. The basis functions for 8-point transforms (which can be obtained from the columns of 8x8 inverse transformation matrix) are shown in Fig. 5.1. Note that the basis functions are discrete-time complex exponential functions. Figs. 5.1(a) and 5.1(b) show the real and imaginary parts of the basis functions, respectively. It is observed that the real part is a sampled cosine function whereas the imaginary part is a sampled sine function. Example 5.1 demonstrated that a signal could be represented by summation of complex basis functions, each weighted by its corresponding DFT coefficients. This is the core principle of the transform-based signal processing. The underlying expectation is that representation by well-known complex exponential functions will help us in the analysis process.

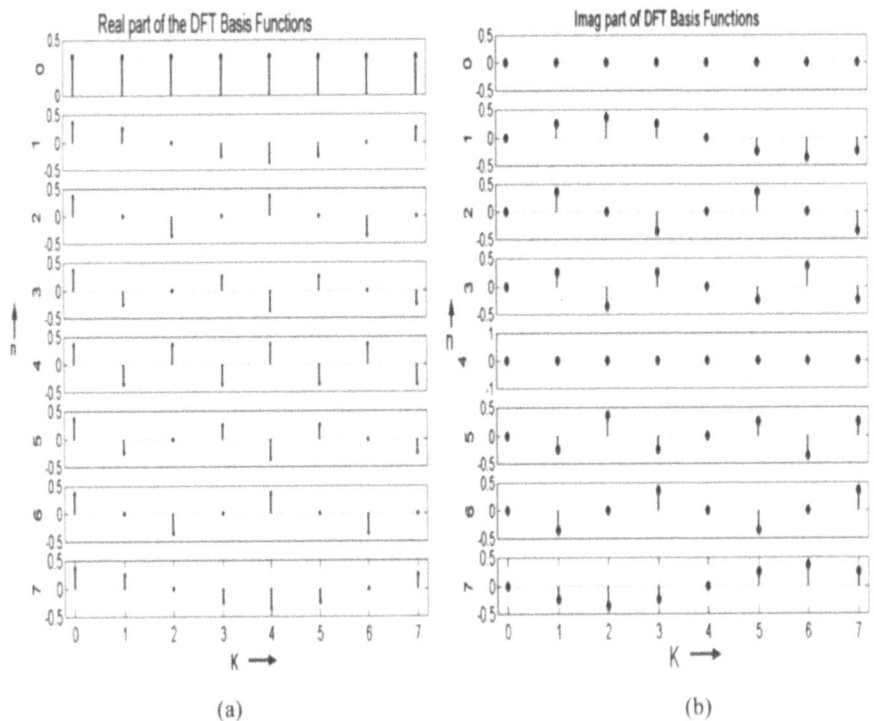

Figure 5.1. The basis vectors of 8-point DFT. a) real part, and b) imaginary parts.

Alternative DFT Transform pair

The DFT definition given in Eqs. (5.7) and (5.8) is not universally used. Many DFT implementations (*e.g.*, MATLAB) use a slightly different definition of the DFT which is as follows:

$$F(k) = \sum_{n=0}^{N-1} f(n)W_N^{kn} \qquad 0 \le k \le N-1 \qquad (5.13)$$

$$f(n) = \frac{1}{N} \sum_{k=0}^{N-1} F(k)W_N^{-kn} \qquad 0 \le n \le N-1 \qquad (5.14)$$

Note that the only difference between the transform pairs {Eqs. (5.7) and (5.8)} and {Eqs. (5.13) and (5.14)} are the scaling factors. Due to the different scaling factors, the transform represented by (Eq. (5.13) and (5.14)) is orthogonal, but not orthonormal or unitary.

Properties of 1-D DFT

DFT is one of the most important transforms in image and signal processing because of its several useful properties. A few selected properties are presented below.

i) Convolution

The DFT of the *circular* (or periodic) convolution [2, 9] of two sequences is equal to the product of their DFTs.

$$f(n) \otimes g(n) \Leftrightarrow F(k)G(k) \quad \text{Time domain Convolution} \quad (5.15)$$

$$f(n)g(n) \Leftrightarrow F(k) \otimes G(k) \quad \text{Frequency domain Convolution} \quad (5.16)$$

where $\otimes$ denotes the circular convolution operation.

ii) Fast Implementation

Note that the N-point DFT coefficients can be calculated by multiplying the 2-D $N \times N$ transformation matrix by 1-D $N \times 1$ matrix. Therefore, the direct computation of 1-D DFT of N-point sequence requires N^2 operations where each operation includes one complex multiplication and one complex addition. However, because of the special property of the DFT transformation matrix, it can be shown that the complexity can be reduced significantly [2]. These algorithms are known as fast Fourier transform (FFT) algorithms. A popular FFT algorithm, known as radix-2 algorithm, has an overall computational complexity of $(N/2)\log_2 N$ *butterfly* operations (for N-point transform) where each *butterfly* consists of one complex multiplication and two complex additions (the term "butterfly" is typically used in FFT literature since the FFT algorithm employs a special data arrangement that looks somewhat like a butterfly). The computational advantage of FFT over direct method is shown in Table 5.1. It is observed that for a 1024-point signal, the computational saving is more than 100 times. The existence of the FFT algorithm is one of the major reasons for the popularity of the DFT.

iii) Energy Conservation and Compaction

The Fourier transform preserves the signal energy, *i.e.*, the energy of a signal in the time or pixel domain is identical to the energy in the frequency domain. In other words, the following relationship is true.

$$\sum_{n=0}^{N-1} |f(n)|^2 = \sum_{k=0}^{N-1} |F(k)|^2 \quad (5.17)$$

Although the total energy does not change in the Fourier domain, the energy however is redistributed among the Fourier coefficients. This is illustrated with the following example.

Table 5.1. Comparison of N^2 versus $N\log_2 N$ for various values of N.

N	N^2	$N\log_2 N$ (FFT)	Computational saving ($N/\log_2 N$)
32	1024	160	6.4
256	65,536	2,048	32
1024	1,048,576	10,240	102
8192	67,108,864	106,496	630

■ Example 5.2

Construct a 1-D signal from the pixel values of horizontal line# 100 of Lena (gray-level) image. In order to avoid a large DC component, make the signal zero mean (by subtracting the average value of the pixels from each pixel amplitude). Calculate the DFT, and the total energy in the first 20 Fourier coefficients. Compare it with the total energy of the signal.

The 1-D signal is plotted in Fig. 5.2(a). The abrupt amplitude fluctuation corresponds to the edges in the horizontal direction. It is observed that the energy of the pixels is distributed across the horizontal direction. The DFT amplitude spectrum of the zero-mean signal is shown in Fig. 5.2(b). The low frequency DFT coefficients lie in the two extreme sides (left side corresponds to positive low frequency and right side corresponds to negative low frequency), whereas the high frequency coefficients lie in the middle region. The total energy of the audio signal (lying across all 512 coefficients) is 11.3×10^5. The first twenty low frequency coefficients together have an energy of 8.6×10^5, which is approximately 76% of the total energy. The remaining 492 coefficients contain only 24% of the total energy. ■

This energy preservation and compaction is the primary reason for the popularity of transform-based image compression techniques. This will be discussed in more detail in Chapter 8.

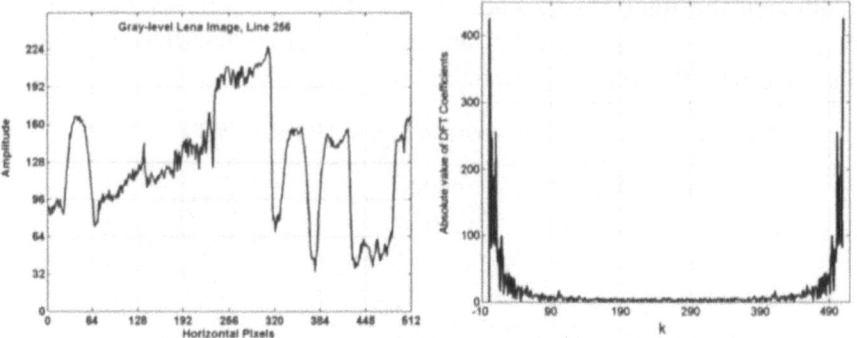

Figure 5.2. Energy compaction with DFT. a) Horizontal line (line# 100) from Lena image, and b) The DFT spectrum of the scanline.

5.3 1-D DISCRETE COSINE TRANSFORM

The one-dimensional discrete cosine transform (DCT) pair for an input sequence $f(n)$ is defined as [3]:

$$F(k) = \alpha(k)\sum_{n=0}^{N-1} f(n)\left[\cos\frac{(2n+1)\pi k}{2N}\right] \qquad \text{for } 0 \le k \le N-1 \qquad (5.18)$$

$$f(n) = \sum_{k=0}^{N-1} \alpha(k)\left[\cos\frac{(2n+1)\pi k}{2N}\right]F(k) \qquad \text{for } 0 \le n \le N-1 \qquad (5.19)$$

where $\alpha(k) = \begin{cases} \sqrt{1/N} & k=0 \\ \sqrt{2/N} & 1 \le k \le N-1 \end{cases}$

Note that the coefficients $\alpha(k)$ are used in Eqs. (5.18) and (5.19) to keep the norm of the basis functions unity. It can be shown that the discrete cosine transform is real and unitary, unlike DFT that requires complex operations.

■ **Example 5.3**

Consider a 4-point sequence f=[2, 5, 7, 6]. Calculate the DCT coefficients. Reconstruct the input sequence from the DCT coefficients.

Substituting $N = 4$ in Eq. (5.18), we obtain the DCT coefficients as follows:

$$\begin{bmatrix} F(0) \\ F(1) \\ F(2) \\ F(3) \end{bmatrix} = \frac{1}{\sqrt{2}} \begin{bmatrix} 1/\sqrt{2} & 1/\sqrt{2} & 1/\sqrt{2} & 1/\sqrt{2} \\ \cos(\pi/8) & \cos(3\pi/8) & \cos(5\pi/8) & \cos(7\pi/8) \\ \cos(2\pi/8) & \cos(6\pi/8) & \cos(10\pi/8) & \cos(14\pi/8) \\ \cos(3\pi/8) & \cos(9\pi/8) & \cos(15\pi/8) & \cos(21\pi/8) \end{bmatrix} \begin{bmatrix} 2 \\ 5 \\ 7 \\ 6 \end{bmatrix} = \begin{bmatrix} 10 \\ -3.15 \\ -2 \\ 0.22 \end{bmatrix} \qquad (5.20)$$

The input sequence can be calculated using the inverse DCT (Eq. (5.19)).

$$\begin{bmatrix} f(0) \\ f(1) \\ f(2) \\ f(3) \end{bmatrix} = \frac{1}{\sqrt{2}} \begin{bmatrix} 1/\sqrt{2} & \cos(\pi/8) & \cos(2\pi/8) & \cos(3\pi/8) \\ 1/\sqrt{2} & \cos(3\pi/8) & \cos(6\pi/8) & \cos(9\pi/8) \\ 1/\sqrt{2} & \cos(5\pi/8) & \cos(10\pi/8) & \cos(15\pi/8) \\ 1/\sqrt{2} & \cos(7\pi/8) & \cos(14\pi/8) & \cos(21\pi/8) \end{bmatrix} \begin{bmatrix} 10 \\ -3.15 \\ -2 \\ 0.22 \end{bmatrix} = \begin{bmatrix} 2 \\ 5 \\ 7 \\ 6 \end{bmatrix} \quad (5.21)$$

∎

Using different values of N in Eq. (5.18), the DCT of arbitrary length sequence can be calculated. The basis vectors corresponding to 8-point DCT is shown in Fig. 5.3(a). The frequency characteristic of the DCT basis vectors is shown in Fig. 5.3(b).

Note that in many signal processing applications, such as audio and image compression, a long sequence of samples is divided into blocks of nonoverlapping samples. The DCT (or DFT) is then calculated for each block of data. This is traditionally known as block transform.

Properties of DCT

i) Energy Compaction

The DCT has an excellent energy compaction performance. This is demonstrated with an example.

∎ Example 5.4

Consider the image scan line in Example 5.2. Calculate the DCT, and the total energy in the first 20 Fourier coefficients. Compare it with the total energy of the signal. Compare the energy compaction performance of DCT and DFT.

Figure 5.4 shows the DCT coefficients of the scan line shown in Fig. 5.2(a). It can be shown that (the MATLAB code is included in the CD) the first 20 DCT coefficients have a total energy of 9.4×10^5 which is about 83% of the total energy. Comparing with the results obtained in Example 5.2, it can be said that the DCT provides a better compaction than the DFT. ∎

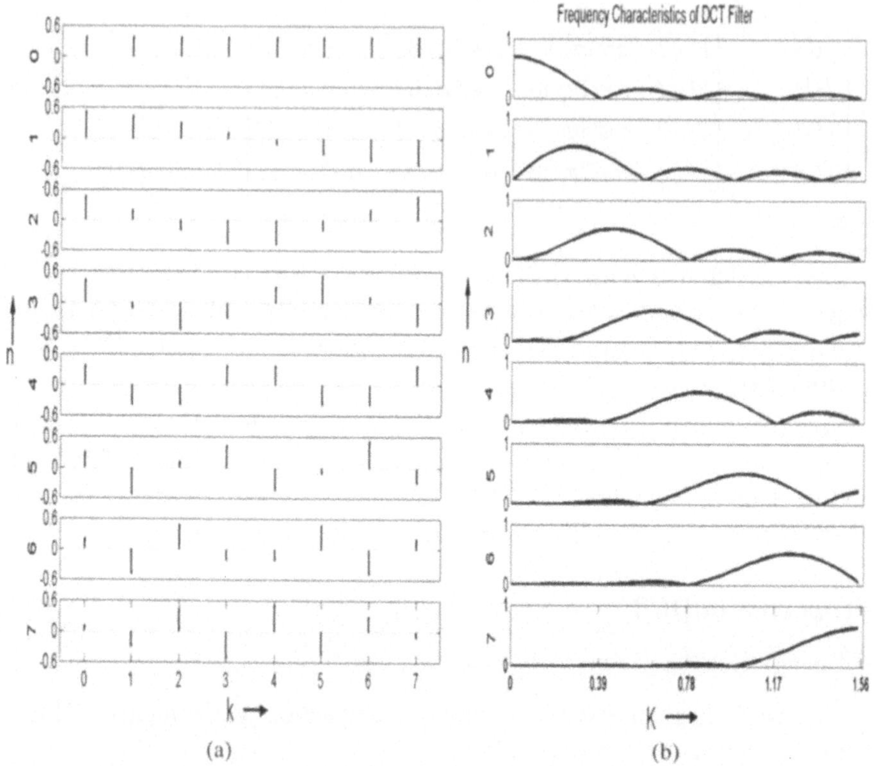

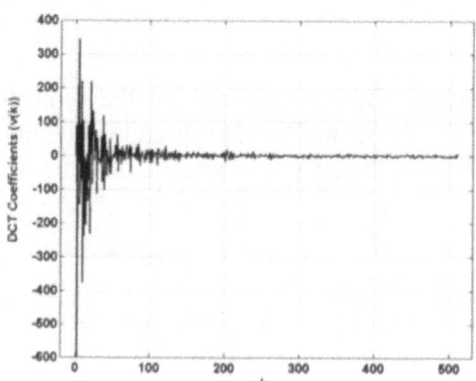

Figure 5.3. Eight-point discrete cosine transform. a) The basis vectors, and
b) the frequency characteristics of each basis function. The frequency axis
is normalized to $[0, \pi / 2]$.

Figure 5.4. DCT coefficients of the 1-D signal shown in Fig. 5.2(a).

ii) Relationship with DFT

Although the DCT basis functions are discrete cosine functions, the DCT is not the real part of the DFT. However, it is related to the DFT. It can be shown that the DCT of an Nx1 sequence $\{f(0), f(1), ...f(N-1)\}$ is related to the DFT of the $2N \times 1$ sequence given as follows.

$$\{f(N-1), f(N-2), ...f(1), f(0), f(0), ...f(N-1)\} \qquad (5.22)$$

The above observation provides us two important corollaries. First, a fast DCT transform exists with the computational complexity $O(N \log_2 N)$ for an N-point input sequence. Secondly, because of the even symmetry of Eq. (5.22), the reconstructed signal from the quantized DCT coefficients will preserve the edge better. The usefulness of the second property in image compression will be shown in Chapter 8.

5.4 DIGITAL FILTERING AND SUBBAND ANALYSIS

The transforms, especially the unitary (or orthogonal) transforms presented in the previous sections, are very useful in analyzing the frequency content of a signal. Two alternative methods of frequency analysis of multimedia signal are digital filtering [9] and subband decomposition [4]. A brief overview of digital filters is presented below. This will be followed by a discussion on subband decomposition.

5.4.1 Digital Filters

Filtering is a process of selecting, or suppressing, certain frequency components of a signal. Digital Filters can also attenuate, or amplify each frequency component of a digital signal by a desired amount. In other words, a digital filter can shape the frequency spectrum of a signal. Digital filters are typically specified in terms of desired attenuation, and permitted deviations from the desired value in their frequency response. The definition of various terms used is given below [9].

Passband: The band of frequency components that are allowed to pass.

Stopband: The band of frequency components that are suppressed.

Transition band: The frequency band between the passband and the stopband where the gain changes from high to low (or low to high).

Ripple: The maximum amount by which gain in the passband or stopband deviates from nominal gain.

Stopband attenuation: The minimum amount by which frequency components in the stopband are attenuated.

The filters can be divided into four basic categories depending on their gain characteristics: lowpass, highpass, bandpass, and bandstop filter. Fig. 5.5 shows typical gain response of these filters. Ideally, the transition band should be zero. However, filters with ideal response are not physically realizable.

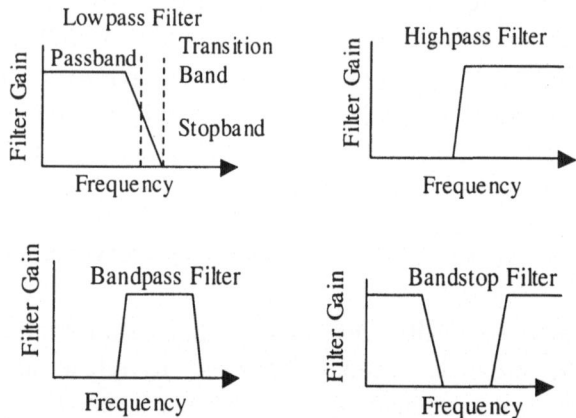

Figure 5.5. Different types of Filter: lowpass, highpass, bandpass filter, and bandstop filter.

Types of Digital Filters

The input $f(n)$ and output $y(n)$ of a digital filter can be expressed by the following relationship.

$$y(n) = \sum_{k=0}^{N} b(k)f(n-k) + \sum_{k=1}^{M} a(k)y(n-k) \qquad (5.23)$$

The coefficients $b(k)$ denote the dependence of the output sample on the current input sample and N-previous input samples. The coefficients $a(k)$ denote the dependence of the output sample on M past output samples. The filters can be divided into two broad categories depending of the values of $a(k)$ and $b(k)$. If all $a(k)$'s are zero, and N is finite, the filter is called a *finite impulse response* (FIR) filter. If $a(k)$'s are not zero for at least one value of k, or N is not finite, the filter is called *infinite impulse response* (IIR) filter. In an FIR filter, the number impulses in the *impulse response* of the filter is finite whereas in IIR filter, the number of impulses is infinite.

The IIR filters can produce a steeper slope with a few coefficients. Therefore, a better frequency response can be obtained with a low implementation cost. However, stability is a prime concern in IIR filtering since there is a feedback element. The FIR filters on the other hand are more robust, and can provide linear phase response. Hence, in practice, most practical systems employ FIR filtering.

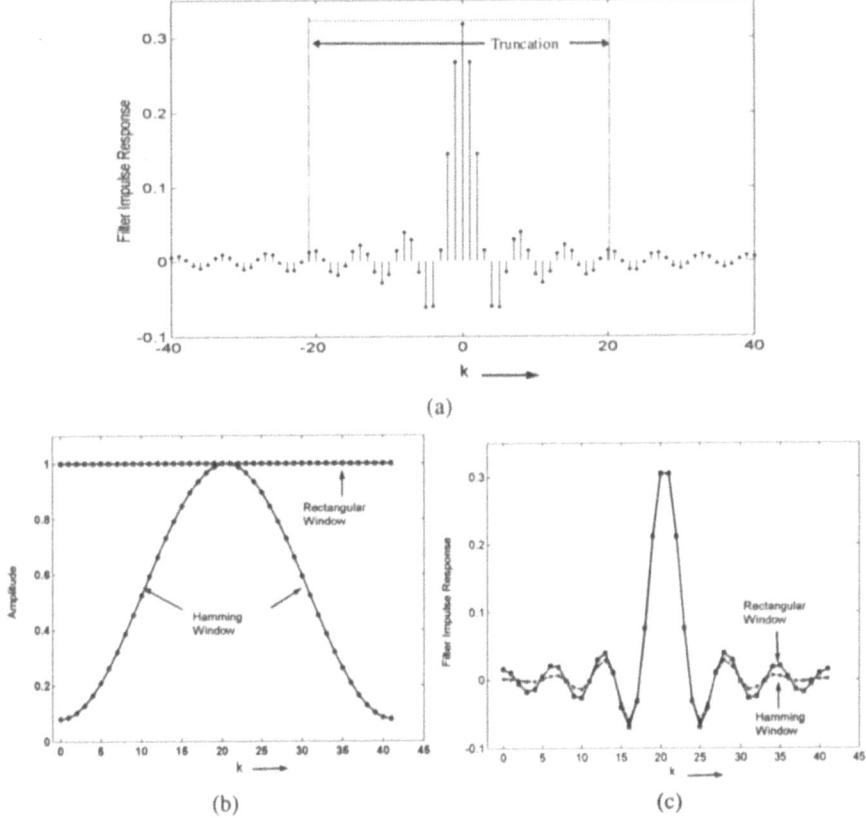

Figure 5.6. Impulse response of lowpass filter. a) ideal lowpass filter impulse response, b) rectangular and Hamming window, c) windowed impulse response. Note that an ideal filter has infinitely long impulse response. In Fig. (a) only 81 impulses are shown.

The coefficients of an FIR filter can be calculated simply by taking the inverse Fourier transform of the desired frequency response. The inverse Fourier transform (of the rectangular pulse frequency response) produces an infinitely long impulse response (which is basically a *sinc* function) as shown in Fig. 5.6(a). It is physically impossible to realize this filter since it will require infinite number of filter taps. Therefore, the filter coefficients are truncated to obtain FIR filter with finite number of taps (or impulses) as shown in Fig. 5.6(a). However, a direct truncation results in a poor frequency response (high ripples in the filter characteristics) due to the Gibbs phenomenon. Hence, the truncation is generally done using a tapered window (*e.g.*, triangular, Hanning, Hamming, Kaiser) [9].

Figure 5.6(b) shows both rectangular (*i.e.*, direct truncation) and hamming windows with 41 taps. The windowed filter impulse responses are shown in Fig. 5.6(c). The frequency characteristics of these windows are shown in Fig. 5.7. It is observed that the ripples near the transition band are significantly smaller for Hamming window. However, the Hamming window increases the width of the transition band (see Fig. 5.7(b)). Typically, Hamming and Kaiser windows (which is an optimal window) are used in FIR filter design.

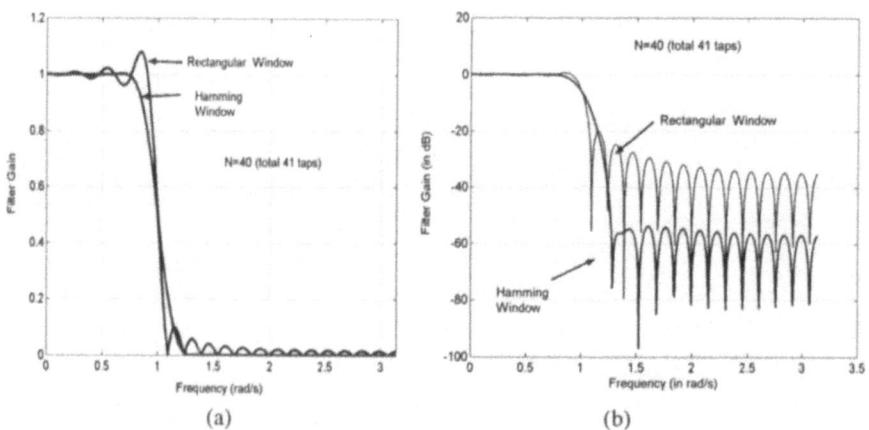

(a)   (b)

Figure 5.7. Gain response of lowpass filters with 41 taps rectangular and Hamming windows. a) Gain response , b) Gain response in dB.

The impulse response of FIR filters with desired cut-off frequency can easily be designed using the MATLAB function "fir1". The *fir1* function uses Hamming window by default. The *fir1* function accepts normalized (with respect to the sampling frequency) cut-off frequency. For example, if we want to design a lowpass filter with a cutoff frequency of 3200 Hz for an audio signal sampled at 8000 samples/sec, the cutoff frequency will be 3200/8000 or 0.4. The following MATLAB code will provide 9-tap lowpass and highpass digital filters with cutoff frequency of 0.4.

```
filter_lowpass = fir1(8,0.4) ;          %8th order, i.e. 9 tap filter
filter_highpass = fir1(8,0.4,'high') ;
```

For bandpass and bandstop filters, two cutoff frequencies (lower and upper) are required. The following code will provide the impulse response of 9-tap bandpass and bandstop filters with cutoff frequencies [0.4, 0.8].

```
filter_bandpass = fir1(8,[0.4 0.8])
filter_bandstop = fir1(8,[0.4 0.8],'stop')
```

The filter coefficients obtained using the above codes are shown in Table 5.2. The corresponding gain characteristics are shown in Fig. 5.8. Note that the gain characteristics are far from ideal, due to their short lengths. If the filter length is increased, a better frequency characteristics can be obtained as shown in Fig. 5.9. However, this will also increase the computational complexity (or, the implementation cost) of the filtering operation.

Table 5.2. Examples of a few 9-tap FIR digital filters. The lowpass and highpass cut-off frequency is 0.4. The bandpass and bandstop cut-off frequencies are [0.4,0.8] and [0.4, 0.8], respectively.

Filter	Coefficients
Lowpass	[-0.0061 -0.0136 0.0512 0.2657 0.4057 0.2657 0.0512 -0.0136 -0.0061]
Highpass	[0.0060 0.0133 -0.0501 -0.2598 0.5951 -0.2598 -0.0501 0.0133 0.0060]
Bandpass	[0.0032 0.0478 -0.1802 -0.1363 0.5450 -0.1363 -0.1802 0.0478 0.0032]
Bandstop	[-0.0023 -0.0354 0.1336 0.1011 0.6061 0.1011 0.1336 -0.0354 -0.0023]

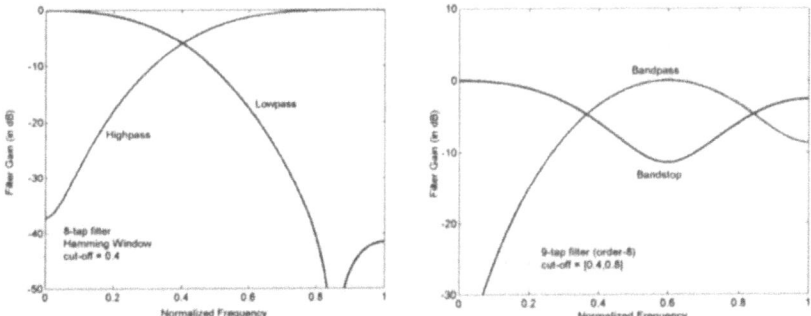

Figure 5.8. Gain response of digital filters whose impulse responses are shown in Table 5.2.

5.4.2 Subband Analysis

A typical schematic of a subband system is shown in Fig. 5.10. The input signal is passed through the analysis section that contains a bank of lowpass, bandpass and highpass filters. The individual outputs produced by the filters contain different frequency band information of the input signal. Hence, the output coefficients corresponding to a particular filter are collectively known as a subband. In other words, an N-band filterbank has N subbands, each representing a part (1/N) of the entire frequency spectrum. The bank of filters is designed in such a way that there is minimum overlap between the passband of the individual filters.

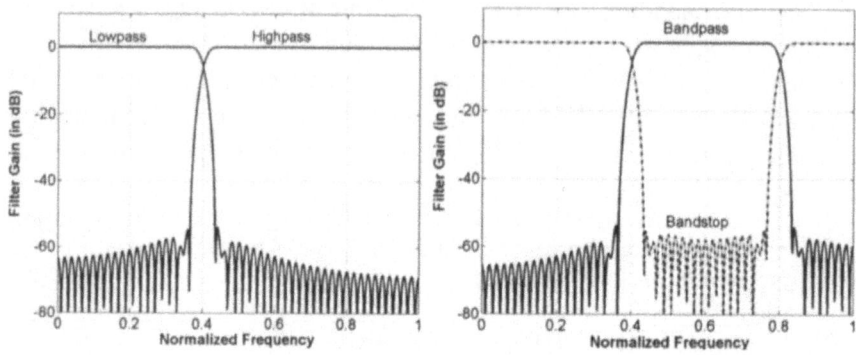

Figure 5.9. Gain response of 101-tap digital filters.

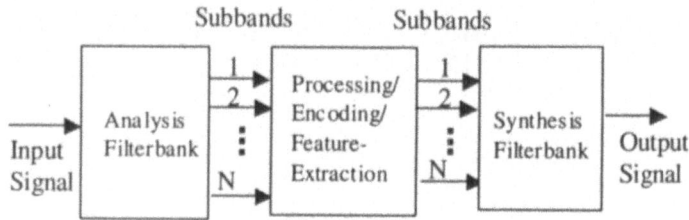

Figure 5.10. A typical subband analysis and synthesis system.

When the analysis section produces different subbands, the subband coefficients are analyzed appropriately depending on the application. The processed subband coefficients can then be passed through the synthesis filterbanks to reconstruct the input signal. The synthesis filterbank again consists of a bank of lowpass, bandpass and highpass filters. The operation of a two-band filterbank is now presented.

The filters in a filterbank can be FIR filters or IIR filters. Because of their design simplicity and better stability, FIR filters are generally used in most applications.

Two-band Filterbank

A two-band FIR filterbank is shown in Fig. 5.11 (a). Here, the input signal $x(n)$ is passed through the lowpass and highpass filters with impulse responses $\tilde{h}(n)$ and $\tilde{g}(n)$, respectively. Typical frequency response of the filters is shown in Fig. 5.11(b). The passband of the lowpass filter is approximately $[0, F_s/4]$, and the highpass filter is approximately $[F_s/4, F_s/2]$ where F_s is the sampling frequency. Because of the filter characteristics, the bandwidths of the intermediate signals $x_0(n)$ and $x_1(n)$ are approximately half of the bandwidth of $x(n)$. Therefore, $x_0(n)$ and $x_1(n)$ can be decimated by 2, without violating the Nyquist criterion, to obtain $v_0(n)$ and $v_1(n)$. The overall data-rate after the decimation is the same as that of the input, but the frequency components have been separated into two bands. After $v_0(n)$ and $v_1(n)$ are generated, they can be decomposed again by another two-band filterbank to obtain a total of four bands. This process can be repeatedly used to obtain a larger number of bands.

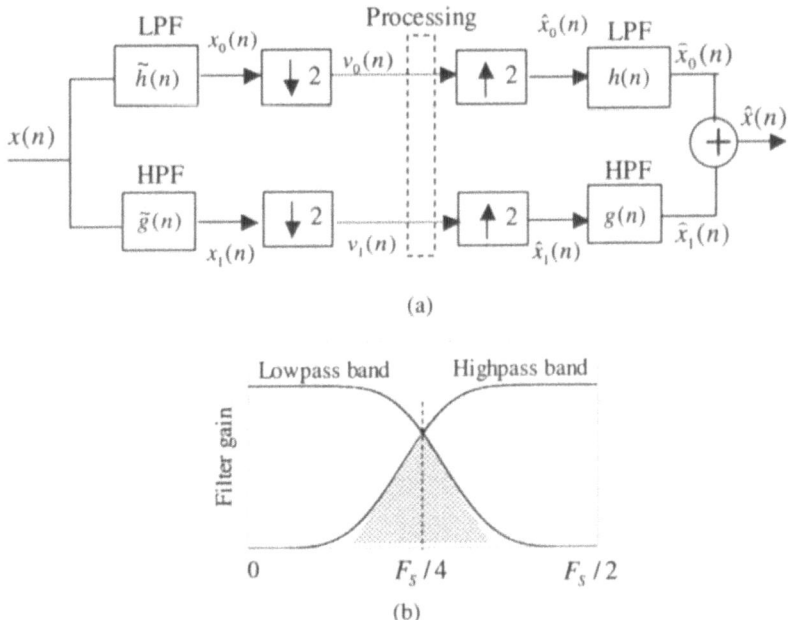

(a)

(b)

Figure 5.11. A two-band filterbank. Multiple two-band filterbanks can be cascaded to form additional subbands.

Once, the decimated outputs of the filterbanks are calculated, the subband data is ready for further processing. For example, the coefficients may be quantized to achieve compression. Assume that the processed signal is represented by $v_0(n)$ and $v_1(n)$ for the lowpass and highpass bands, respectively. The reconstruction of $x(n)$ is as follows. The signals $v_0(n)$ and $v_1(n)$ are upsampled by 2, and $\hat{x}_0(n)$ and $\hat{x}_1(n)$ are obtained. Note that *upsampling* (by 2) operation inserts a zero between every two samples. The signals $\hat{x}_0(n)$ and $\hat{x}_1(n)$ are then passed through the $h(n)$ and $g(n)$ filters, respectively. The outputs $\hat{x}_0[n]$ and $\hat{x}_1[n]$ are added to obtain $\hat{x}[n]$.

Figure 5.11(b) shows the frequency response of a special class of filters, where the lowpass and highpass filters have responses symmetric to the frequency $F_s/4$. Hence, these filters are known as the *quadrature mirror filter* (QMF). The lowpass and highpass filters do not strictly have a bandwidth of $F_s/4$. Hence, there will be aliasing energy in $v_0(n)$ and $v_1(n)$. However, these filters have a special property: if the original subband coefficients (*i.e.*, with no quantization) are passed through the synthesis filterbank, the aliasing in the synthesis bank will cancel the aliasing energy in $v_0(n)$ and $v_1(n)$, and hence reconstruct the input signal $x[n]$ without any distortion.

The filterbanks can be divided into several categories. They can be perfect reconstruction or reversible if the input sequence can be perfectly reconstructed using the unquantized subband coefficients and the synthesis filterbanks. Otherwise, the filterbank is called non-perfect reconstruction or irreversible. Another important category is the *paraunitary filterbank* where the forward transformation matrix is the inverse of the inverse transformation matrix (similar to an orthonormal transform).

An important question in the filterbank analysis is how do we design a filterbank, and especially how do we ensure that it is paraunitary? It can be shown that a two-band filterbank will be paraunitary if Eqs. (5.24)-(5.27) are satisfied (see Ref [10] for more details).

$$\sum_{n=0}^{N-1} h(n)h(n-2k) = \delta(k) \tag{5.24}$$

$$h(n) = (-1)^{n+1}\tilde{g}(n) \tag{5.25}$$

$$\tilde{g}(n) = (-1)^{n+1}\tilde{h}(N-1-n) \tag{5.26}$$

$$g(n) = (-1)^{n}\tilde{h}(n) \tag{5.27}$$

The first condition satisfies the paraunitary property. The other three conditions provide a convenient way to calculate the three other filters from a given filter. The above conditions state that if we can find a filter $h(n)$ such that it together with its derivatives (i.e., $\tilde{h}(n)$, $\tilde{h}(n)$, $\tilde{h}(n)$) satisfy the four above equations, we are sure that the filterbank is paraunitary. The following example illustrates how to obtain the other filters from $h(n)$.

■ **Example 5.5**

Consider the following $h(n)$.
$$h(n) = [0.4830, 0.8365, 0.2241, -0.1294] \qquad (5.28)$$
Using the conditions given in Eqs. (5.25)-(5.27), determine the other filters of the filterbank.

It can be easily verified that Eq. (5.24) is satisfied for the $h(n)$ given in Eq. (5.28). As per the convention used in Fig. 5.11, $h(n)$ is the lowpass filter in the synthesis section. Eq. (5.25) states that $h(0) = -\tilde{g}(0)$, $h(1) = \tilde{g}(1)$, $h(2) = -\tilde{g}(2)$, and $h(3) = \tilde{g}(3)$. In other words:

$$\tilde{g}(n) = [-0.4830, 0.8365, -0.2241, -0.1294]$$

According to Eq. (5.26), $\tilde{g}(0) = -\tilde{h}(N-1) = -\tilde{h}(3)$, $\tilde{g}(1) = \tilde{h}(2)$, $\tilde{g}(2) = -\tilde{h}(1)$, and $\tilde{g}(3) = \tilde{h}(0)$. Using the $\tilde{g}(n)$ values obtained in the last step, we get
$$\tilde{h}(n) = [-0.1294, 0.2241, 0.8365, 0.4830]$$
From Eq. (5.27), we get $g(0) = \tilde{h}(0)$, $g(1) = -\tilde{h}(1)$, $g(2) = \tilde{h}(2)$, $g(3) = -\tilde{h}(3)$. Therefore,
$$g(n) = [-0.1294, -0.2241, 0.8365, -0.4830] \quad ■$$

■ **Example 5.6**

Consider the lowpass filter:
$$h(n) = \left[1/\sqrt{2} \quad 1/\sqrt{2}\right] \approx [0.7071, 0.7071] \qquad (5.29)$$

Using an approach similar to Example 5.4, the other filters can be obtained as follows:
$$\tilde{g}(n) = [0.7071, -0.7071]$$
$$\tilde{h}(n) = [0.7071, 0.7071]$$
$$g(n) = [-0.7071, 0.7071] \quad ■$$

■ **Example 5.7**

Consider the filterbank in Example 5.6. Calculate the output at various stages of the filterbank for the input $x(n) = [1, 2, 5, 7, 8, 1, 4, 5]$

After the first stage filters:

The outputs of the filter $x_0(n)$ and $x_1(n)$ can be calculated by convolving the input with $\tilde{h}(n)$ and $\tilde{g}(n)$. Since the input signal has a finite width, circular convolution should be performed in order to achieve the perfect reconstruction at the output of the synthesis filterbank.

$x_0(n) = [2.1213 \quad 4.9497 \quad 8.4853 \quad 10.6066 \quad 6.3640 \quad 3.5355 \quad 6.3640 \quad 4.2426]$

$x_1(n) = [-0.7071 \quad -2.1213 \quad -1.4142 \quad -0.7071 \quad 4.9497 \quad -2.1213 \quad -0.7071 \quad 2.8284]$

Note that the wrapped around convolution has been pushed to the end of $x_0(n)$ and $x_1(n)$. For example, $x_0(0) = 1 \times 0.7071 + 2 \times 0.7071 = 2.1213$ whereas $x_0(7) = 5 \times 0.7071 + 1 \times 0.7071 = 4.2426$.

After the Decimation:

$v_0(n)$ and $v_1(n)$ are obtained by decimating $x_0(n)$ and $x_1(n)$ by 2 (in other words, keep one sample, drop the next one).

$v_0(n) = [2.1213 \quad 8.4853 \quad 6.3640 \quad 6.3640]$

$v_1(n) = [-0.7071 \quad -1.4142 \quad 4.9497 \quad -0.7071]$

After Upsampling:

Assume that the original coefficients are fed to the synthesis filterbank. The upsampled (insert one zero after every sample) output $\hat{x}_0(n)$ and $\hat{x}_1(n)$ will be:

$\hat{x}_0(n) = [2.1213 \quad 0.0 \quad 8.4853 \quad 0.0 \quad 6.3640 \quad 0.0 \quad 6.3640 \quad 0.0]$

$\hat{x}_1(n) = [-0.7071 \quad 0.0 \quad -1.4142 \quad 0.0 \quad 4.9497 \quad 0.0 \quad -0.7071 \quad 0.0]$

After Synthesis Filtering

The outputs of the filter $\tilde{x}_0(n)$ and $\tilde{x}_1(n)$ can be calculated by (circular) convolving the $\hat{x}_0(n)$, $\hat{x}_1(n)$ with $h(n)$ and $g(n)$, respectively.

$\tilde{x}_0(n) = [1.500 \quad 1.500 \quad 6.000 \quad 6.000 \quad 4.500 \quad 4.500 \quad 4.500 \quad 4.500]$

$\tilde{x}_1(n) = [-0.50 \quad 0.50 \quad -1.00 \quad 1.00 \quad 3.500 \quad -3.500 \quad -0.500 \quad 0.500]$

$[\tilde{x}_0(0) = 0.0 \times 0.7071 + 2.1213 \times 0.7071 = 1.5, \tilde{x}_0(1) = 2.1213 \times 0.7071 + 0.0 \times 0.7071 = 1.5]$

Reconstructed Output

The reconstructed input can be calculated by adding $\tilde{x}_0(n)$ and $\tilde{x}_1(n)$.

Reconstructed signal $\hat{x}(n) = [1, 2, 5, 7, 8, 1, 4, 5]$

■

Note that the signal has been reconstructed perfectly as the original output of the analysis filterbank was fed to the synthesis filterbank. Even if a single coefficient is changed marginally, the output will not be identical to the input.

5.4.3 Transforms and Digital Filtering

We have seen that the unitary transforms provide us with coefficients corresponding to different frequencies. On the other hand, subband coefficients also provides us coefficients corresponding to different subbands, which are non-overlapping (in the ideal case) in the frequency domain. It can be easily demonstrated that block transforms are special class of filterbanks. An N-point transform can be calculated using an N-band filterbank where each filter corresponds to the complex conjugate of a basis function (for a real transform, the complex conjugation operation is not required). The output of each filter is then decimated by N, so that for each set of N input samples there would be one output from each of the N filters. These outputs are basically the transform coefficients. This is illustrated with an example.

■ **Example 5.8**

Consider a 12-point sequence $f=[\{2, 5, 7, 6\}, \{1, 3, 9, 4\}, \{6, 8, 5, 7\}]$. Calculate the 2nd DCT coefficients of each block of 4 coefficients. Can we implement the DCT as a filter bank?

Using Eq. (5.18), the 2nd coefficients of each block can be calculated as: {-3.1543, -3.5834, -0.6069, 0.1585}. Yes, the DCT coefficients can also be implemented using a filter bank. The coefficients can be calculated by passing the input sequence through a digital filter with impulse response h = [cos(pi/8), cos(3*pi/8), cos(5*pi/8), cos(7*pi/8)] which is the 2nd basis function (see Example 5.3). Note that when the digital filter is in operation, each block of four input samples will produce one sample at the output of the 2nd filter. ■

The filter characteristics of 8-point DCT was shown in Fig. 5.3(b). It was observed that the neighboring filters have substantial overlap, and hence may not be efficient in many applications.

5.5 1-D DISCRETE WAVELET TRANSFORM

Although the Fourier transform and its derivatives (such as DCT) have been used extensively in signal processing, these transforms have several limitations. First, the basis functions have a long width. For example, for 256-point DFT, there are 256 basis functions, each with a width of 256 impulses. To represent a small rectangular pulse, a large number of frequency components are required, which may be inefficient in many applications, such as signal compression. Secondly, Fourier transform does not provide a good performance for nonstationary signal analysis. Thirdly, it is difficult to estimate the time-domain (for audio) or spatial domain (for images) characteristics from the Fourier or DCT amplitude spectrum. Finally, they do not provide good filter characteristics.

The subband decomposition provides efficient separation of the frequency spectrum. However, it is suitable only for analysis of discrete data. In addition, the unitary properties of the transforms are not readily available.

In recent years, wavelet transform (WT) has emerged as a powerful mathematical tool [5] in many areas of science and engineering. Wavelet transform provides the best of the transform and subband decomposition techniques. The WT can be applied to both continuous and discrete signal, and can be considered as a transform as well as a subband decomposition. Because of the space constraint, we briefly introduce the discrete wavelet transform (DWT) in the following. Interested readers can find a detailed step-by-step development of continuous and discrete wavelet transform in [7], a copy of which is provided in the accompanying CD in PDF format.

The basis functions of Fourier transform are complex exponentials, which are unique. The basis functions of DCT are discrete cosine functions which are once again unique. However, there is no unique set of basis functions for the DWT. One can design a set of basis functions most suitable for a given application. Here, we consider a special class of DWT, known as dyadic wavelets.

Forward Transform (Analysis)

A given dyadic wavelet transform is uniquely defined by a FIR lowpass filter $h[.]$ and a FIR highpass filter $g[.]$. Consider a discrete N-point sequence $c_{0,k}$, $0 \le k \le N-1$. The DWT coefficients are calculated recursively using the following equations.

$$c_{j+1,k} = \sum_m c_{j,m} h[m-2k] \qquad 0 \le k \le N/2^{j+1} - 1 \qquad (5.30)$$

$$d_{j+1,k} = \sum_m c_{j,m} g[m - 2k] \qquad 0 \le k \le N/2^{j+1} - 1 \qquad (5.31)$$

where

$c_{p,q}$ = Lowpass coefficient of p th scale at q th location.

$d_{p,q}$ = Highpass coefficient of p th scale at q th location.

There are several points to be noted:

i) The input sequence can be considered as the DWT coefficients at 0th scale.

ii) The lowpass DWT coefficients $c_{1,k}$ can be obtained by convolving $c_{0,m}$ and $h[-n]$, and decimating the convolution output by a factor of 2 (because of the "$2k$" factor). Similarly, $d_{1,k}$ can be obtained by convolving $c_{0,m}$ and $g[-n]$, and decimating the convolution output by a factor of 2. The numbers of $c_{1,k}$ and $d_{1,k}$ coefficients are N/2 each.

iii) $c_{2,k}$ and $d_{2,k}$ can be obtained from $c_{1,k}$ using a step similar to step (ii). The numbers of $c_{2,k}$ and $d_{2,k}$ coefficients are N/4 each. Higher scale coefficients can be obtained in a similar manner.

iv) Only the lowpass coefficients output of a given stage is used for further decomposition (there is no $d_{j,m}$ in the right hand side of Eqs. (5.30) and (5.31)).

Eqs. (5.30) and (5.31) provide a simple algorithm of calculating $j+1$ scale wavelet coefficients from j scale coefficients. We can use the same set of equations to calculate $j+2$ scale coefficients from $j+1$ scale scaling coefficients. Figure 5.12(a) shows the schematic of computing the DWT coefficients of lower resolution scales from the wavelet coefficients of a given scale by passing it through a lowpass filter (LPF) and a highpass filter (HPF), and by decimating the filters' outputs by a factor of two. This recursive calculation of DWT coefficient is known as the *tree* algorithm [6].

Inverse Transform (Synthesis)

Using an approach similar to the wavelet decomposition, the wavelet coefficients of a given scale can be reconstructed using the coefficients of higher scales. The reconstructed coefficients are expressed as follows:

$$c_{j,m} = \sum_k c_{j+1,k} h[m - 2k] + \sum_l d_{j+1,l} g[m - 2l] \qquad (5.32)$$

For inverse wavelet transform, the wavelet coefficients of a given scale can be upsampled by 2 (*i.e.*, just insert a zero between every consecutive coefficients), and passed through a set of lowpass and highpass filters, and then added to obtain the wavelet coefficients corresponding to the next higher resolution scale. The synthesis process is shown in Fig. 5.12(b).

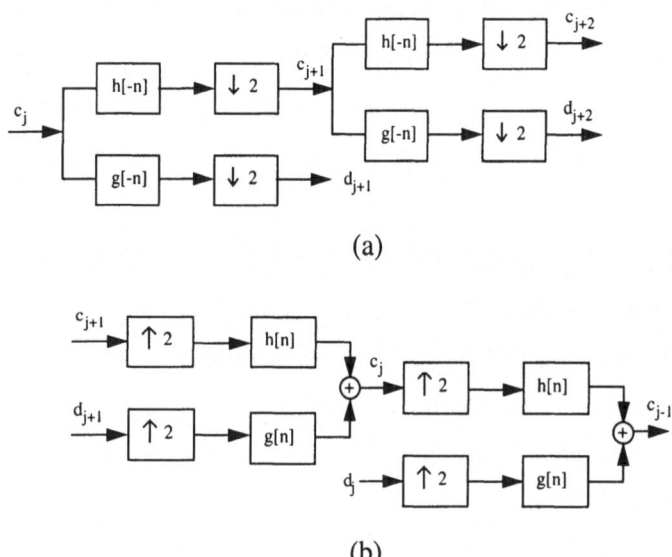

Figure 5.12. Calculation of wavelet transform coefficients using the *tree* algorithm [6]. a) Signal decomposition using analysis filters, and b) signal reconstruction using synthesis filters.

The decomposition can also be done using a matrix multiplication approach. Assume an 8-point discrete time input sequence $c_{j,k}$ ($0 \leq k \leq 7$) and 4-tap lowpass and highpass filters. Further assume that the wavelet is unitary. The $h[n]$ and $g[n]$ are then related according to Eqs. (5.25)-(5.27), which can be expressed as:

$$g[n] = (-1)^n h[N - 1 - n] \qquad (5.33)$$

$$
\begin{bmatrix}
c_{j+1,0} \\
d_{j+1,0} \\
c_{j+1,1} \\
d_{j+1,1} \\
c_{j+1,2} \\
d_{j+1,2} \\
c_{j+1,3} \\
d_{j+1,3}
\end{bmatrix}
=
\begin{bmatrix}
h[0] & h[1] & h[2] & h[3] & & & & \\
h[3] & -h[2] & h[1] & -h[0] & & & & \\
& & h[0] & h[1] & h[2] & h[3] & & \\
& & h[3] & -h[2] & h[1] & -h[0] & & \\
& & & & h[0] & h[1] & h[2] & h[3] \\
& & & & h[3] & -h[2] & h[1] & -h[0] \\
h[2] & h[3] & & & & & h[0] & h[1] \\
h[1] & -h[0] & & & & & h[3] & -h[2]
\end{bmatrix}
\begin{bmatrix}
c_{j,0} \\
c_{j,1} \\
c_{j,2} \\
c_{j,3} \\
c_{j,4} \\
c_{j,5} \\
c_{j,6} \\
c_{j,7}
\end{bmatrix}
\quad (5.34)
$$

The 8×8 H matrix in the right side of Eq. (5.34) can be considered as the forward transformation matrix of the wavelet transform. This representation is comparable to other transforms, such as Fourier (Eq. (5.7)) and DCT (Eq. (5.18)). If the transform is unitary, the H matrix should be unitary, *i.e.*, $H^{-1} = H^T$.

Table 5.3 shows four orthogonal wavelets from a wavelet family first constructed by Daubechies [7]. The first wavelet is also known as Haar wavelet. Since the wavelets are orthogonal, the corresponding $g[n]$ can be calculated using Eq. (5.33).

Table 5.3. Minimum phase Daubechies wavelets for N=2, 4, 8, and 12 taps.

	n	h[n]		n	h[n]
N=2	0	0.70710678	N=12	0	0.111540743350000
	1	0.70710678		1	0.494623890398000
N=4	0	0.482962913144		2	0.751133908021000
	1	0.836516303737		3	0.315250351709000
	2	0.224143868042		4	-0.226264693965000
	3	-0.129409522551		5	-0.129766867567000
N=8	0	0.230377813308		6	0.097501605587000
	1	0.714846570552		7	0.027522865530000
	2	0.630880767939		8	-0.031582039318000
	3	-0.027983769416		9	0.000553842201000
	4	-0.18703481171		10	0.004777257511000
	5	0.030841381835		11	-0.001077301085000
	6	0.032883011666			
	7	-0.010597401785			

■ **Example 5.9**

Consider a 4-point sequence f=[2, 5, 7, 6]. Decompose the sequence using 2-tap wavelet given in Table 5.3, which is also known as Haar wavelet. Reconstruct the input sequence from the Haar coefficients.

The filter coefficients of Haar wavelet are h=[0.707 0.707]). The 1st stage (*i.e.*, scale 1) Haar coefficients can be calculated as:

$$
\begin{bmatrix} c_{1,0} \\ d_{1,0} \\ c_{1,1} \\ d_{1,1} \end{bmatrix} = \begin{bmatrix} h[0] & h[1] & 0 & 0 \\ h[1] & -h[0] & 0 & 0 \\ 0 & 0 & h[0] & h[1] \\ 0 & 0 & h[1] & -h[0] \end{bmatrix} \begin{bmatrix} c_{0,0} \\ c_{0,1} \\ c_{0,2} \\ c_{0,3} \end{bmatrix} = \begin{bmatrix} 0.707 & 0.707 & 0 & 0 \\ 0.707 & -0.707 & 0 & 0 \\ 0 & 0 & 0.707 & 0.707 \\ 0 & 0 & 0.707 & -0.707 \end{bmatrix} \begin{bmatrix} 2 \\ 5 \\ 7 \\ 6 \end{bmatrix} = \begin{bmatrix} 4.95 \\ -2.12 \\ 9.19 \\ 0.71 \end{bmatrix}
$$

In the second recursion, only two lowpass coefficients will be used.

$$
\begin{bmatrix} c_{2,0} \\ d_{2,0} \end{bmatrix} = \begin{bmatrix} h[0] & h[1] \\ h[1] & -h[0] \end{bmatrix} \begin{bmatrix} c_{1,0} \\ c_{1,1} \end{bmatrix} = \begin{bmatrix} 0.707 & 0.707 \\ 0.707 & -0.707 \end{bmatrix} \begin{bmatrix} 4.95 \\ 9.19 \end{bmatrix} = \begin{bmatrix} 10 \\ -3 \end{bmatrix}
$$

We now have only one lowpass coefficient, and further decomposition cannot be done. After rearrangement, the four Haar coefficients becomes $[c_{2,0}, d_{2,0}, d_{1,0}, d_{1,1}] = [10 \ -3 \ -2.12 \ 0.71]$.

The signal reconstruction is performed starting at the highest scale (*i.e.*, lowest pass DWT coefficients. The first stage synthesis is done as follows.

$$
\begin{bmatrix} c_{1,0} \\ c_{1,1} \end{bmatrix} = \begin{bmatrix} h[0] & h[1] \\ h[1] & -h[0] \end{bmatrix} \begin{bmatrix} c_{2,0} \\ d_{2,0} \end{bmatrix} = \begin{bmatrix} 0.707 & 0.707 \\ 0.707 & -0.707 \end{bmatrix} \begin{bmatrix} 10 \\ -3 \end{bmatrix} = \begin{bmatrix} 4.95 \\ 9.19 \end{bmatrix}
$$

$$
\begin{bmatrix} c_{0,0} \\ c_{0,1} \\ c_{0,2} \\ c_{0,3} \end{bmatrix} = \begin{bmatrix} h[0] & h[1] & 0 & 0 \\ h[1] & -h[0] & 0 & 0 \\ 0 & 0 & h[0] & h[1] \\ 0 & 0 & h[1] & -h[0] \end{bmatrix} \begin{bmatrix} c_{1,0} \\ d_{1,0} \\ c_{1,1} \\ d_{1,1} \end{bmatrix} = \begin{bmatrix} 0.707 & 0.707 & 0 & 0 \\ 0.707 & -0.707 & 0 & 0 \\ 0 & 0 & 0.707 & 0.707 \\ 0 & 0 & 0.707 & -0.707 \end{bmatrix} \begin{bmatrix} 4.95 \\ -2.12 \\ 9.19 \\ 0.71 \end{bmatrix} = \begin{bmatrix} 2 \\ 5 \\ 7 \\ 6 \end{bmatrix}
$$

■

Figure 5.13(a) shows the basis functions corresponding to 8-point Haar wavelets (h=[0.707 0.707]). The frequency response of these basis functions is shown in Fig. 5.13(b). It is observed that four basis functions corresponding to n=4, 5, 6, and 7 have support of only two-samples, and hence can be considered as compact in the time-domain. These basis functions are the shifted version of each other, and have identical gain response. They correspond to the highest frequency subband with highest time resolution but poorest frequency resolution. Next consider the basis functions for n=2 and 3. These functions have a support-width of 4 samples. These basis functions have moderate time and frequency resolution. Finally, the basis functions corresponding to n=0 and 1, have the support width of 8 pixels. These functions have poorer time resolution, but better frequency resolution. This adaptive time-frequency is useful in nonstationary signal (such as audio and image) analysis.

5.6 2-D UNITARY TRANSFORM

The 1-D transform is useful for 1-D signal analysis. However, for two-dimensional signals, such as an image, 2-D transforms are required. We will now consider the extension of the 1-D concepts discussed already to calculate 2-D transform.

Let an $N \times N$ image be denoted by

$$I = [i(m,n)], \qquad\qquad 0 \le m,n \le N-1$$

The forward and inverse transforms are defined as

$$\theta(k,l) = \sum_{m=0}^{N-1}\sum_{n=0}^{N-1} \omega(k,l;m,n)i(m,n) \qquad 0 \le k,l \le N-1 \qquad (5.35)$$

$$i(m,n) = \sum_{k=0}^{N-1}\sum_{l=0}^{N-1} \upsilon(m,n;k,l)\theta(k,l) \qquad 0 \le m,n \le N-1 \qquad (5.36)$$

where $\omega(.)$ and $\upsilon(.)$ are the forward and inverse transform kernels.

Properties of 2-D Unitary transform

In this section, three important properties of 2-D unitary transform are discussed. These properties are useful for signal (such as audio and image) processing and compression (see Chapter 8).

i) Energy Conservation

It can be shown that for 2-D unitary transforms satisfy Parseval's relation *i.e.* the total energy in the frequency domain is equal to that of the spatial domain.

$$\sum_{m=0}^{N-1}\sum_{n=0}^{N-1}|i(m,n)|^2 = \sum_{k=0}^{N-1}\sum_{l=0}^{N-1}|\theta(k,l)|^2 , \text{ i.e., } \|I\|^2 = \|\Theta\|^2 \qquad (5.37)$$

Eq. (5.37) states that a unitary transform preserves the signal energy or, equivalently the length of the vector I in N^2 dimensional vector space. A unitary transformation can be considered simply as a rotation of the vector I in the N^2 dimensional vector space. In other words, a unitary transform rotates the basis coordinates, and the components of Θ (*i.e.*, the transform coefficients) are the projections of I on the new basis coordinates.

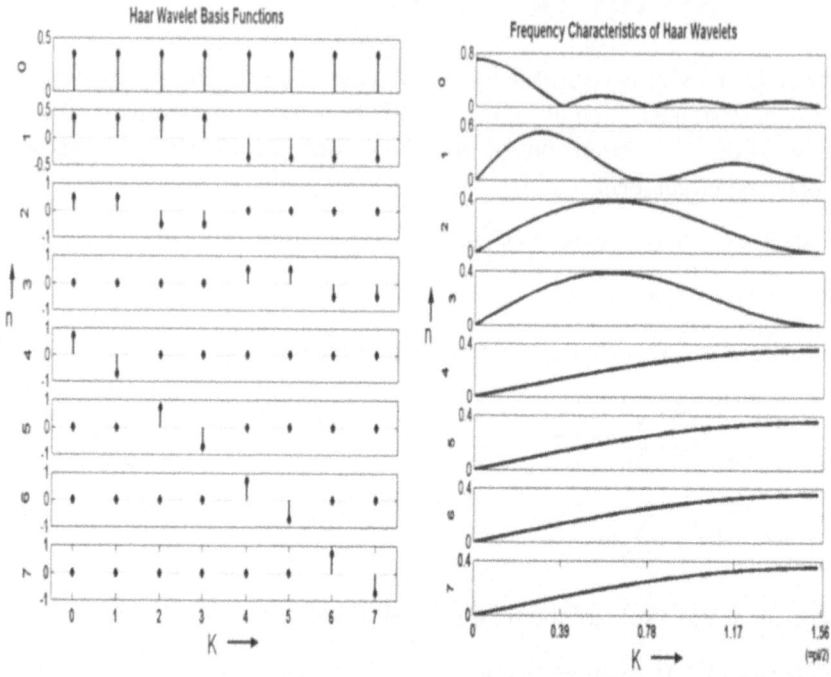

Figure 5.13. Basis functions of 8-point Haar Wavelet.

ii) Sum of Squared Error

Eq. (5.36) shows how $i(m,n)$ can be reconstructed perfectly using $N \times N$ (i.e., N^2) transform coefficients. Assume that during processing, the value of some of the transform coefficients change from $\theta(k,l)$ to $\hat{\theta}(k,l)$. This change might occur due to transmission error or during data compression. If we reconstruct the image by calculating the inverse transform using the coefficients $\hat{\theta}(k,l)$, the reconstructed image will be different from the original image. Let the image be denoted as $\hat{i}(m,n)$. It can be shown that the following relationship always hold true for unitary transform.

$$\sum_{m=0}^{N-1}\sum_{n=0}^{N-1}\left|i(m,n)-\hat{i}(m,n)\right|^2 = \sum_{k=0}^{N-1}\sum_{l=0}^{N-1}\left|\theta(k,l)-\hat{\theta}(k,l)\right|^2 \tag{5.38}$$

The left side of Eq. (5.38) is the total square error between the original and the reconstructed image whereas the right side is the total square error between the original and noisy coefficients. The above relationship basically states that the total squared reconstruction error in the pixel domain is equal to the total squared quantization error. The squared error will be zero if and only if all N^2 original coefficients are used for the reconstruction. This property has been used extensively to design transform based data compression.

iii) Separable Basis Images

In most cases of practical interest, the 2-D kernels are separable and symmetric. Therefore, the 2-D kernel can be expressed as the product of two 1-D orthogonal basis functions. If the 1-D transform operator is denoted by Φ, the forward and inverse transformations can be expressed as

$$\Theta = \Phi^* I \Phi^T \tag{5.39}$$

$$I = \Phi^T \Theta \Phi^* \tag{5.40}$$

The above formulations reveal that the image transformation can be done in two stages: by taking the unitary transform Φ of each row of the image array, and then by applying transformation Φ^* to each column of the intermediate result.

5.7 2-D DISCRETE FOURIER TRANSFORM

The two-dimensional unitary DFT pair is defined as

$$\theta(k,l) = \frac{1}{N} \sum_{m=0}^{N-1} \sum_{n=0}^{N-1} i(m,n) W_N^{km} W_N^{ln}, \qquad 0 \le k,l \le N-1 \tag{5.41}$$

$$i(m,n) = \frac{1}{N} \sum_{k=0}^{N-1} \sum_{l=0}^{N-1} \theta(k,l) W_N^{-km} W_N^{-ln}, \qquad 0 \le m,n \le N-1 \tag{5.42}$$

■ **Example 5.10**

Calculate the DFT, and plot the amplitude spectrum of the following 32x32 image.

$$i(m,n) = \begin{cases} 1 & 15 \le 16 \le 17 \& 8 \le n \le 24 \\ 0 & \text{otherwise} \end{cases}$$

The image is shown in Fig. 5.14(a), which is basically a rectangular parallelepiped. The DFT is calculated using MATLAB, and the amplitude spectrum is shown in Fig. 5.14(b). The spectrum has been made centralized (i.e., the DC frequency is in the middle) for clarity. It is observed that the amplitude spectrum is a 2-D *sinc* pattern, which is expected. Note that the rectangle has a larger width in the horizontal direction compared to the vertical direction. This is reflected in the spectrum. The spectrum has better frequency resolution in the horizontal direction compared to the vertical direction.

Note that the width of the rectangle is larger in the horizontal direction than in the vertical direction. It has an opposite effect in the frequency domain: the *sinc* pattern is narrower in the horizontal frequency axis, and wider in the vertical frequency axis. This is generally true for all transforms due to the uncertainty principle: an improved time resolution degrades the frequency resolution, and vice versa. ∎

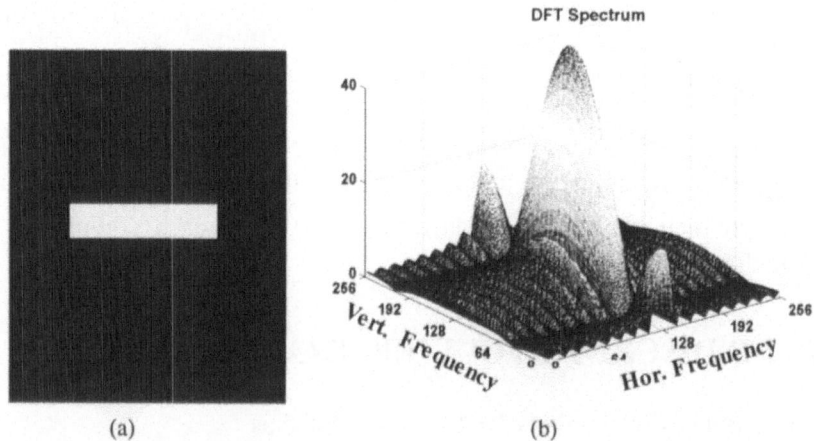

(a)   (b)

Figure 5.14. Discrete Fourier transform of a rectangle. (a) The 2-D input pattern, and (b) it's centralized amplitude spectrum. Note that the spectrum has been centralized (the DC frequency is in the middle), and zero padding has been applied to make the spectrum smooth.

Properties of 2-D DFT

Most of the properties of 1-D DFT can be extended to 2-D DFT [8], such as *convolution, correlation, energy conservation*, and *minimum sum of squared error*. We present a few selected properties of the 2-D DFT below.

i) Separability

It can be easily shown that the 2-D DFT is separable. Therefore, the 2-D DFT calculation of an image can be done in two simple steps:

 a) Calculate the 1-D DFT of each row of the image. The row is substituted by its DFT coefficients.

b) Calculate the 1-D DFT of each column. The column is replaced by its DFT coefficients

The 2-D output matrix represents the 2-D DFT coefficients of the input image. The block schematic of separable DFT calculation is shown in Fig. 5.15. Due to the separability property, complexity of $N \times N$ transform is equivalent to the complexity of $2N$ N-point 1-D DFT.

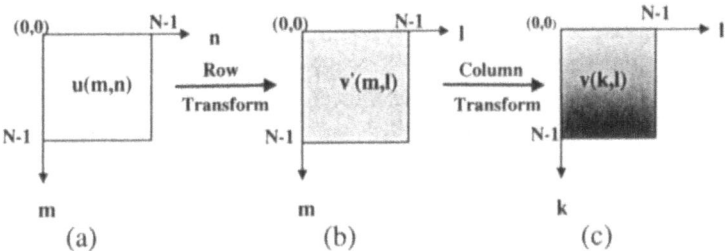

Figure 5.15. 2-D Fourier transform calculation using separable approach.

■ Example 5.11

Using separable approach, calculate the 2-D DFT of the image:

$$\begin{bmatrix} 2 & 3 & -4 & 7 \\ 1 & 5 & 2 & 8 \\ 2 & 4 & 9 & 3 \\ 2 & 7 & 1 & 4 \end{bmatrix}$$

To calculate the 2-D DFT, the 1-D transform of each row is first calculated. The 1-D DFT of [2 3 -4 7] is [4, 3+j2, -6, 3-j2]. Similarly, 1-D DFT of other rows can be calculated, and the coefficients matrix after row transform will be:

$$\frac{1}{2}\begin{bmatrix} 8 & 6+j4 & -12 & 6-j4 \\ 16 & -1+j3 & -10 & -1-j3 \\ 18 & -7-j & 4 & -7+j \\ 14 & 1-j3 & 8 & 1+j3 \end{bmatrix}$$

After the column transform, the final DFT coefficient matrix will be

$$F(k,l) = \frac{1}{4}\begin{bmatrix} 56 & -1+3j & -26 & -1-3j \\ -10-2j & 19+7j & -16+2j & 7-3j \\ -4 & -1+3j & 10 & -1-3j \\ -10+2j & 7+3j & -16-2j & 19-7j \end{bmatrix}$$

The results can be verified easily by direct calculation using Eq. (5.41). ■

ii) Conjugate Symmetry

If the image $i(m,n)$ is real, then the DFT coefficients $\theta(k,l)$ satisfy the following relationship.

$$\theta(k,l) \Leftrightarrow \theta^*(N-k, N-l) \tag{5.43}$$

Consider the 2-D image in Example 5.11. The symmetry property can easily be verified from the DFT coefficient matrix. For example, a 4x4 input image and its 2-D DFT are shown below. It is observed that the DFT coefficients satisfy the symmetry properties. Since natural images have real-valued pixels, the above property is applicable most of the time.

iii) Energy Compaction

The DFT provides significant energy compaction for most images. Figure 5.16 shows Lena image and its DFT amplitude spectrum. It is observed that most energy in the spectrum is concentrated in the four corners that represent the lower frequencies.

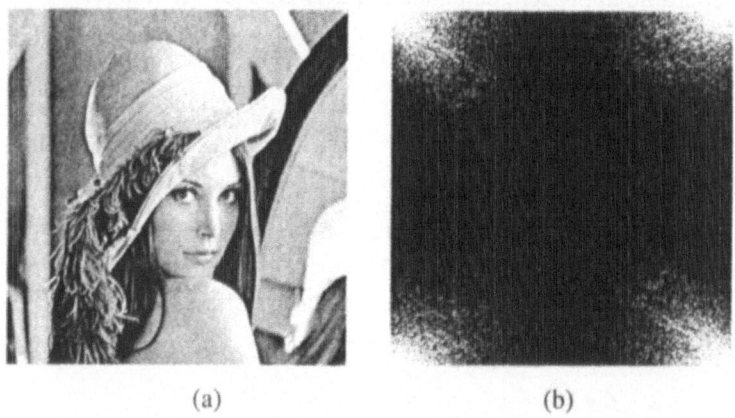

(a)   (b)

Figure 5.16. Fourier spectrum of image data. a) Lena image and b) it's amplitude spectrum. The white pixels in Fig (b) represents higher amplitude coefficients.

5.8   2-D DISCRETE COSINE TRANSFORM

The $N \times N$ two-dimensional DCT transform pair is defined as follows.

$$\theta(k,l) = \alpha(k)\alpha(l)\sum_{m=0}^{N-1}\sum_{n=0}^{N-1} i(m,n)\left[\cos\frac{(2m+1)k\pi}{2N}\right]\left[\cos\frac{(2n+1)l\pi}{2N}\right] \tag{5.44}$$

$$i(m,n) = \sum_{k=0}^{N-1}\sum_{l=0}^{N-1} \alpha(k)\alpha(l)\left[\cos\frac{(2m+1)k\pi}{2N}\right]\left[\cos\frac{(2n+1)l\pi}{2N}\right]\theta(k,l) \qquad (5.45)$$

The basis images (totaling 64) corresponding to 8x8 DCT is shown in Fig. 5.17. It is observed that the DC basis images represent the average intensity of the image. When we move horizontally in a row towards right, the basis images contain more vertical bars. Hence, the rightmost basis images represent mostly vertical edges. On the other hand, the bottom-most basis images represent mostly horizontal edges of an image.

The properties of 1-D DCT can be readily extended to 2-D DCT. First, the 2-D DCT is a separable transform. Hence, the coefficients can be calculated using row and column transforms.

■ Example 5.12

Using separable approach, calculate the 2-D DCT of the image:

$$\begin{bmatrix} 2 & 3 & -4 & 7 \\ 1 & 5 & 2 & 8 \\ 2 & 4 & 9 & 3 \\ 2 & 7 & 1 & 4 \end{bmatrix}$$

The coefficient matrix after row transform will be:

$$\begin{bmatrix} 4 & -1.37 & 5 & -5.92 \\ 8 & -3.76 & 1 & -3.85 \\ 9 & -2 & -4 & 2.99 \\ 7 & 0.32 & -1 & -4.4 \end{bmatrix}$$

After the column transform, the final DFT coefficient matrix will be

$$\Theta(k,l) = \begin{bmatrix} 14 & -3.41 & 0.5 & -5.62 \\ -2.23 & -1.58 & 5.27 & -2.81 \\ -3 & 2.35 & 3.5 & -4.76 \\ -0.16 & 0.69 & -1.64 & 4.08 \end{bmatrix}$$

The results can be verified easily by direct calculation using Eq. (5.44). ■

As in the 1-D case, the 2-D DCT can compact energy of typical images very efficiently. Fig. 5.18 shows the DCT spectrum of the Lena image. It is observed that most of the energy is concentrated in the low frequency region (upper left corner). It has been shown that DCT provides near optimal energy compaction performance [3] for most natural images. As a result, it has been accepted as the transform kernel for most existing image and video coding standard. More details about the DCT-based coding will be provided in Chapters 7 and 8.

Figure 5.17. 8x8 DCT Basis images   Figure 5.18. DCT spectrum of Lena image

5.9 2-D DISCRETE WAVELET TRANSFORM

We have seen in section 4 that there is no unique transformation kernel for 1-D discrete wavelet transforms. This is also the case for 2-D DWT. There are 2-D DWT's that are unitary, and there are others that are non-unitary. Although unitary DWT's are widely used, there is a special class of non-unitary DWT, known as bi-orthogonal wavelets, which are very popular in image compression application.

On the other hand, there are 2-D DWT that are separable, and there are also DWTs that are not separable. Most 2-D DWTs of interest are separable dyadic DWT's. In 1-D DWT, each level of decomposition produces two bands corresponding to low and high-resolution data (see Fig. 5.12). In case of 2-D wavelet transform, each level of decomposition produces four bands of data, one corresponding to lowpass band and three others corresponding to *horizontal*, *vertical* and *diagonal* highpass bands. The filtering can be done on "rows" and "columns" in the two-dimensional array corresponding to horizontal and vertical directions in images. The separable transform is shown in Fig. 5.19.

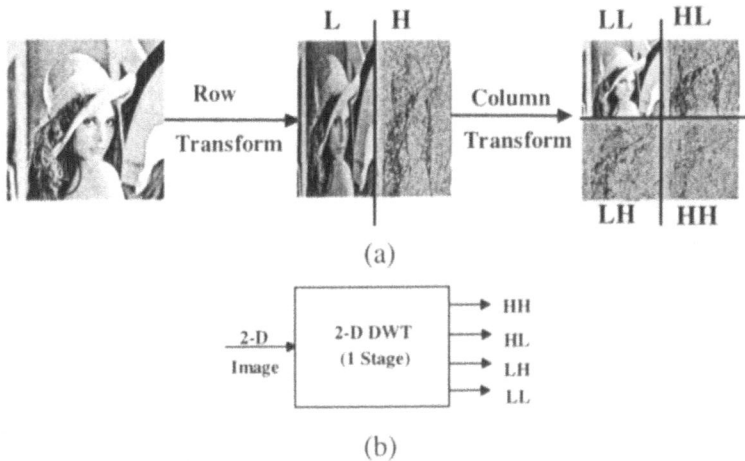

(a)

(b)

Figure 5.19. Two-dimensional wavelet transform. a) 2-D DWT by 1-D row and column transforms, b) Equivalent block schematic. L: output of lowpass filter after decimation, H: output of highpass filter after decimation.

■ **Example 5.13**

Consider the Lena image in Example 5.2. i) Calculate two-stage wavelet transform of the image using Daub-4 tap wavelet, ii) display the transform coefficients, iii) calculate the root mean square energy (RMSE) of the pixels of different bands, and iv) calculate the percentage of total energy compacted in the LL band.

As in the 1-D case (see Fig. 5.12), multi-stage 2-D wavelet decomposition can be performed by decomposing the LL band recursively. The two stage DWT calculation program is provided in the accompanying CD. Fig. 5.20(a) shows the 2-level wavelet decomposition of the Lena image. The RMSE of different bands are shown in Fig. 5.20(b). It can be easily shown that about 99.5% of the total energy is compacted in the lowpass subband.

■

Several features can be noted in Fig. 5.20. First, all subbands provide spatial and frequency information simultaneously. The spatial structure of the image is visible in all subbands. A low resolution image can be observed in the lowest resolution lowpass band. The highpass bands provide the detailed (*i.e.*, edge) information about the image at various scales. Secondly, the wavelet decomposition provides a multiresolution representation of the image. A coarse image is already available in the lowest resolution band. A higher resolution image can be obtained by calculating inverse transform of the four low resolution subbands. The full resolution image can be obtained by calculating inverse transform of all seven subbands. Thirdly, the coefficients of high pass subbands have small magnitude. Therefore, superior compression of images can be achieved by quantizing the wavelet coefficients.

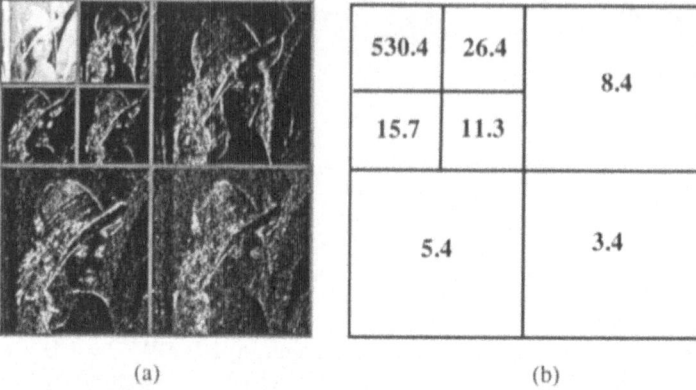

(a)	(b)

Figure 5.20. Wavelet decomposition of Lena Image. a) Magnitude of wavelet coefficients, and b) root means square energy of different wavelet bands. The high pass coefficients have been magnified to show the details.

The inverse wavelet transform is calculated in the reverse manner, *i.e.* starting from the lowest resolution subband, the higher resolution images are calculated recursively. We note that at each stage, the size of the LL subband becomes one-fourth. For typical images, the LL subband become very small after 5 or 6 levels of decompositions, and further decomposition does not provide any improvement in the image analysis.

REFERENCES

1. A. K. Jain, *Fundamentals of Digital Image Processing*, Prentice Hall, 1989.

2. E. O. Brigham, *The Fast Fourier Transform*, Prentice Hall, 1974.

3. K. R. Rao and R. C. Yip, *The Transform and Data Compression Handbook*, CRC Press, New York, 2000.

4. P. P. Vaidyanathan, *Multirate Systems and Filterbanks*, Prentice Hall, 1992.

5. C. S. Burrus, R. A. Gopinath, and H. Guo, *Introduction to Wavelets and Wavelet Transforms*, Prentice Hall, 1998.

6. S. G. Mallat, "A theory for multiresolution signal representation: the wavelet decomposition," *IEEE Trans. on Pattern Analysis and Machine Intelligence*, Vol. 11, pp. 674-693, July 1989.

7. M. K. Mandal, *Wavelet Theory and Implementation*, Chapter 3 of M.A.Sc Thesis, *Wavelets for Image Compression*, University of Ottawa, 1995 (Included in the CD).

8. R. C. Gonzalez and Richard E. Woods, *Digital Image Processing*, Addison Wesley, 1993.

9. B. P. Lathi, *Signal Processing and Linear Systems*, Berkeley Cambridge Press, 1998.

10. A. N. Akansu and R. A. Haddad, *Multiresolution Signal Decomposition: Transforms, Subbands, and Wavelets*, 2nd Edition, Academic Press, San Diego, 2001.

QUESTIONS

1. Show that the 4x4 forward DFT transformation matrix given in Eq. (5.11) is unitary.

2. Show that the 4x4 forward DCT matrix given in Eq. (5.20) is unitary.

3. Show that the the 4x4 forward DWT (Haar) transformation matrix given in Example-5.9 is unitary.

4. Calculate (by hand or using calculator) the DFT, DCT and Haar coefficients of the signal $f(n) = [1, 1, 0, 0]$. Calculate the total energy in the signal domain, and in the frequency (*i.e.*, DFT, DCT, and Haar) domain. Show that the total energies in all three transform domains are individually identical to the total signal energy.

5. Calculate the DCT of the sequence $f(n) = [0.5, -0.5, -0.5, 0.5]$. You will find that only one DCT coefficient is non-zero. Can you explain why is it so?

   [Hints: show that the signal $f(n)$ is one of the basis vectors.]

6. Using MATLAB, calculate the DFT coefficients of the following signal

   $$f(n) = \begin{cases} 1 & 0 \le n \le 4 \\ 0 & 5 \le n \le 15 \end{cases}$$

   Plot the amplitude spectrum (in this particular case you can verify that the spectrum follows a *sinc* pattern). Show that the DFT preserves the signal energy.

7. Repeat the above problem for the DCT.

8. Repeat the above problem for 2-stage Daub-4 wavelet.

9. Calculate the total energy of the 4 largest (absolute value) coefficients in each of the above three cases. Compare the energy compaction performance of the three transforms: DFT, DCT, and DWT in this particular case.

10. Can we consider the block transforms as a special case of filterbank?

11. Compare and contrast orthogonal transform and subband decomposition.

12. Show that the Daub-4 wavelet filter coefficients satisfy Eq. (5.24).

13. Write down 8x8 forward transform matrix for Haar wavelet. Show that the matrix is unitary.

14. Using separable approach, calculate the DFT and DCT coefficients of the following 3x3 image.

    $$\begin{bmatrix} 4 & 9 & 5 \\ 2 & 7 & 3 \\ 5 & 6 & 2 \end{bmatrix}$$

    Show that the energy is preserved in the transform domain.

15. Wavelet transform is said to have multiresolution property. Can you justify it from Fig. 5.20?

Chapter 6

Text Representation and Compression

The characteristics of audio and image media were considered in Chapters 2 and 3. The text is another important medium for representing multimedia information. Indeed, it has been the oldest medium to store and convey information, and most of today's information is still stored in text form. In this chapter, a brief introduction to digital text representation and compression techniques is presented.

6.1 TEXT REPRESENTATION

Unlike audio and visual data, text data is truly digital data. There is no time-domain sampling or quantization. Text data can be assumed to be created by a discrete information source that emits symbols representing letters corresponding to an alphabet. The representations are different for various languages due to their different alphabets. Here, we mainly focus on the digital representation of the English text.

The most popular English character set representation is the American Standard Code for Information Interchange (ASCII) that was published in 1968 as ANSI X3.4. The standard ASCII character set [1] uses 7 bits for each character, which includes letters, punctuation marks, and a few other symbols associated with computer keyboards. The 7-bits ASCII codes are shown in Table 6.1.

Although ASCII code was very popular, it became clear that 7-bit ASCII, with only 128 distinct characters, is not enough. Hence, several larger character sets have been developed to use 8 bits (*i.e.*, 256 characters). These 8-bits codes have 128 characters in addition to the 128 ASCII codes. The extra characters represent non-English characters, graphics symbols, and mathematical symbols. The DOS operating system uses a superset of ASCII called *extended ASCII* or *high ASCII*. Character sets used today in the US are generally 8-bit sets with 256 different characters that are extensions of ASCII.

Table 6.1. ASCII character set. DE: Decimal Equivalent, and CH: Character

DE	CH	DE	CH	DE	CH	DE	CH	DE	CH
0	NUL	26	SUB	52	4	78	N	104	h
1	SOH	27	ESC	53	5	79	O	105	I
2	STX	28	FS	54	6	80	P	106	j
3	ETX	29	GS	55	7	81	Q	107	k
4	EOT	30	RS	56	8	82	R	108	l
5	ENQ	31	US	57	9	83	S	109	m
6	ACK	32	Space	58	:	84	T	110	n
7	BEL	33	!	59	;	85	U	111	o
8	BS	34	"	60	<	86	V	112	p
9	TAB	35	#	61	=	87	W	113	q
10	LF	36	$	62	>	88	X	114	r
11	VT	37	%	63	?	89	Y	115	s
12	FF	38	&	64	@	90	Z	116	t
13	CR	39	'	65	A	91	[	117	u
14	SO	40	(	66	B	92	\	118	v
15	SI	41	)	67	C	93	]	119	w
16	DLE	42	*	68	D	94	^	120	x
17	DC1	43	+	69	E	95	_	121	y
18	DC2	44	,	70	F	96	'	122	z
19	DC3	45	-	71	G	97	a	123	{
20	DC4	46	.	72	H	98	b	124		
21	NAK	47	/	73	I	99	c	125	}
22	SYN	48	0	74	J	100	d	126	~
23	ETB	49	1	75	K	101	e	127	DEL
24	CAN	50	2	76	L	102	f		
25	EM	51	3	77	M	103	g		

The 128 ASCII characters are adequate to exchange information in English. However, other languages that use the Roman alphabet need additional symbols not covered by ASCII, such as β (German), and $\mathring{a}$ (Swedish and other Nordic languages). Hence, ISO 8859 sought to remedy this problem by extending 7-bit ASCII to eight bits, allowing positions for another 128 characters [2]. Unfortunately, more characters were needed than could fit in a single 8-bit character encoding, and hence several extensions were developed. All the character encodings, however, encode the first 128 positions (from 0 to 127) using ASCII code. The last 128 (from 128 to 255) code points of each ISO 8859 encoding are different. Table 6.2 shows the ISO Latin 1 character set, which is used by several operating systems, as well as by different Web browsers.

The ISO 8859 encoding provides the diacritic marks required for various European languages. They also provide non-Roman alphabets: Greek, Cyrillic (used by Russian, Bulgarian, and other languages), Hebrew, and Arabic. However, the standard makes no provision for the scripts of East

Asian languages such as Chinese or Japanese, as these highly ideographic writing systems require many thousands of code points, many more than can be placed in a single 8-bit plane.

Table 6.2: ISO 8859(Latin1) to ASCII character set. The codes represented by EMP (empty) has not been defined, and left for future expansion.

128	EMP	154	EMP	180	´	206	Î	232	è
129	EMP	155	EMP	181	µ	207	Ï	233	é
130	EMP	156	EMP	182	¶	208	Ð	234	ê
131	EMP	157	EMP	183	·	209	Ñ	235	ë
132	EMP	158	EMP	184	¸	210	Ò	236	ì
133	EMP	159	EMP	185	¹	211	Ó	237	í
134	EMP	160	EMP	186	º	212	Ô	238	î
135	EMP	161	¡	187	»	213	Õ	239	ï
136	EMP	162	¢	188	¼	214	Ö	240	ð
137	EMP	163	£	189	½	215	×	241	ñ
138	EMP	164	¤	190	¾	216	Ø	242	ò
139	EMP	165	¥	191	¿	217	Ù	243	ó
140	EMP	166	¦	192	À	218	Ú	244	ô
141	EMP	167	§	193	Á	219	Û	245	õ
142	EMP	168	¨	194	Â	220	Ü	246	ö
143	EMP	169	©	195	Ã	221	Ý	247	÷
144	EMP	170	ª	196	Ä	222	Þ	248	ø
145	EMP	171	«	197	Å	223	ß	249	ù
146	EMP	172	¬	198	Æ	224	à	250	ú
147	EMP	173		199	Ç	225	á	251	û
148	EMP	174	®	200	È	226	â	252	ü
149	EMP	175	¯	201	É	227	ã	253	ý
150	EMP	176	°	202	Ê	228	ä	254	þ
151	EMP	177	±	203	Ë	229	å	255	ÿ
152	EMP	178	²	204	Ì	230	æ		
153	EMP	179	³	205	Í	231	ç		

Storage Space Requirements

Since, text data is used extensively to represent many different types of documents such as books, periodicals, and newspaper articles, it is important to represent them efficiently in order to reduce the storage space requirement.

■ **Example 6.1**

Consider a book that consists of 800 pages. Each page has 40 lines, and each line has on average 80 spaces (including spaces). If the book is stored in digital form, how much storage space is required?

The book contains approximately 800x40x80 or 2.56 million characters. Assuming that each character is represented by eight bits (or one byte), the book will require 2.56 million bytes or 2.44 Mbytes. ■

In the above example, it was observed that a typical document might require several Mbytes of storage space. The important question: is it possible to store the same document using a smaller space? This is the topic of the next section. It will be shown that the text can be compressed without losing any information. In fact, most readers are likely to be aware of various tools such as WinZip (for Windows OS), compress, and gzip [3] (for Unix OS) that employ compression techniques to save storage space.

6.2 PRINCIPLES OF TEXT COMPRESSION

Typical text data have redundant information. The text compression techniques reduce or eliminate this redundancy, and achieve compression. There are primarily two types of redundancies present in text data [4]:

Statistical Redundancy: It refers to the non-uniform probabilities of the occurrences of the characters (or symbols). The symbols that occur more frequently should be given a larger number of bits than the less probable symbols.

Knowledge redundancy: When the text to be coded is limited in its scope, a common knowledge can be associated with it at both the encoder and decoder. The encoder can then transmit only the necessary information required for reconstruction.

In the following sections, we will illustrate the above two types of redundancies and present techniques to achieve compression by exploiting the redundancies.

6.3 STATISTICAL REDUNDANCY

Statistical redundancy reduction methods are based on a rich body of literature collectively known as source coding theory. The source coding theory treats each data sample (*e.g.*, a character) as a symbol generated by an information source. The collection of all possible different symbols is called the alphabet. Source coding theory deals with the compression of data generated by an information source that emits a sequence of symbols chosen from a finite alphabet.

In source coding theory, *entropy* and *rate-distortion* functions are the two most fundamental concepts. Entropy provides a measure for information contained in the source data and, therefore, determines the minimum average bit rate required for perfect reconstruction of the source symbols. Rate distortion theory provides a lower bound on the average bit rate for a given distortion in the reconstructed symbols. The concepts of entropy and rate distortion are detailed in the following.

6.3.1 Probability Density Function and Entropy

Consider a discrete memoryless information source S producing characters that are statistically independent. Further assume that the source has an alphabet $S = \{s_1, s_2, \ldots s_K\}$ with K symbols. A character can be represented by any one of the symbols in the alphabet. Let a character sequence consisting of N sample be represented by $x[i], 1 \leq i \leq N$.

The *histogram* of the random variable x is a discrete function $h[k] = n_k$, where n_k is the number of characters represented by the symbol s_k. The function

$$p[k] = h[k]/N \tag{6.1}$$

provides an estimate of the probability of occurrence of the symbol s_k. Therefore, p is known as the *probability density function* (pdf) of x. Note that

$$\sum_{k=0}^{K-1} p[k] = 1 \tag{6.2}$$

The above equation states that the summation of the probabilities of occurrence of all symbols of an information source is unity.

■ Example 6.2

Consider the string of characters: X = "*aaabbbbbbbbccaaabbcbbbb*". Determine the alphabet; calculate the histogram and probability density function of the characters. Show that relation (6.2) is satisfied.

The alphabet of the above source contains three characters $\{a,b,c\}$. In the character string, there are 6 a's, 13 b's, and 3 c's. Hence, the histogram will be h(a)=6, h(b)=13, and h(c)=3. Since there are a total of 22 characters, p(a)=6/21=0.27, p(b)=0.59, and p(c)=0.14. It can be easily verified that the summation of all probabilities is 1. ∎

Although, an information source may have K symbols, the information provided by each symbol is not equally important. The average information carried by a symbol is related to the reciprocal of its probability and is given by

$$I(s_k) = \log\left(\frac{1}{p[k]}\right)$$ (6.3)

Eq. (6.3) states that the information carried by a symbol is inversely logarithmic proportional to its probability of occurrence. If a symbol has a very small probability, the information carried by the symbol is very large, and vice versa. Consider this typical observation: the newspaper reports the unusual events (*i.e.*, events with a low probability) with greater emphasis, and ignores the events that are considered normal (*i.e.*, events with a higher probability).

The *entropy* of a source is defined as the average amount of information per source symbol, which can be expressed as

$$H(S) = \sum_{k=1}^{K} p[k]I(s_k) = -\sum_{k=1}^{K} p[k]\log_2 p[k]$$ (6.4)

In Eq. (6.4), the base of the logarithm has been taken as 2 in order to express the information in bits per symbol.

Although each symbol can be analyzed individually, it is often useful to deal with large blocks of symbols, rather than individual symbols. The output of a source s can be grouped into blocks of N symbols, and the symbols can be assumed to be generated by a source S_N with an alphabet of size K^N. The source S_N is called the N-th extension of the source s. A total of K^N possible audio patterns can be generated by such a source. Let the probability of a specific character string $\mathbf{s}$ be given by $p(\mathbf{s})$. The entropy, *i.e.*, the average information of the source can be calculated as

$$H(S) = -\sum_{all\ \mathbf{s}} p(\mathbf{s})\log_2 p(\mathbf{s}) \quad bits/source$$ (6.5)

$$= -\frac{1}{N}\sum_{all\ \mathbf{s}} p(\mathbf{s})\log_2 p(\mathbf{s}) \quad bits\ /\ sample$$

The true entropy of an information source can be theoretically estimated using Eq. (6.5). However, in order to do the calculation, the probability function $p(s)$ has to be determined, which is a very difficult task. Furthermore, a large number of audio patterns s has to be generated which may not be feasible, or meaningful. Hence, in practice, the first order entropy (FOE), which is defined on a sample-by-sample basis is often used. The calculation of the FOE is performed using Eq. (6.4) where $p[k]$ is simply the probability of the occurrence of the kth symbol. The FOE is often called *memoryless entropy* as it provides the minimum bit rate required for lossless reproduction of a discrete data sequence.

■ **Example 6.3**

Consider the source in Example 6.2. Calculate the first and second order entropy of the source.

Since p(a)=0.27, p(b)=0.59, and p(c)=0.14, the first order entropy can be calculated as follows:

$$H_1 = -(0.27 \times \log_2 0.27 + 0.59 \times \log_2 0.59 + 0.14 \times \log_2 0.14) = 1.36 \text{ bits/symbol}$$

In order to calculate the second order probability, two characters have to be taken into consideration. The alphabet will now contain nine symbols $\{aa,bb,cc,ab,ba,ac, ca,bc,cb\}$. Since there are 22 symbols, the string can be considered to contain eleven sets of two symbols: $\{aa,ab,bb,bb,bb,cc,aa,ab, bc,bb,bb\}$. The probability of different symbols are p(aa)=2/11=0.18, p(ab)=2/11=0.18, p(bb)=5/11=0.45, p(cc)=1/11=0.09, and p(bc)=1/11=0.09. The second order entropy can now be calculated as follows.

$$H_2 = -(2 \times 0.18 \times \log_2 0.18 + 0.45 \times \log_2 0.45 + 2 \times 0.09 \times \log_2 0.09) = 2.03$$

$$= 2.03 \text{ bits/symbol} \qquad (1 \text{ symbol} \equiv 2 \text{ characters}) \quad ■$$

6.3.2 Shannon's Noiseless Source Coding Theorem

The probability density function and entropy of an information source can be used to design an efficient text compressor. However, before the compressor design technique is demonstrated, an important theorem, known as Shannon's noiseless source coding theorem [5], must be presented.

Let S be a text source with an alphabet of size K and entropy $H(S)$. Consider encoding blocks of N source symbols at a time into binary code words. Shannon's noiseless source coding theorem states: for any $\delta > 0$ it is possible, by choosing N large enough, to construct a code in such a way that the average number of bits per original source symbol $\bar{R}$ satisfying the following relationship.

$$H(S) \leq \overline{R} < H(S) + \delta \qquad (6.6)$$

In other words, a source can be coded losslessly with an average bit-rate close to its entropy, but the average bit-rate cannot be less than the entropy. However, to achieve a bit-rate close to entropy, one has to encode higher extensions of the source.

It can be shown that the source entropy is bounded by 0 and $\log_2 K$. In other words,

$$0 \leq H(S) \leq \log_2 K \qquad (6.7)$$

The left side equality holds if $p[k]$ is zero for all source symbols s_k except one, in which case the source is totally predictable. The right side equality holds when every source symbol s_k has the same probability. The redundancy of the source is defined as:

$$redundancy = \log_2 K - H(S) \qquad (6.8)$$

Eq. (6.7) states that if a source has an alphabet of size K, the maximum entropy the source can have is $\log_2 K$. If the entropy is equal to $\log_2 K$, the source is said to have zero redundancy. In most cases, the information contains dome redundancy.

■ **Example 6.4**

Consider an information source with the alphabet $\{a,b,c,d\}$. The symbols have equal probabilities of occurrence. Calculate the entropy of the source, and redundancy of the source. What is the average bit-rate required to transmit the symbols generated by the source? Design a suitable coder to encode the symbols.

Since p(a)=p(b)=p(c)=p(d)=0.25, the entropy of the source will be equal to $-4*0.25*\log_2 0.25$ or 2 bits/symbol. The redundancy is zero since the entropy is equal to $\log_2 K$. The average bit-rate required to transmit the symbols is 2 bits/symbol. The coder design in this case is trivial. Assign a:00, b:01, c:10, and d:11.

Note that it might be possible to encode a four-symbol source using less than 2 bits/symbol if the entropy is less than 2 bits/symbol. ■

■ **Example 6.5**

What is the average bit-rate required for a three-symbol source? What is the minimum bit-rate required to encode the three-symbol source considered in Example 6.2? Calculate the redundancy in the source. When is the redundancy of a source zero?

The bit-rate required for an arbitrary three-symbol source is $\log_2 3$ or 1.59 bits/symbol. However, it might be possible to encode a three-symbol source using less than 1.59 bits/symbol. The source in Example 6.2 has entropy of 1.36 bits/symbol. Hence, it is possible to assign codes to the symbols of the source such that the average bit-rate is 1.36 bits/symbol. However, the information source should generate the symbols according to the probability density function mentioned in the example. Otherwise, the entropy will change, and the required bit-rate will also be changed.

The redundancy in the 3-symbol source in Example 6.2 is (1.59-1.36) or 0.23 bits/symbol.

The redundancy of a source is zero when the entropy is equal to $\log_2 K$. This will happen only if all the symbols have equal probability (*i.e.* 1/K). ∎

Significance of Noiseless Coding Theorem

The source-coding theorem, as specified in Eq. (6.6), states that the average number of bits/symbol must be greater than or equal to the entropy of the source. A pertinent question is: what is the significance of the term δ ?

Note that Shannon's theorem provides a lower bound of the bit-rate requirement. However, the theorem does not tell us how to achieve it. In practice, it is very difficult to achieve a bit-rate equal to the entropy. Consider Example 6.2 again. If we want to design a three-symbol coder, we can easily obtain a coder by assigning 2 bits/symbol. The penalty of the simplest design is the bit-rate 0.64 (2-1.36) bits/symbol more than the entropy level, i.e., $\delta = 0.64$. If we do a non-trivial coder design (see Problem 9), we can achieve a bit-rate of 1.41 bits/symbol (corresponding to $\delta = 0.05$). As the δ is reduced, the coder complexity is expected to increase significantly.

6.3.3 Huffman Coding

According to Shannon's noiseless coding theorem, the average bit rate $\overline{R}$ for encoding an information source is lower bounded by the entropy of the source. Shannon's theorem, however, does not state the design procedure of an encoder that will encode the information source at an average bit rate $\overline{R}$. Huffman has provided a practical method [6] to design an encoder that will provide a bit-rate close to the entropy. This method designs a variable length code (VLC) for each source symbol such that the number of bits in the code is approximately inversely proportional to the probability of occurrence of that symbol.

Table 6.3 shows the alphabet of a source along with the probabilities of individual symbols. Since there are six symbols, a straightforward pulse code modulation (PCM) will require 3 bits. However, if 3 bits, which can represent 8 symbols, is used in this case, the bits are under-utilized. An information source with six symbols requires at most $\log_2 6$ or 2.58 bits. Since the probability of occurrences of the symbols are not uniform, there is a further scope of improvement. The entropy of the source can be calculated using Eq. (6.4).

$$H = -(0.3 * \log_2 0.3 + 2 * 0.2 * \log_2 0.2 + 3 * 0.1 * \log_2 0.1) = 2.44 bits / symbol$$

In other words, it is possible to design an encoder that will spend an average of 2.44 bits/symbol.

Table 6.3. Example of a fixed model for alphabet {$a, b, c, d, e, !$}

Symbol	Probability	Huffman code
a	0.2	10
b	0.1	011
c	0.2	11
d	0.1	0100
e	0.3	00
$!$	0.1	0101

The Huffman code for an information source is generated in two steps: source reduction and code assignment, which are discussed below.

Source reduction

The source reduction process for the information source corresponding to Table 6.3 is explained in Fig. 6.1. There are several steps that are as follows:

Step 1: The symbols are arranged with decreasing probability of occurrences as shown in the first two columns of Fig. 6.1. In case there is a tie, these symbols can be listed in any order. For example, in this case, p(b)=p(d)=p(!)=0.1. We chose the order "b, d, !", but we could also choose "d, b, !".

Step 2: Once the symbols are arranged, a new symbol is created by combining the two least probable symbols. In this example, we have merged d and $!$, and hence the new symbol has a probability of occurrence of 0.2 (=0.1+0.1). It can be considered that the source now has only four symbols instead of five. The probabilities of these four symbols are rearranged in descending order as shown in column-3.

Step 3: The two least probable symbols are again merged, and the new probabilities are rearranged. This process is repeated until the newest source has only two symbols. In this example, the last two probabilities are 0.6 and 0.4.

Original Source		Reduction of Source			
Symbol	Probability	1	2	3	4
e	0.3	0.3	0.3	0.4	0.6
a	0.2	0.2	0.3	0.3	0.4
c	0.2	0.2	0.2	0.3	
b	0.1	0.2	0.2		
d	0.1	0.1			
!	0.1				

Figure 6.1. Huffman Source Reduction Process

Code Assignment

Step 1: After the source reduction process is over, the Huffman code assignment process is started (see Fig. 6.2). The first two symbols are assigned the trivial codewords "0" and "1". In this case, we assign "0" to the symbol with probability 0.6, and "1" to the symbol with probability 0.4.

Step 2: The probability "0.6" was obtained by merging two probabilities 0.3 and 0.3. Hence, the code assigned to these probabilities are "0 followed by 0", and "0 followed by 1". After this assignment, there are three symbols with probabilities 0.4, 0.3, 0.3 with codes 1, 00, and 01, respectively.

Step 3: The probability 0.4 was obtained by merging two probabilities 0.2 and 0.2. In this step, the code assigned to these probabilities (0.2 and 0.2) are "1 followed by 0", and "1 followed by 1." After this assignment, there are four symbols with probabilities 0.3, 0.3, 0.2 and 0.2 with codes 00, 01, 10, and 11, respectively.

Step 4: The second probability 0.3 was obtained by merging two probabilities 0.2 and 0.1. In this step, the codes assigned to these probabilities (0.2 and 0.1) are "01 followed by 0," and "01 followed by 1." After this assignment, there are five symbols with probabilities 0.3, 0.2, 0.2 and 0.2, and 0.1 with codes 00, 10, 11, 010, and 011, respectively.

Step 5: The fourth probability 0.2 was obtained by merging two probabilities 0.1 and 0.1. In this step, the code assigned to these probabilities (0.1 and 0.1) are "010 followed by 0," and "010 followed by 1." After this assignment, there are six symbols with probabilities 0.3, 0.2, 0.2, 0.1, 0.1, and 0.1 with codes 00, 10, 11, 011, 0100, and 0101, respectively.

At this point, we have the Huffman codes for all the original source symbols. The codes are shown along with the assigned symbols in the third column of Table 6.3.

Original Source			Code Assignment			
Sym.	Prob.	Code	1	2	3	4
e	0.3	00	0.3 00	0.3 00	0.4 1	0.6 0
a	0.2	10	0.2 10	0.3 01	0.3 00	0.4 1
c	0.2	11	0.2 11	0.2 10	0.3 01	
b	0.1	011	0.2 010	0.2 11		
d	0.1	0100	0.1 011			
!	0.1	0101				

Figure 6.2. Huffman Code Assignment Process

The average length of the Huffman code is, $\overline{R} = (0.3+0.2+0.2)*2+0.1*3 +0.2*4 = 2.5\ bits\ /\ symbol$, which is close to the first order entropy (=2.44 bits/symbol) of the source.

■ **Example 6.6**

i) Using the Huffman table shown in Table 6.3, calculate the Huffman code for the text string "baecedeac!".

ii) A string of characters was Huffman coded, and the code is "0001100010011000010111". Determine the text string. Is the decoding unique?

The Huffman code for text string "baecedeac!" is "0111000110001000 010110101". Note that 10 symbols (or characters) could be compressed using 25 bits. So, an average of 2.5 bits is used to represent each symbol.

The code can arranged as "00,011,00,0100,11,00,00,0101,11". In other words, the decoded string is "ebedcee!c". Decoding of the Huffman code is unique for a given Huffman table. One cannot obtain a different output string. ■

Note that the Huffman code has been used extensively since its discovery by Huffman in 1952. However, it has several limitations. First, it requires at least one bit to represent the occurrence of each symbol. Therefore, we cannot design a Huffman code where we assign 0.5 bits to a symbol. Hence, if the entropy of a source is less than 1 bit/symbol, Huffman coding is not efficient. Even if the entropy is greater than one, in most cases the Huffman code requires a higher average bit-rate than the entropy (see Problem 16). Second, it cannot efficiently adapt to changing source statistics. Although dynamic Huffman coding schemes have been developed to address these issues, these schemes are difficult to implement.

6.3.4 Arithmetic Coding

Arithmetic coding is another entropy coding technique that achieves a compression ratio higher than Huffman coding. Unlike Huffman coding, arithmetic coding can provide a bit-rate less than one. In most cases, the average bit-rate approaches the theoretical entropy bound for any arbitrary source.

Arithmetic coding principle is significantly different from the Huffman coding. In Huffman coding, each symbol is assigned a variable length codeword. While encoding, the symbols are encoded independently. In arithmetic coding, a codeword is assigned to the entire input message. In other words, consider a text string of 100000 symbols. The Huffman code will calculate the codeword for each symbol, and concatenate them to produce the overall code. Arithmetic coding, however, just produces one codeword (which may be very long, say 2000 bits!!) for the entire string. How is this done? Before explaining further, consider this brief example.

■ Example 6.7

Consider the English alphabet with 26 letters. Calculate the maximum number of different text strings with length *i*) 1, *ii*) 2, and *iii*) 100.

Examples of length-1 strings are "a", "b", "c", ...and "z". Hence, the maximum number of possible strings is 26.

Examples of length-2 strings are "aa", "ab",...,"az", "ba", "bb", "bc",.... "bz",......,"zz". Hence, the maximum number of distinct strings are 26^2 or 676.

Similarly, it can be shown that there are 26^{100} or 3.14×10^{141} distinct 100-character strings. ■

The above example demonstrates that if a source has an alphabet of size K, there are K^m different possible messages of length *m*. If the probability density function of the individual symbols is known, the probability of the occurrence of a given message s_m of length m, can be calculated. Since there are K^m distinctly possible messages, one can theoretically calculate the probability ($p(s_m)$) of occurrences of these individual messages where as per the probability theory:

$$\sum_{m=0}^{K^m} p(s_m) = 1 \qquad\qquad (6.9)$$

If one wishes to encode an entire message at once, a code table (using the Huffman coding technique) can be generated. The length of the codeword will be approximately equal to $-\log_2 p(s_m)$. Since there is a large number of code words, the individual $p(s_m)$'s will be very small. Therefore, the length of the code words is likely to be significantly longer than the example shown in Table 6.3.

Encoding an entire message at once increases the efficiency of the coder. Even the Huffman code will provide an average bit-rate close to the entropy even if it is much less than one bit/symbol. However, generating a Huffman table of K^m symbols would be very difficult, if not almost impossible. The arithmetic coding provides a convenient way to achieve this goal, and provides optimal performance. The coding principle is explained below.

Assume that the $p(s_m)$ of different possible messages are known. In Eq. (6.9), it has been shown that their overall sum is equal to 1. Divide the interval $[0,1)$ in the real axis into K^m sub-intervals, such that the width of each sub-interval corresponds exactly to the probability of occurrence of a message. Then assume that the sub-intervals are $[L_l, R_l)$, $(l = 1,2,.....K^m)$. Once the subintervals are determined, any real number in that sub-interval will correspond to a unique message, and hence the entire input message can be decoded. Since the sub-intervals are non-overlapping, a codeword for s_m can be constructed by expanding any point in the interval in binary form and retaining only the $n_l = \lceil -\log_2 p(s_m) \rceil$ bits after the decimal point. Consequently, the number of bits required to represent the message can differ from the source entropy by a maximum of one bit.

The above procedure might seem too difficult to implement. However, a simple recursive technique can be employed to implement such a scheme. To explain the coding technique with a concrete example, consider the information source given in Table 6.3.

Step 1: List the symbols and their probabilities (not in any particular order) as shown in Table 6.4.

Step 2: Divide the interval [0,1) into 6 (=K) sub-intervals since there are six symbols. The width of each sub-interval is equal to the probability of the individual symbol occurrences.

Step 3: Now suppose we want to transmit the symbol *baad!*. Since the first symbol is *b*, we chose the interval [0.2,0.3). This is shown in Table 6.5, 3rd row, as well as in Fig. 6.3. Once the sub-interval is chosen, it is divided into six second-level sub-intervals. Note that the widths of these sub-intervals are smaller than the first-level sub-intervals. This is shown in Fig. 6.3.

Step 4: The second symbol is "*a*". Therefore, we choose the sub-interval corresponding to "a" in the interval [0.2,0.3), which is [0.2,0.22) (see Fig. 6.3). The sub-interval [0.2,0.22) is further divided into six third-level sub-intervals.

Step 5: The third symbol is again "*a*". Therefore, the sub-interval [0.2,0.204) is chosen. This sub-interval is again divided into six fourth-level sub-intervals.

Step 6: The fourth symbol is "d". Therefore, the sub-interval (see Fig. 6.3) [0.2020, 0.2024) is chosen. This sub-interval is again divided into six fifth-level sub-intervals.

Step 7: The fifth symbol is "!". Hence, the sub-interval [0.20236,0.2024) is chosen.

Since there is no other symbol, the encoding is now complete. The transmitter can send an arbitrary number in the interval [0.20236, 0.2024) and the decoder will be able to reconstruct the input sequence (of course, the decoder has to know the information provided in Table 6.4).

Table 6.4. Example of a fixed model for alphabet {*a, b, c, d, e, !*}

Symbol	Probability	Range (for Arithmetic coder)
a	0.2	[0, 0.2)
b	0.1	[0.2, 0.3)
c	0.2	[0.3, 0.5)
d	0.1	[0.5,0.6)
e	0.3	[0.6,0.9)
!	0.1	[0.9,1.0)

Table 6.5. Generation of bitstream in Arithmetic Coding

		Sub-interval	Binary-interval	Bitstream Generated
Starting		[0,1.0)		
after seeing	b	[0.2,0.3)	[0.0,0.5)	0
	a	[0.2,0.22)	[0,0.25)	0
	a	[0.2,0.204)	[0.125,0.25)	1
	d	[0.2020,0.2024)	[0.1875,0.25)	1
	!	[0.20236,0.2024)	[0.1875,0.21875)	0

However, to stop decoding process, the decoder will face the problem of detecting the end of the message. For example, the single number 0.20236 could represent *baad!*, *baad!a*, or *baad!aa*. To resolve this ambiguity, normally a special EOF (end of file) symbol is used. This EOF marker is known to both the encoder and the decoder and thus the ambiguity can be resolved. In this case, "!" can be used as the EOF marker.

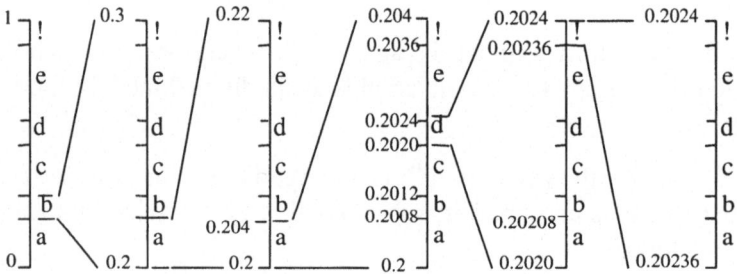

Figure 6.3. Arithmetic Encoding Process

The procedure described so far produces the final real number that can be sent to the decoder. The real number is generally represented by a stream of bits, which can be generated once we know the final real number. However, we do not need to wait until the final real number is calculated. The bitstream can be generated as the input symbols are coming in, and the sub-intervals are being segmented. This is shown in Table 6.5. After seeing the first symbol *b*, the encoder narrows the range to [0.2, 0.3). As soon as it is determined that the interval is [0.0,0.5), the encoder knows that the first symbol is "0". Hence, it can start sending the output bits. After the second symbol, the encoder narrows the range further to [0.2, 0.22). At this moment the encoder knows that the second symbol is "0" because the interval falls within [0,0.25). Similarly, the third symbol will be "1" as the interval falls within [0.125,0.25). The fourth symbol is 1 as the interval.

The decoding scheme is very simple. Suppose the number 0.20238 is sent. The decoder knows that the first symbol is *b*, as the number lies in [0.2,0.3).

The second symbol is *a,* since the number lies in the first one-fifth of the range, *i.e.*, within [0.2, 0.22). Thus, the process continues.

In the above example, a fixed model has been used, which is not suitable for coding nonstationary sources where the source statistics change with time. In these cases, an adaptive model should be employed to achieve a superior performance. Witten *et al.* [7] have presented an efficient implementation of adaptive arithmetic coding that provides a bit-rate very close to the entropy of the source.

6.4 DICTIONARY-BASED COMPRESSION

The entropy coding methods use a statistical model to encode symbols contained in the input data. The compression is achieved by removing the statistical redundancy present in the data. This is done by encoding symbols with codes that use fewer bits for frequently occurring symbols.

Dictionary-based compression is another popular data-encoding method that exploits the knowledge redundancy present in the data. These methods do not encode single symbols as variable length code. Instead, they encode several symbols as a single token or word. Dictionary-based compression is easier to understand because it follows a familiar strategy that large amount of storage can be retrieved using small indices from a database.

Assume that the sentence "Canada is a beautiful country" is to be encoded. There are 25 characters (ignoring the spaces between the words). A straightforward ASCII code will require 25 bytes to store the sentence. However, if sentence is coded word-by-word, less storage space may be required.

For example, consider an English Dictionary with 2000 pages, with each page containing about 256 words. Assume that the words of the above sentence are encoded by their page number and the word number in the given page. For example, assume that the word Canada is the 10th word on page 40 of the Dictionary. The sentence may be represented by the following code:

40/10   440/30   1/1   20/89   43/55   (page# / word#)

Since the Dictionary has 2000 pages, it would require at most 12 bits to encode each page number. Similarly, it would require 8 bits to encode each word number. Hence, each word would effectively require an average 20 bits for representation. The above-mentioned sentence contains 5 words, which can be represented with 100 bits or (14 bytes). In this case, a compression ratio of more than 40% is achieved with the dictionary scheme.

Static Dictionary Based Approach

The previous example uses a static dictionary. One of the biggest advantages is that the static dictionary can be tuned to fit the data. But static dictionary schemes have the problem of how to pass the dictionary from the encoder to the decoder. The compression schemes using static dictionaries are mostly ad hoc, implementation dependent, and not very popular.

Adaptive Dictionary Based Approach

Most well-known dictionary algorithms use an adaptive dictionary scheme. In this scheme the compression begins either with no dictionary or with default baseline dictionary. As the compression proceeds, the algorithms add new phrases to be used later as encoded tokens.

The basic principle of the adaptive dictionary-based approach is parsing the input text stream into fragments for testing against the dictionary and testing the fragments against the dictionary; if the fragments does not match in the dictionary it is added as a new word and then encoded. The decompression program does not need to parse the input stream into fragments and test fragments against the dictionary. Instead, it has to decode the input stream into either dictionary indices or plain text and add new phrases to the dictionary. Also, the program has to convert the dictionary indices into phrases and display them as plain text.

6.4.1 LZ77 Technique

LZ77 is a dictionary-based compression scheme proposed [8] by Lempel and Ziv in 1977. It employs a text window that is divided into two parts. The first part is a dictionary or a sliding window that can hold large blocks of text. The second part is a look-ahead buffer that can hold only a small amount of text data. The text data in the sliding window is used as the dictionary to compress the remainder of the input source to find existing patterns. If a pattern in the input source is already in the dictionary, this pattern is replaced with the reference to the position in the dictionary and the length of the pattern. As the compression progresses, the dictionary is updated by shifting in more bytes from the input source, and subsequently forcing the earlier entries out. The dictionary is referred to as the sliding window since it maintains a constantly updated input stream into the window.

The working of the LZ77 technique is illustrated with the following example.

■ Example 6.8

Consider a 20-character dictionary or sliding window and an 8-character look-ahead window. The input source string is

RYYXXZERTYGJJJASDERRXXZERFEERZXYURPP

Calculate the output of the text compressor.

Step 1

The sliding window is loaded with the first 20 characters from the input source, with the symbol or data byte "R" in window position 0, and the symbol R in window position 19. The remainder of the input file consists of symbols starting with XXZERFEE that are loaded into the look-ahead window.

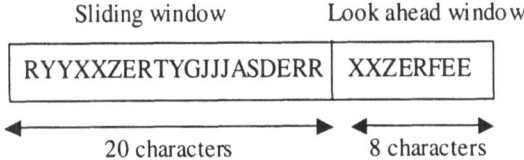

Step 2

Search for a sequence in the sliding window that begins with the character in the look-ahead position 0 (*i.e.*, 'X'). The sequence of five characters starts at sliding window index 5 ("XXZER") and matches the pattern in the look ahead buffer. These five characters can be replaced by a (offset, length) record as shown below:

RYYXXZERTYGJJJASDERR | (4,5)

Step 3

The sliding window is then shifted over five characters. The five characters "RRYYX" are moved out of the sliding window. The new 5 characters "RXXYU "are moved into the look ahead window.

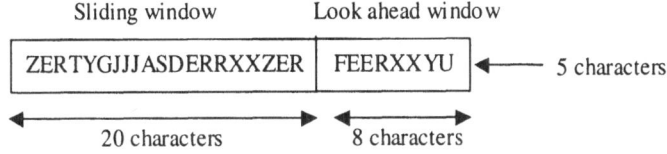

Step 4

There is no sequence match in the first three characters of the look ahead window and the sliding window. Hence, the sliding window is shifted by 3 characters, and three new characters come into the look ahead window. Again, search for a sequence in the sliding window that matches the look-ahead window sequence is carried. There is a sequence of 3 characters starting at sliding window index 12 ("RXX") that matches the pattern in the look-ahead window. Hence these 3 characters are replaced by (12,3).

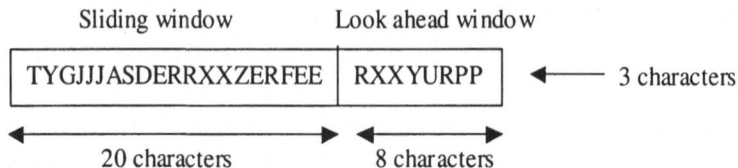

The above procedure is followed for the new symbols, which are coming into the look-ahead window. The final compressed output ignoring the first 20 characters is [(4,5), FEE, (12,3),........]. ∎

During the decompression, a sliding window of identical size is required. However, the look-ahead window is not required. The uncompressed data is put into the sliding window. When a record (offset, length) is detected, the decompressor points to the position of the offset, and begins to copy the specified number of symbols and shifts them into the same sliding window.

6.4.2 LZ78 Technique

The LZ78 is another technique [9] that forms the basis for several highly efficient text encoders. It uses a different approach to build and maintain the dictionary. Instead of using a fixed-size sliding window as in LZ77, LZ78 builds a string-based dictionary with previously matched substrings during compression. There is no restriction on how far back the match can be with previous strings. Hence, matching with previous strings in the long past is possible in this technique. This increases the probability of finding a match with the previous string compared to the LZ77 technique. In this approach, the length of the matched string is not limited by window size. There is no need for the encoder to send the dictionary to the decoder. The decoder can automatically regenerate the dictionary without error. The LZ78 technique is illustrated with the following example.

■ **Example 6.9**

Assume that the text string to be encoded is "PUP PUB PUPPY...". Show how the dictionary is built up, and the first few compressor output symbols.

Step 1

At first, the encoder starts with a dictionary having the null string only. The first character 'P' is read in from the source, and there is no match in the dictionary, and hence it can be considered as matching the null string. Thus the encoder outputs the corresponding dictionary index '0', which corresponds to the null string and character output as 'P', and then adds string to the dictionary as entry 1 (see Fig. 6.4). The second character 'U' is read from the source; there is no match in the dictionary. The encoder outputs again dictionary index '0' and character output " U " and adds it to the dictionary as entry 2.

Step 2

The next character 'P' is read in. It matches an existing dictionary phrase, and hence the next character ' ' is read in which creates a new phrase with no match in the dictionary. The encoder outputs the dictionary index '1' and the character output as ' '. The string is then added to the dictionary as entry "3".

DICTIONARY		CODED OUTPUT	
Dict. entry	**String added to dict**	**Index Output**	**Charcter Output**
0	" "		
1	P	0	P
2	U	0	U
3	"P "	1	' '
4	PU	1	U
5	B	0	B
6	" "	0	' '
7	PUP	4	P
8	PY	1	Y

Figure 6.4. LZ78 compression of a text string.

Step 3:

The next character 'P' is read in. It matches an existing dictionary phrase, so the next character "U" is read which creates a new phrase with no match in the dictionary. The encoder outputs the dictionary index '1' and the character output as "U". The string is the added to the dictionary as entry "4".

Step 4:

The next character 'B' is read in, there is no match in the dictionary. The encoder outputs the corresponding dictionary index '0', which is for the null string and character output as 'B' and then adds string to the dictionary as entry "5". The next character is ' ' read which does not the match the dictionary. Again, the encoder outputs the Index output as '0' and character index as ' '.

Step 5:

When the next character 'P' is read in, it matches the dictionary phrases. So the next character is read in; it again matches the dictionary phrases. When the next character 'P' is read, there is no match in the dictionary. The encoder outputs the dictionary index '4' which matches "PU" and the character output 'P'.

The encoding continues in this procedure. The encoded string for this example is "0P0U1' '1U0B0' '4P1Y........". ∎

Decoding of the Encoded Stream

The decoding procedure is simply the inverse of the encoding process. It only includes interpretation of index output and then the addition of new dictionary entries, which is an easy task to perform. In this example decoding process takes place in the following way.

Step 1

The decoder reads the encoded data. The first index '0' is read, which indicates that it matches the null index in the dictionary and next character should be added to the dictionary. The decoder reads the next character 'P' and adds the character 'P' as dictionary entry '1', and outputs the character 'P'.

Step 2

The decoder reads the next character '0', which again indicates that the character read next should be added to the dictionary. The decoder reads the next character 'P', and adds the character 'U' as dictionary entry '2' and outputs the character 'U'.

Step 3

The decoder reads next character '1' which indicates the dictionary index 1 and the decoder outputs the corresponding phrase to the dictionary index '1' as 'P' and the next character ' 'is read and is added to the dictionary as

"P " entry "3" in the dictionary and 'P 'is sent as decoder output. The string decode is 'PUP '. The same procedure used for decoding the remaining encoded symbols and building the dictionary.

Although the LZ technique provides good compression, Welch proposed an efficient implementation and improvement of LZ technique, known as LZW technique [10]. This technique has been used in many compression schemes, including the popular GIF image format.

6.5 SUMMARY

Several lossless coding techniques have been presented in this Chapter. The first set of techniques is based on entropy-based approaches. Although these techniques are presented for encoding text data, these techniques are also used extensively for audio and image coding (more details in Chapters 7 and 8). The second set of techniques is based on dictionary-based approaches. Most text-compression scheme employs these approaches to achieve high compression.

REFERENCES

1.   ASCII Table and Description, http://www.asciitable.com/

2.   Unicode in the Unix Environment, http://czyborra.com/

3.   P. Deutsch, *GZIP file format specification version 4.3*, RFC 1952, May 1996. Can be downloaded from http://www.ietf.org/rfc/rfc1952.txt.

4.   M. Nelson, *The Data Compression Book*, 2nd Edition, M&T Books, New York, 1996.

5.   C. E. Shannon and W. Weaver, *The Mathematical Theory of Communication*, University of Illinois Press, Urbana, IL, 1949.

6.   D. A. Huffman, "A method for the construction of minimum redundancy codes," *Proc. of the IRE*, Vol. 40, Sept. 1952.

7.   I. H. Witten, R. M. Neal and J. G. Cleary, "Arithmetic coding for data compression," *Communications of the ACM*, Vol 30, pp 520-540, June 1987.

8.   J. Ziv and A. Lempel, "A universal algorithm for sequential data compression," *IEEE Trans. on Information Theory*, Vol. 23, No. 3, pp. 337-343, 1977.

9.   J. Ziv and A. Lempel, "Compression of individual sequences via variable-rate coding," *IEEE Trans. on Information Theory*, Vol. 24, No. 5, pp. 530-536, 1978.

10. T. A. Welch, "A technique for high-performance data compression," *IEEE computer*, Vol. 17, pp. 8-19, June 1984.

QUESTIONS

1.   How many bits/character are generally required for simple text files?

2.   Why are there so many extensions of the 7-bit ASCII character set?

3. A book has 900 pages. Assume that each page contains on average 40 lines, and each line contains 75 characters. What would be the file size if the book is stored in digital form?

4. What are the main principles of the text compression techniques?

5. Justify the inverse logarithmic relationship between the information contained in a symbol and its probability of occurrence.

6. Why is entropy a key concept in coding theory? What is the significance of Shannon's noiseless coding theorem?

7. Calculate the entropy of a 5-symbol source with symbol probabilities {0.25, 0.2, 0.35, 0.12, 0.08}.

8. Show that the first order entropy of a source with alphabet-size K is equal to $\log_2 K$ only when all the symbols are equi-probable.

9. Design a Huffman table for the information source given in Example 6.2. Show that the average bit-rate is 1.41 bits/symbol.

10. How many unique sets of Huffman codes are possible for a three-symbol source? Construct them.

11. An information source has generated a string "AbGOODbDOG" where the symbol "b" corresponds to a blank space. Determine the symbols you need to represent the string. Calculate the entropy of the information source (with the limited information given by the string). Design a Huffman code to encode the symbols. How many bits does the Huffman coder need on average to encode a symbol?

12. How many distinct text files are possible that contain 2000 English letters? Assume an alphabet size of 26.

13. Consider the information source given in Example 6.2. Determine the sub-interval for arithmetic coding of the string "abbbc".

14. Repeat the above experiment with an arithmetic coder.

15. Why do we need a special EOF symbol in arithmetic coding? Do we need one such symbol in Huffman coding?

16. Show that Huffman coding performs optimally (*i.e.*, it provides the bit-rate identical to entropy of the source) only if all the symbol probabilities are integral powers of ½.

17. Explain the principle of dictionary-based compression. What are the different approaches in this compression method?

18. Explain sliding window-based compression technique. Consider the text string "PJRKTYLLLMNPPRKLLLMRYKMNPPRLMRY". Compress the file using the sliding window of length 16 and a look-ahead window of length 5.

19. Explain the LZ78 compression method. Encode the sentence "DOG EAT DOGS".

20. Compare the LZ77 and LZ78 compression technique?

Chapter 7

Digital Audio Compression

Audio data requires a large number of bits for representation. For example, CD quality stereo audio requires 176.4 Kbytes/sec data rate for transmission or storage. This bandwidth is too large for many applications such as voice transmission over the Internet. Even when there is no live audio transmission, the storage cost may be high. An audio CD only typically contains up to 74 minutes of audio. It has been found that if audio data is compressed carefully, excellent quality audio can be stored or transmitted at a much lower bit-rate. In this Chapter, we present the basic principles of audio compression techniques, followed by brief discussions on a few selected audio compression standards.

7.1 AUDIO COMPRESSION PRINCIPLES

Audio data generally requires a large bandwidth that may not be always available. Hence, compression techniques are used to represent the audio at a lower bit-rate without degrading the quality significantly. The following example demonstrates audio transmission with various constraints.

■ **Example 7.1**

We want to transmit a stereo audio signal through a 56 kbps channel in real time. Consider the following scenarios:

i)   We are using a sampling frequency of 44.1 KHz. What is the maximum average number of bits we can use to represent an audio sample?

      Bit-rate = 56000 bits/s, Sample rate = 44100 samples/sec
      Average bit-rate = (56/44.1) = 1.26 bit/stereo sample
                       = 0.63 bit/channels/sample

ii) We want to use 16 bit/sample/channel representation. What is the maximum sampling frequency? What should we do in order to avoid aliasing?

$$Sampling Frequency = \frac{56000}{2 \times 16} Hz = 1750 Hz$$

In order to avoid aliasing, the audio signal has to be passed through a lowpass filter with a cut-off frequency of 875 Hz or lower.

iii) We want to use a sampling frequency of 44.1 KHz, as well as 16 bits/sample/channel representation. What is the minimum compression ratio we need in order to transmit the audio signal?

Uncompressed bit-rate = $44100 \times 16 \times 2 \, bit / s = 1.41 \times 10^6$ bit/s

Minimum compression factor = $1.41 \times 10^6 / 56100 = 25.2$ ∎

The above example demonstrates that low bit-rate audio can be obtained by reducing the sampling bit resolution of each audio sample, or by reducing the sampling frequency. Although, these methods can reduce the bit-rate, the quality may be degraded significantly. In this Chapter, we will present techniques that achieve superior compression performance (good quality at low bit-rate) by removing redundancies in an audio signal. There are primarily three types of redundancies:

Statistical Redundancy: The audio samples do not have equal probability of occurrence resulting in statistical redundancy. The sample values that occur more frequently are given a larger number of bits than the less probable sample values.

Temporal redundancy: The audio signal is time varying. However, there is often a strong correlation between the neighboring sample values. This inter-sample redundancy is typically removed by employing compression techniques such as predictive coding, and transform coding.

Knowledge redundancy: When the signal to be coded is limited in its scope, a common knowledge can be associated with it at both the encoder and decoder. The encoder can then transmit only the necessary information (*i.e.* the change) required for reconstruction. For example, an orchestra can be encoded by storing the names of the instruments, and how they are played. The MIDI files exploit knowledge redundancy, and provide excellent quality music at a very low bit-rate.

In addition to the above techniques, the human auditory system properties can be exploited to improve the subjective quality of audio signal.

In this Chapter, several audio compression techniques, which exploit the above-mentioned redundancies, will be presented. Unlike text data, which typically employs lossless compression, the audio signal is generally compressed using lossy technique. A brief overview of rate-distortion theory for lossy coding is presented below.

7.1.1 Rate Distortion Function

Shannon's noiseless source coding theorem was discussed in Chapter 6. It was shown that the theorem presents a fundamental limit on error free compression. Text data is typically encoded using lossless methods, and noiseless coding theorem is suitable for this application. For most natural audio sources, however, error free compression cannot provide a compression ratio greater than $2:1$. Thus, in practice, lossy compression techniques are often used to get a higher compression ratio.

Given a random source vector S, the objective of lossy coding is to design an encoder that operates at a given rate, and which minimizes the average distortion given by

$$d_{avg} = E\{d(S,\hat{S})\} \qquad (7.1)$$

where $\hat{S}$ is the reconstructed vector and $d(S,\hat{S})$ represents the distortion between S and $\hat{S}$, and $E\{x\}$ is the expected value of the variable x. $R(d_{avg})$ is the minimum value of this information over all transition distributions yielding an average distortion d_{avg}.

Shannon's source coding theorem [1] states that for a given distortion d_{avg}, there exists a rate-distortion function $R(d_{avg})$ corresponding to an information source, which is the minimum bit-rate required to transmit the signals coming out from the source with distortion equal to (or less than) d_{avg}.

The significance of the above theorem is that it presents a limit to the performance of an encoder. The theorem states that no encoder, regardless of the form or complexity, can yield an average distortion less than d_{avg} at an average transmission rate of $R(d_{avg})$, but it is possible to design an encoder of sufficiently high complexity to yield an average distortion d_{avg} at a rate arbitrarily close to $R(d_{avg})$. It can be shown that $R(d_{avg})$ is a convex hull, continuous and strictly decreasing function of d_{avg}. A typical rate-distortion curve is shown in Fig. 7.1. This $R(d_{avg})$ is a theoretical limit and can be approached asymptotically by increasing the number of data samples, which is never achievable in practice. However, it provides a good benchmark against which practical encoding systems can be compared.

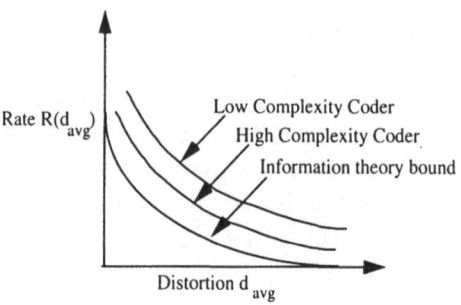

Figure 7.1. A typical rate-distortion function

7.2 STATISTICAL REDUNDANCY

It was demonstrated in Chapter 6 that efficient text compression could be achieved using entropy coding methods that rely on the statistical redundancy of the text symbols. The same approach can be applied to audio coding.

■ **Example 7.2**

Consider an audio acquisition system that has 10,000 mono audio samples with 3-bit resolution with sample levels 0 to 7. The number of occurrences of the eight samples levels is [700, 900, 1500, 3000, 1700, 1100, 800, 300]. Calculate and plot the probability density function of each symbol. Calculate the entropy of the source.

The probability of the occurrence of each sample level is given by

$$p[0] = 700/10000 = 0.07$$
$$p[1] = 900/10000 = 0.09$$
$$p[2] = 1500/10000 = 0.15$$
$$p[3] = 3000/10000 = 0.30$$
$$p[4] = 1700/10000 = 0.17$$
$$p[5] = 1100/10000 = 0.11$$
$$p[6] = 800/10000 = 0.08$$
$$p[7] = 300/10000 = 0.03$$

The *pdf* is plotted in Fig. 7.2.

$$H = -(0.07 * \log_2 0.07 + 0.09 * \log_2 0.09 + 0.15 * \log_2 0.15 + 0.30 * \log_2 0.30 +$$
$$0.17 * \log_2 0.17 + 0.11 * \log_2 0.11 + 0.08 * \log_2 0.08 + 0.03 * \log_2 0.03)$$
$$= 1.88 \ bits \ / \ sample$$

Therefore, the entropy of the source is 1.88 bits/audio sample. ■

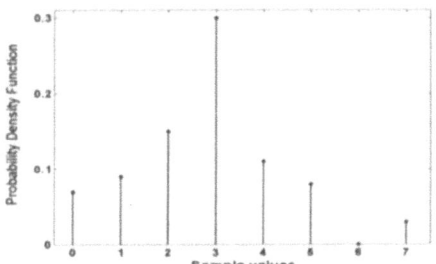

Figure 7.2. Probability density function of an information source.

The waveform corresponding to the audio signal "chord.wav" was shown in Fig. 4.14 (a). If the signal is quantized with 8 bits/sample, the dynamic range of the sample values will be [-128,127]. The *pdf* of the sample values is shown in Fig. 7.3. The entropy of the signal is 3.95 bits/sample. As a result, a direct application of entropy coding will produce a compression ratio of about 2:1 for this signal without any loss.

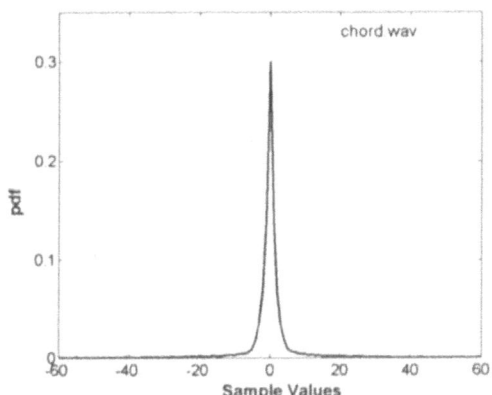

Figure 7.3. Probability density function of the audio signal "chord". It is observed that most of the sample values lie close to zero.

The entropy coding method generally does not provide a high compression ratio for most audio signals. However, these coding methods provide good performance when applied on the transform coefficients of the audio samples. Hence, MPEG-1 audio compression standard employs entropy coding, which will be discussed in a later section.

7.2.1 Companding and Expanding

It was observed in Fig. 7.3 that most of the audio samples have values close to zero. In that case, uniform quantization was employed to obtain the samples, which is not very efficient since the smaller values (that occur frequently) are quantized with the same step-size that is used to quantize larger amplitude samples. A superior performance can be obtained using a nonuniform quantizer that quantizes frequently occurring sample values more precisely than others. However, designing a nonuniform quantizer may be a tedious process.

A performance close to the nonuniform quantization can be achieved with a nonlinear transformation followed by uniform quantization. Here, the signal undergoes a fixed nonlinear transformation (known as *companding*) as shown in Fig. 4.14. A uniform quantization is then applied on the transformed signal, and the digitized sample values are stored. During the playback, we dequantize the signal, and perform the inverse nonlinear transformation (known as *expanding*). The original audio, along with some quantization noise, is obtained. Implementing such a system is easier than implementing a non-uniform quantizer.

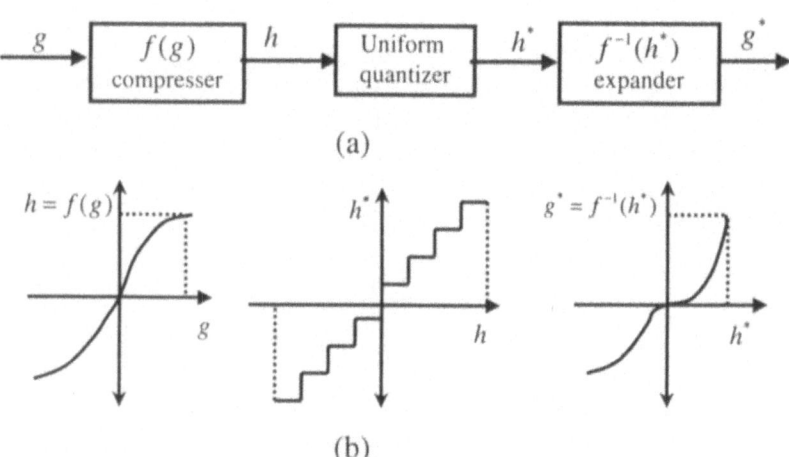

Figure 7.4. Companding and Expanding. a) block schematic of companding and expanding , b) the compressor and expander functions

Telephone systems in North America employ companding and expanding, which is known as μ-*law* encoding. The input-output relationship for the companding is as follows:

$$h = \frac{\ln(1+\mu|g|)}{\ln(1+\mu)} \text{sgn}(g)$$

where g and h are the input and output of the compander, respectively. After the quantization, the signal can be reconstructed using the inverse relationship which is as follows:

$$g^* = \frac{(1+\mu)^{|h^*|} - 1}{\mu} \text{sgn}(h^*)$$

The input-output characteristics of compander and expander with $\mu = 255$ are shown in Fig. 7.5.

■ **Example 7.3**

Consider the audio signal *chord*. Quantize the signal uniformly with 8 bits after companding using $\mu = 255$. Expand the signal, and calculate the SNR. Compare the SNR with the corresponding SNR obtained in Example 4.6.

When the audio signal is quantized with companding, the SNR obtained is approximately 44 dB. In example 4.6, it was observed that the SNR corresponding to 8 bit/sample was approximately 32 dB. Hence, in this experiment, an improvement of about 12 dB is obtained with companding.

■

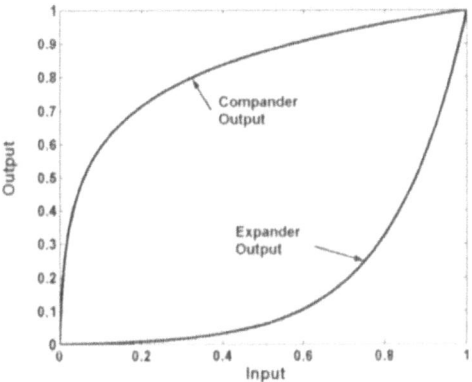

Figure 7.5. Input-output characteristics of compander and expander with $\mu = 255$.

Studies have revealed that an improvement in the range of 25 dB is generally achieved for telephone signals. In other words, even though the telephone signals are quantized with 8 bits/sample, the effective quality with companding is comparable to that of 12 bits with uniform quantization. European telephone systems use a slightly different scheme, known as *A*-law companding.

7.3 TEMPORAL REDUNDANCY

In audio sampling with pulse-code modulation (PCM), the samples are encoded independent of other samples. Most audio signals exhibit a strong correlation among neighbor signals. This is especially true if the signal is sampled above Nyquist frequency. This is demonstrated in Fig. 7.6, which shows the blow-up of the audio signal shown in Fig. 4.14(a). It is observed that the neighboring samples are reasonably well correlated. Hence, the value of an audio signal at any time instance k can be predicted from the values of the signal at time $(k-m)$ where m is a small integer. This property can be exploited to achieve a compression ratio of 2 to 4 without degrading the audio quality significantly. Differential pulse code modulation (DPCM) [2, 3] is one of the simplest but effective techniques in this category, which is presented below.

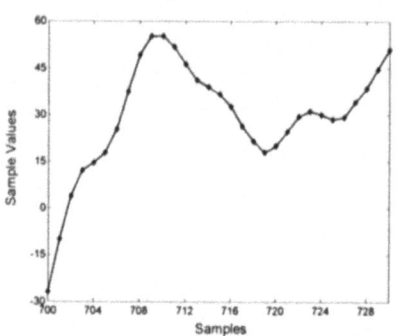

Figure 7.6. Blow up of "chord" signal representing 30 samples (700-730).

Differential PCM (DPCM)

In DPCM, an audio sample is predicted from the previous samples. The block diagrams of DPCM encoding and decoding systems are shown in Fig. 7.7. In most cases, the predicted value $\hat{s}_n$ will be close to the s_n (the actual sample value at time instance n), but not identical. In the linear predictive coding (LPC) technique, the predicted value $\hat{s}_n$ is calculated as follows:

$$\hat{s}_n = \sum_{i=1}^{M} \alpha_i s'_{n-i} \qquad (7.2)$$

where s'_{n-i} is the reconstructed sample at the i-th previous instance, and α_i, $i=1, 2,...,M$, are the coefficients of an M-th order predictor. The prediction error e_n (*i.e.*, the difference between the original sample and its predicted value) is calculated as follows:

$$e_n = s_n - \hat{s}_n \qquad (7.3)$$

In order to achieve a good prediction, the predictor coefficients can be chosen appropriately to minimize the variance or energy of the error sequence e_n.

The optimal set of LPC coefficients $\alpha_{i,opt}$ can be obtained by solving the following set of simultaneous equations (known as *Yule-Walker* or *normal* equations):

$$\sum_{i=1}^{M} \alpha_{i,opt} R(j-i) = R(j) \tag{7.4}$$

where $R(j), j = 0, \pm 1, \pm 2, \ldots$ is the autocorrelation function of the input data samples, defined as

$$R(j) = \sum_{m=1}^{N} s_m * s_{m+j} \tag{7.5}$$

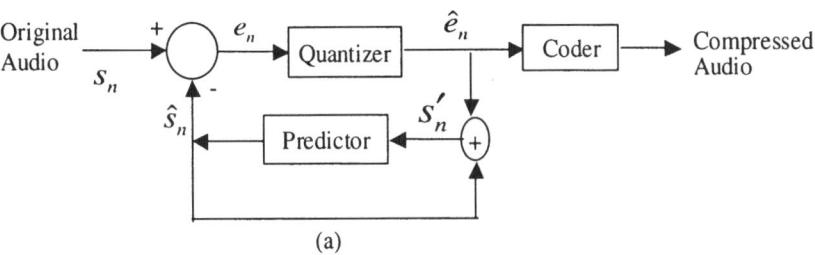

(a)

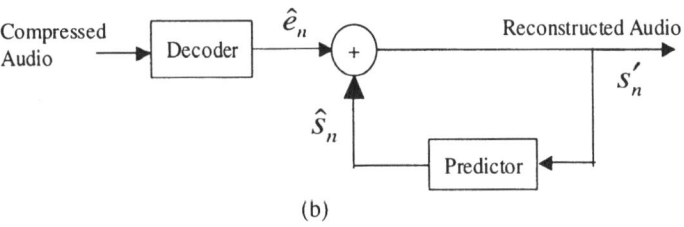

(b)

Figure 7.7. Schematic of a DPCM (a) Encoder and (b) decoder. s_n: original input signal, $\hat{s}_n$: predicted input, s'_n: reconstructed output, e_n: prediction error, $\hat{e}_n$: prediction error.

■ Example 7.4

Consider the *chord* audio signal. Determine the optimal set of 1st order, 2nd order and third order optimal prediction coefficients.

The normalized autocorrelation function of the *chord* signal is as follows:

$$R = [1.0000 \quad 0.9708 \quad 0.8872 \quad 0.7587 \quad 0.5977 \quad 0.4167 \quad 0.2276]$$

1st order predictor

From Eq. (7.4), we obtain:

$$R(0)\alpha_1 = R(1)$$

Hence, $\alpha_1 = R(1)/R(0) = 0.97$.

2nd order predictor

$$\begin{bmatrix} R(0) & R(1) \\ R(-1) & R(0) \end{bmatrix} \begin{bmatrix} \alpha_1 \\ \alpha_2 \end{bmatrix} = \begin{bmatrix} R(1) \\ R(2) \end{bmatrix} \text{ or, } \begin{bmatrix} 1 & 0.97 \\ 0.97 & 1 \end{bmatrix} \begin{bmatrix} \alpha_1 \\ \alpha_2 \end{bmatrix} = \begin{bmatrix} 0.97 \\ 0.89 \end{bmatrix}$$

or, $\alpha_1 = 1.81$, and $\alpha_2 = -0.86$

3rd order predictor

$$\begin{bmatrix} R(0) & R(1) & R(2) \\ R(-1) & R(0) & R(1) \\ R(-2) & R(-1) & R(0) \end{bmatrix} \begin{bmatrix} \alpha_1 \\ \alpha_2 \\ \alpha_3 \end{bmatrix} = \begin{bmatrix} R(1) \\ R(2) \\ R(3) \end{bmatrix}, \text{ or, } \begin{bmatrix} 1 & 0.97 & 0.89 \\ 0.97 & 1 & 0.97 \\ 0.89 & 0.97 & 1 \end{bmatrix} \begin{bmatrix} \alpha_1 \\ \alpha_2 \\ \alpha_3 \end{bmatrix} = \begin{bmatrix} 0.97 \\ 0.89 \\ 0.76 \end{bmatrix}$$

Hence, $\alpha_1 = 1.16$, $\alpha_1 = 0.49$, $\alpha_3 = -0.75$

The optimal predictors (of different orders) for the chord signal are

$$\hat{s}_n = 0.97 s'_{n-1} \qquad\qquad 1^{st} \text{ order predictor} \qquad (7.6)$$

$$\hat{s}_n = 1.81 s'_{n-1} - 0.86 s'_{n-2} \qquad 2^{nd} \text{ order predictor} \qquad (7.7)$$

$$\hat{s}_n = 1.16 s'_{n-1} + 0.49 s'_{n-2} - 0.75 s'_{n-3} \qquad 3^{rd} \text{ order predictor} \qquad (7.8)$$

The first predictor produces an error energy of 2.79, whereas the second order predictor produces an error energy of 0.25. To verify the validity of the optimal coefficients, total error energies were calculated in the neighborhood of the optimal coefficients, and are shown in Fig. 7.8. It is observed that the first order predictor provides the smallest error energy. For the second order case, however, it is observed that the error energy is minimum (or very close) for a large number of coefficients. These coefficients generally follow the relationship: $\alpha_1 + \alpha_2 \approx 0.95$. In other words, so long as the above relationship is satisfied, a good prediction is expected. ∎

After the prediction error sequence e_n has been obtained, it has to be encoded to reconstruct the signal perfectly. In lossy coding, however, a reasonably good reconstruction quality is often acceptable. As a result, the error sequence is typically quantized using the scheme shown in Fig. 7.7(a). Note that the quantization is the only operation in the DPCM scheme that introduces noise in the reconstructed signal. The quantized error sequence e_n can be entropy coded to achieve further compression. The quantized difference value is in turn used to predict the next data sample s_{n+1}. The following example illustrates various steps of the DPCM coding.

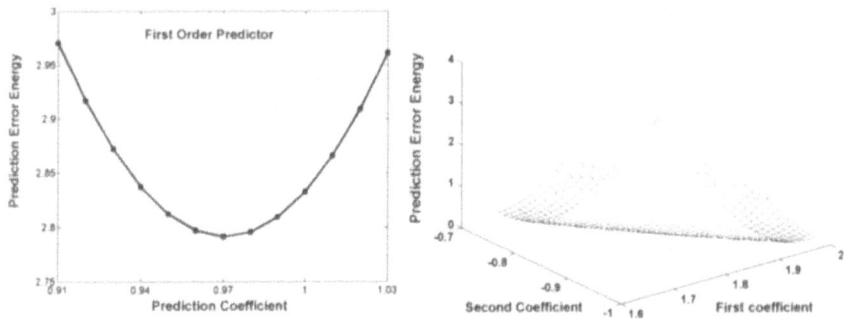

Figure 7.8. Prediction error for different predictors. a) first order predictor, b) second order predictors.

■ Example 7.5

The first four samples of a digital audio sequence are [70, 75, 80, 82,...]. It would require at least seven bits to encode each of the audio samples. The audio samples are to be encoded using DPCM with a first order predictor as given in Eq. (7.6). The prediction error coefficients are quantized by 2 and rounded to nearest integer, and stored losslessly. Determine the number of approximate bits required to represent each sample and the reconstructed error at each sampling instance.

The various steps at different sampling instances are shown in Table 7.1. It is observed that the first audio sample is encoded as a reference sample. From the second sample onwards, the prediction is employed. With a smaller quantization step size of 2, the reconstruction error is very small (magnitude less than 1), and tolerable. However, the number of bits required to encode the quantized error signal is significantly reduced (3, 2, and 2 bits for coding the 2nd, 3rd and 4th predicted error samples, respectively). ■

■ Example 7.6

Consider again the *chord* audio signal. The signal is encoded using a DPCM coder using the predictors given in Eqs. (7.6) and (7.7). The error sequence is quantized with step-sizes {1, 2, 3, 4, 5}. It is assumed that the quantized error coefficients would be encoded using an entropy-coding method. Hence, the entropy is considered as the bit-rate.

The rate-distortion performance of the DPCM coder is shown in Fig. 7.9. It is observed that as the bit-rate increases, the SNR also improves. The second predictor is seen to provide a margibal improvement over the first order predictor. ■

Table 7.1. Various steps of DPCM coding of the sequence [70, 75, 80, 82,....].
$\alpha_0 = 0.97$. The error sequence is quantized by 2 and rounded to the nearest integer.

	Sampling Instances			
	0	1	2	3
Original signal, S_n	70	75	80	82
Error signal, e_n	0	75-67.9=7.1	80-73.6=6.4	82-77.2=4.8
Quantized Error Signal	0	7.1/2 = 4	6.4/2=3	4.8/2=2
Reconstructed Error	0	4*2=8	3*2=6	2*2=4
Reconstructed Signal, s'_n	70	67.9+8=75.9	73.6+6=79.6	77.2+4=81.2
Predicted Signal for the next sample, $\hat{S}_n$	70*0.97 =67.9	75.9*0.97 =73.6	79.6*0.97 =77.2	81.2*0.97 =78.8
Reconstruction Error	0	-0.9	0.4	0.8
No. of bits required	7	3	2	2

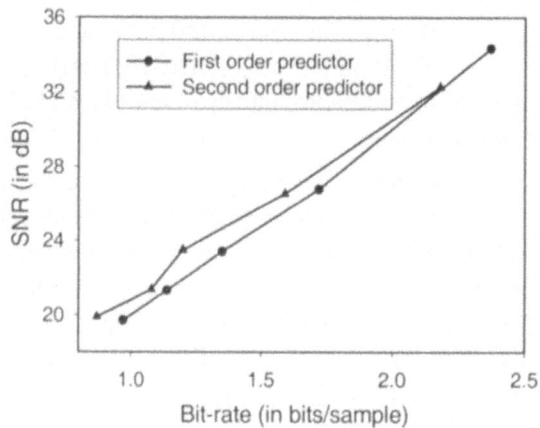

Figure 7.9. Performance of the DPCM coder.

7.4 PERCEPTUAL AUDIO CODING

It was explained in Chapter 2 that the sensitivity of the ear is not uniform throughout the entire audio spectrum (20 Hz-20KHz). It was observed the ear is most sensitive in the 1-5 KHz frequency range. The sensitivity decreases at lower and higher frequencies. In addition, a strong tonal signal at a given frequency can mask the weaker signals corresponding to the neighboring frequencies. When the two properties are combined, the threshold of hearing as shown in Fig. 2.5 (in Chapter 2) is obtained. Any audio signal whose amplitude is below the masking threshold is inaudible to the human ear. This characteristic of the auditory system can be exploited to achieve superior audio compression.

In order to employ the psychoacoustics model, the audio signal is divided into data-blocks with short duration. The spectrum of each audio block is then estimated using the Fourier transform, and masking thresholds at different audio frequencies are obtained. The perceptual bit allocation is then performed. The bit allocation process is illustrated in Fig. 7.6, with a signal containing several tones. Due to the masking threshold, only tones Q, S, and U would be heard, and the other tones (P, R, T, and V) would be masked by the stronger tones. Since the tones P, R, T, and V would not be heard, these signals need not be encoded at all. Therefore, the bits are allocated to the signals corresponding to Q, S, and U tones. The bits allocated would be proportional to the signal level above the masking threshold, as stronger signals are generally more important, and need to be encoded more precisely.

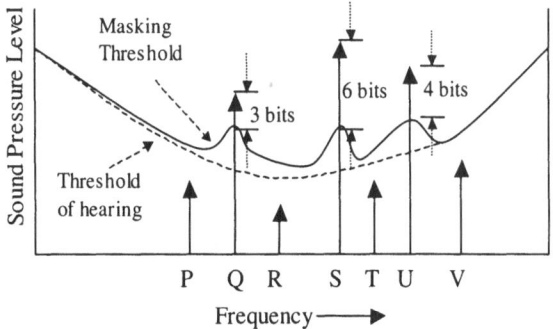

Figure 7.6. Perceptual bit allocation. The bits are allocated only to tones that are above the masking levels.

The perceptual coding technique can be implemented efficiently in the subband domain [4]. Here, an audio signal is passed through a digital filterbank consisting of a bank of lowpass and bandpass filters (see Fig. 5.10 in Chapter 5). The filterbank divides the signal into multiple bandlimited channels to approximate the critical band response of the human ear. The samples in each subband are analyzed and compared to a psychoacoustics model. Then the coder adaptively quantizes the samples in each subband based on the masking threshold in that subband. Figure 7.7 shows the operation of a typical subband coder.

Each subband is coded independently with greater or fewer bits allocated to the samples in the subband. The level of quantization noise introduced in each subband is determined by the number of bits used (more allocated bits produce less noise). Bit allocation is determined by the psychoacoustic model and analysis of the signal itself. These operations are repeated for each subband for every new block of data. The samples are dynamically quantized according to the audibility of signals, and the entire process is highly flexible. The decoder dequantizes the quantized data, which is then passed through the synthesis filterbank resulting in the reconstructed audio output, which can then be played back.

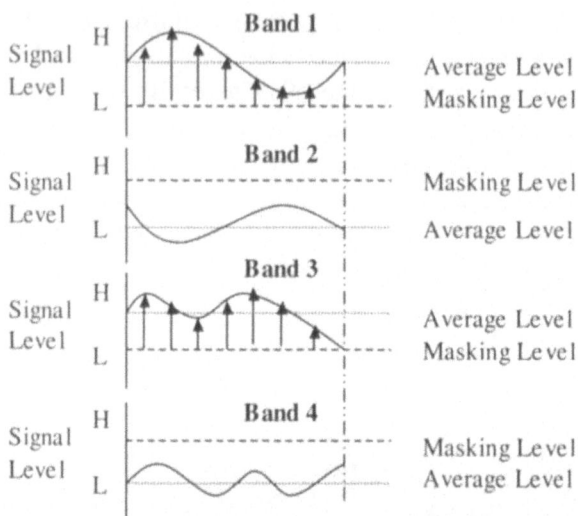

Figure 7.6. Perceptual coding of different subbands. Bands 2 and 4 are perceptually unimportant since the signal levels are below the masking levels. Bands 1 and 3 are important since the signal levels are above the masking levels. The bits are allocated according to the peak level above the masking threshold corresponding to the respective subbands.

7.5 AUDIO COMPRESSION STANDARDS

Several audio coding standards have been developed in last 15 years [5]. These can be broadly divided into two categories: low bit-rate audio coders for telephony, and general-purpose high fidelity audio coders.

ITU-T Recommendation G.711 [6] is the ISDN default standard for coded audio. The audio is sampled at 8000 samples/sec with a precision of 8bits/sample. It uses μ-law and A-law companding for better efficiency. For higher fidelity audio, ITU-T G.722 [7] can be used, which encodes 7 KHz audio within 64 kbits/s. Further coding efficiency can be achieved in G.729 that can encode speech at 8 kbits/s using a conjugate structure algebraic-code-excited linear-prediction algorithm [8].

Adopted in 1992, the MPEG-1 audio compression algorithm [9, 10] is the first international standard for compressing high fidelity digital audio. It is a generic audio compression standard that can compress audio signals coming from a wide variety of sources. MPEG-2 [11, 12] audio codec is based on MPEG-1 codec, and includes several enhancements. The MPEG-4 audio codec [13] can compress both natural (e.g., speech, music) and synthesized sound using structural description. The synthesized sound can be represented by text or by the description of musical instruments together with different effects. The AC-2 and AC-3 [14] codecs have been developed by Dolby Digital Laboratories [15] for multichannel high fidelity audio coding.

A brief overview of MPEG-1, MPEG-2 and AC-3 standards is presented in the following.

7.6 MPEG-1 AUDIO COMPRESSION STANDARD

In MPEG-1 audio compression algorithm, the psychoacoustics properties of the human ear are exploited to achieve high compression. Typically, the audio sampling rate in MPEG-1 is 32, 44.1 or 48 KHz. The encoder can compress both mono and stereo (i.e., 2 channels) audio. Also, the compressed bitstream can have a predefined bit-rate in the range 32-192 kbits/s/ch as well as a free bit-rate mode. The algorithm offers three independent layers of compression that provide a wide range of tradeoffs among coder complexity, compression ratio and audio quality. Layer 1 is the simplest layer, and best suited for bit-rates above 128 kbits/s/ch. Layer 2 has moderate complexity, and is targeted for bit-rates around 128 kbits/s/ch. Layer 3, also known as *mp3* standard, is the most complex among the three layers. It is targeted for bit-rates around 64 kbits/s/ch, and offers the best audio quality. This layer is well suited for audio transmission over ISDN.

Coding Algorithm

The MPEG-1 generic encoder (for all three layers) is shown in Fig. 7.8(a). The input audio signal is passed through a filterbank that represents the signal in multiple subbands. The input signal is also passed through a psychoacoustics model that determines the masking threshold for each subband. Two pyschacoustics model are used: Model 1 is typically used for Layers 1 and 2, whereas Model 2 is used for Layer 3. The coefficients in a subband are then quantized according to its masking threshold. Then the quantized coefficients from all subbands are formatted and some ancillary data are added to form the encoded bitstream.

The MPEG-1 decoder is the inverse of the encoder, and is shown in Fig. 7.8(b). Here, the encoded bitstream is unpacked. The quantized subband coefficients are dequantized, passed through the synthesis subband, and added to obtain the reconstructed audio.

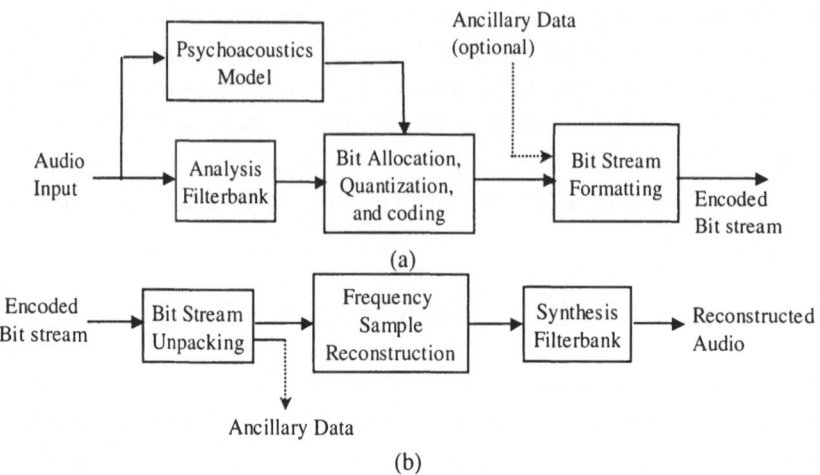

Figure 7.8. MPEG-1 audio coder. a) Encoder, b) Decoder

Filter Bank

In order to achieve superior performance, the bandwidths of individual filters in the analysis filterbank should match the critical bands (Table 2.2, Chapter 2) of the ears. Therefore, the bandwidth of the subbands should be small in the low frequency range, and large in the high frequency range. However, in order to simplify the codec design, the MPEG-1 filterbank contains 32 banks of equal width. The filters are relatively simple and provide a good time resolution with a reasonable frequency resolution. Note that the filterbank is not reversible, *i.e.*, even if the subband coefficients are not quantized, the reconstructed audio will not be identical to the original audio.

Layer Coding

Since there are 32 filters, the individual subband filters in the filterbank produces 1 sample for each set of 32 input samples (as shown in Fig. 7.9). The samples are then grouped together for quantization, although grouped differently in the three layers [9].

The Layer 1 algorithm encodes the audio in frames of 384 audio samples that is obtained by grouping 12 samples from each of the 32 subbands. Each group of 12 samples gets a bit allocation, ranging from 0-15 bits depending on the masking level. If the bit allocation is not zero, each group also gets a scale factor (which is represented by 6 bits) that sizes the samples to make full use of the range of the quantizer. Together the bit allocation and scale factor can provide up to 20 bits resolution, resulting in a SNR of 120 dB. The decoder multiplies the scale factor with the dequantized output to reconstruct the subband samples.

The Layer 2 encoder uses a frame size of 1152 samples per audio channel, grouping 36 samples from each subband. There is one bit allocation for each channel, and up to three scale factors for each trio of 12 samples [9]. The different scale factors are used only to avoid the audible distortion. In this layer, three consecutive quantized values may be coded with a single codeword.

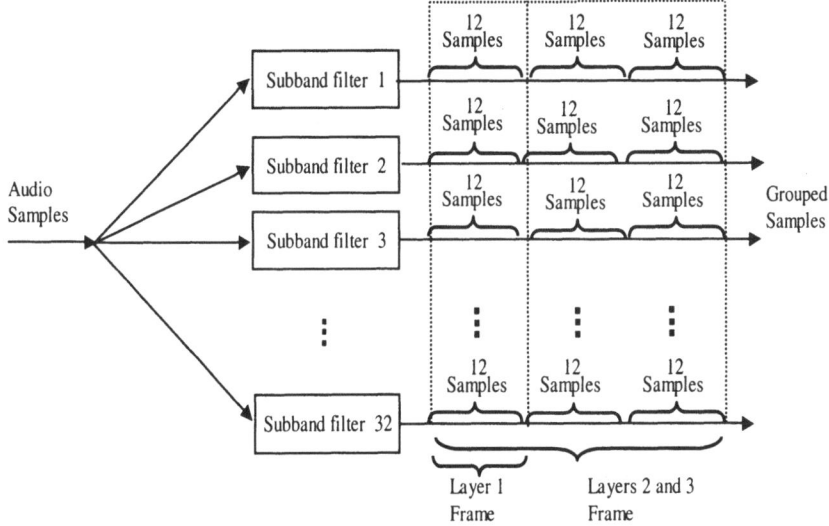

Figure 7.9. Grouping of subband samples in MPEG-1 layers 1, 2, and 3.

To achieve superior compression performance, the Layer 3 employs a much more refined algorithm than Layers 1 and 2. The schematic of Layer 3 is shown in Fig. 7.10. It employs modified discrete cosine transform (MDCT). The MDCT is calculated for a long block of 18 samples or a short block of 6 samples. There is 50% overlap between successive transform windows so that the window size is 36 or 12, respectively. The long block length provides a better frequency resolution, whereas the short block length provides a better time resolution. However, the switch between the long and short blocks is not instantaneous. A long block with a specialized long-to-short or short-to-long data window is used to serve as the transition between long and short blocks.

Note that since the MDCT employs a block of 36 or 12 samples, any quantization error of the MDCT coefficients will be spread over a large time window compared to Layers 1 and 2, resulting in audible distortion. Therefore, it employs an alias reduction mechanism that processes MDCT values to remove the artifacts caused by the overlapping bands of the polyphase filterbank. In addition to the MDCT, the Layer 3 includes several enhancements. It employs non-uniform quantization to provide a more consistent SNR over the range of quantizer values. Scalefactor bands are used in Layer 3 to color the quantization noise to fit the varying frequency contours of the masking threshold. Layer 3 also employs a variable length Huffman code to encode the quantized samples, and to achieve higher compression.

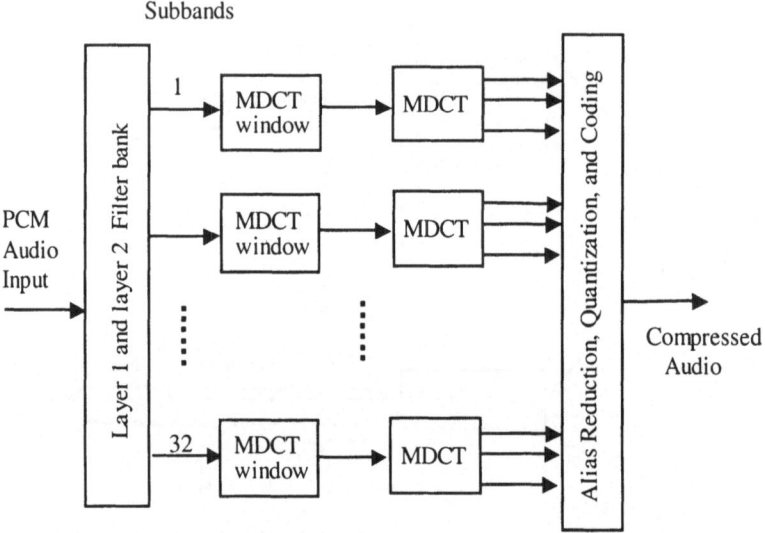

Figure 7.10. MPEG Audio Layer 3 Processing. The MDCT window can be long or short. Alias reduction is applied only to long blocks.

7.7 MPEG-2 AUDIO COMPRESSION STANDARD

MPEG-2 defines two audio compression standards: MPEG-2 backward-compatible (BC) standard and MPEG-2 advanced audio coding (AAC) standard. The MPEG-2 BC is an extension of the MPEG-1 audio standard, which includes multichannel and multilingual coding. In addition, coding at sampling frequencies lower than 32 KHz is allowed. On the other hand, the MPEG-2 Advanced Audio coding (AAC) is a highly advanced audio encoder with superior compression performance. The MPEG-AAC algorithm is briefly presented in the following.

MPEG-2 AAC

The AAC has three modes or profiles. The *low complexity* (LC) mode is used for applications where processing speed and memory are the bottlenecks. The *main* mode is used when processing power and memory are readily available. The scalable sampling rate (SSR) mode is used for applications that required scalable decoding. The block schematic of the

MPEG-2 AAC encoder is shown in Fig. 7.11. A brief description of the basic tools is provided below.

Filterbank: A modified DCT (MDCT) is used. The MDCT output consists of 1024 or 128 frequency lines. The window shape is selected between two alternative window shapes.

Perceptual Model: A psychoacoustics model similar to the MPEG-1 model 2 calculates the masking threshold.

Temporal Noise Shaping (TNS): Controls the temporal shape of the quantization noise within each window of the transform.

Prediction: Reduces the redundancy for stationary signals. A second order backward-adaptive predictor is used.

Scale-factor: The audio spectrum is divided into several groups of spectral coefficients, known as the *scale factor bands*, which share one scale factor. A *scale factor* represents a gain value that changes the amplitude of the spectral coefficients of a scale-factor band.

Quantization: A nonuniform quantizer is used with a step-size of 1.5 dB.

Noiseless Coding: Huffman coding is applied for the quantized subband coefficients, the differential scale factors, and the directional information. Up to twelve static Huffman codebooks are employed to encode pairs or quadruples of spectral values.

7.8 AC AUDIO COMPRESSION STANDARDS

Dolby Laboratory has developed several high performance audio codecs [14]. Here, we mainly focus on the AC family of audio coders. There are three codecs in the AC family: AC-1, AC-2 and AC-3. The AC-1 codec employs adaptive delta modulation along with analog companding. It is not a perceptual coder. The AC-2 coder is more sophisticated. It is a perceptual coder that uses a low complexity TDAC (time domain alias cancellation) transform. A series of modified DCT and DST (sine transform) are employed. An optimal window function based on the Kaiser-Bessel kernel is used in designing the AC-2 windows. AC-3 is a much more sophisticated and popular coder for compression of digital audio. It was originally developed for use in digital cinema application, and was introduced in 1991. Later (in 1993), it was selected as the audio coding system for the North American high definition TV (HDTV) standard, as well as for the consumer DVD movies in North America. Because of its widespread use, we will discuss the AC-3 coder in more details.

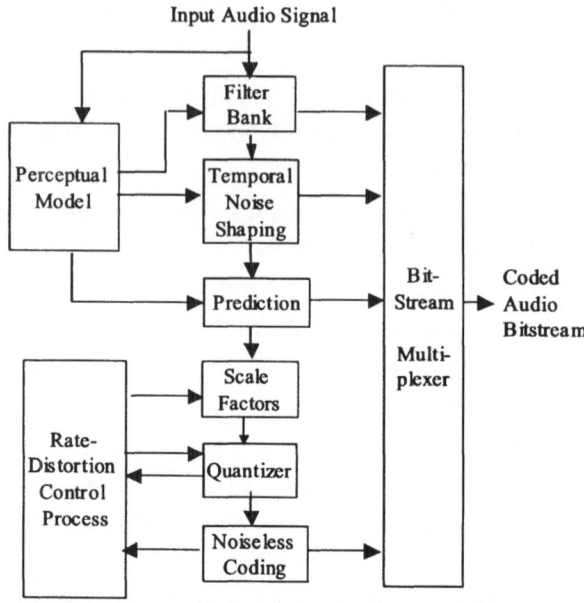

Figure 7.11. A simplified schematic of the MPEG-2 AAC encoder.

The schematic of the AC-3 encoder is shown in Fig. 7.12. It employs TDAC transform that allows tradeoffs between the coding delay and the bit-rate. Here, a series of modified DCT and DST (sine transform) are applied on audio data blocks overlapped 50% in the time domain. The transform coefficients are then sub-sampled by a factor of 2 (*i.e.*, drop every other coefficients) to obtain critically sampled transform. The default transform block length is 256 samples (actually 512 samples with 50% overlap), providing 93.75 Hz of frequency resolution at 48 KHz sampling rate. The input audio is constantly monitored to detect the presence of transients. If any transients are present, the block length is halved to improve the temporal resolution (as a consequence, the frequency resolution will degrade).

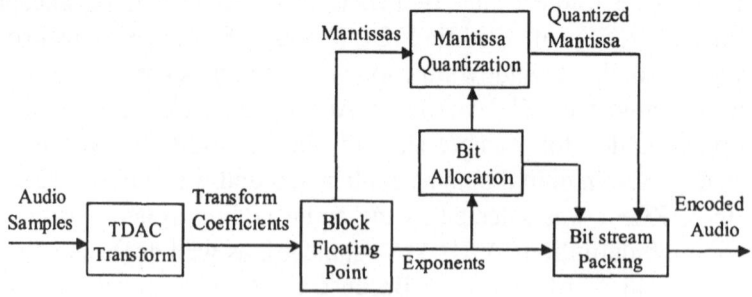

Figure 7.12. A simplified schematic of the AC-3 audio encoder.

The transform coefficients are appropriately combined to form subbands that approximate the critical bands. The subband coefficients are normalized, and expressed in floating point notations with exponents and mantissas ($s = x * 2^{-exp}$). Note that an exponent is the number of leading zeros in the binary representation of a frequency coefficient. These exponents are employed to drive the perceptual model that performs bit allocation. The mantissas are then quantized according to the bit allocation. Exponent values are allowed to range from 0 to 24 (hence 5 bits are required for the representation). It has been found that the exponents of the subband coefficients are highly redundant. Therefore, to achieve further compression, the exponents are differentially encoded. These different exponents are combined into groups in the audio block. The "exponent strategy" defines how different exponents are encoded.

The AC-3 decoding is the inverse of the encoding process. Figure 7.13 shows the schematic of the decoder operation.

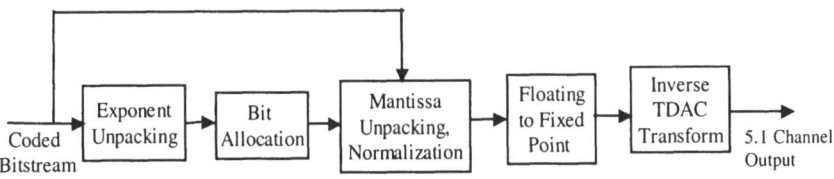

Figure 7.13. A simplified schematic of the AC-3 audio decoder.

7.9 COMPARISON OF COMPRESSION ALGORITHMS

Since a large number of audio coding standards have been developed for different applications, it is important to know their relative performance. Table 7.2 shows the comparison of MPEG-1 codec at various layers, and AC-3 codec. MPEG Layer 1 provides mild compression, but has a lower complexity, and can be used where cost is an issue. Layer 2 provides better performance at a reasonable complexity, and is used in many existing applications. Layer 3, known as mp3, provides excellent compression, but has a high complexity. The AC-3 coder is best for good quality multichannel applications. The MPEG-2 AAC provides a superior compression performance compared to most existing coding algorithms including layer 3 and AC-3 coder. Experiments have demonstrated that the MPEG-2 AAC coder at 320 kb/s outperforms MPEG-2 Layer 2 at 640 kb/s [16].

Table 7.2. Comparison of MPEG-1 and AC-3 codecs [16]

	Bit-rate (in kb/s)	Quality	Applications	Available since
MPEG-1 L-1	32-448	Good quality at 192 kb/s/ch	DCC	1991
MPEG-1 L-2	32-384	Good quality at 256 kb/s/ch	Digital audio bradcasting, CD-I, DVD	1991
MPEG-1 L-3	32-320	Good quality at 96 kb/s/ch		1993
Dolby AC-3	32-640	Good quality at 384 kb/s/5.1 ch	HDTV, Cable, DVD	1991

7.10 AUDIO FORMATS

Digital audio data is available in a wide variety of formats [17]. A few selected formats are shown in Table 7.3. Note that some formats represent audio as PCM data (raw samples), while other formats use some form of compression.

Table 7.3. Common file types used for storing sounds

File Type/ Extension	Comments
aiff	Uncompressed audio, typically 16 bits/sample. Other bit precision is also allowed. Used mostly in Mac/Unix platforms.
au	Uses μ-law companding. Typically used in SUN workstations.
mov	Quick Time Video
mpa/mp2	MPEG audio format. Various layers are used to trade-off complexity and performance.
mp3 (*)	MPEG audio layer 3.
qt (*)	Quick time format. Propriety format by Apple Computer.
ra, ram (*)	Real audio format. Proprietary format by Real Networks. Supports live audio over Internet
wav	Uncompressed audio, typically 16 bits/sample.
wma (*)	Windows media audio. Propriety format by Microsoft.

*Common file types that can be used for audio streaming through the web.

REFERENCES

1. C. E. Shannon, "A Mathematical theory of communication," *Bell Systems Technical Journal*, Vol. XXVII, No. 3, pp. 379-423, 1948.

2. N. S. Jayant and P. Noll, *Digital Coding of Waveforms: Principles and Applications to Speech and Video*, Prentice-Hall, New Jersey, 1984.

3. A. Gersho, "Advances in speech and audio compression," *Proc. of the IEEE*, Vol. 82, No. 6, p 900-918, Jun 1994.

4. B. Tang, A. Shen, A. Alwan, G. Pottie, "Perceptually based embedded subband speech coder," *IEEE Transactions on Speech and Audio Processing*, Vol. 5, No. 2, pp. 131-140, Mar 1997.

5. K. C. Pohlmann, *Principles of Digital Audio*, McGraw-Hill, New York 2000.

6. ITU-T Recommendation G.711, *Pulse Code Modulation (PCM) of Voice Frequencies*, 1988.

7. ITU-T Recommendation G.722, *7 KHz audio coding within 64 kbits/s*, 1988.

8. ITU-T Recommendation G.729, *Coding of speech at 8 kbits/s using conjugate structure algebraic-code-excited linear-prediction (CS-ACELP)*, 1996.

9. D. Pan, "An overview of the MPEG/audio compression algorithm," *Proc. of SPIE - Digital Video Compression on Personal computers: Algorithms and Technologies*, Vol. 2187, San Jose, February 1994.

10. ISO/IEC JTC1/SC29/WG11 MPEG, International Standard IS 11172-3, "Coding of Moving Pictures and Associated Audio for Digital Storage Media at up to about 1.5 Mbits/s, Part 3: Audio", 1992.

11. ISO/IEC JTC1/SC29/WG11 MPEG, International Standard IS 13818-3, "Information Technology – Generic Coding of Moving Pictures and Associated Audio, Part 3: Audio"

12. ISO/IEC JTC1/SC29/WG11 Doc. N1430, "DIS 13818-7 (MPEG-2 Advanced Audio Coding)" (1996).

13. ISO/IEC JTC1/SC29/WG11 Doc. N0937, "MPEG-4 Proposal Package Description (PPD)" (1995).

14. Dolby Laboratory, http://www.dolby.com/.

15. S. Vernon, "Design and implementation of AC-3 coders," *IEEE Trans. on Consumer Electronics*, Vol. 41, No. 3, August 1995.

16. K. Brandenburg and M. Dosi, "Overview of MPEG audio: current and future standards for low-bit-rate audio coding," *Journal of Audio Engineering Society*, Vol. 45, No. 1/2, pp. 4-21, Jan/Feb 1997.

17. Chris Bagwell, *Audio File Formats FAQ*, http://home.attbi.com/~chris.bagwell/AudioFormats.html.

QUESTIONS

1. What are the main principles of the audio compression techniques?

2. You are quantizing an analog audio signal. How much quantization noise (in *dB*) do you expect if you use 8, 12, 16 and 24 bits/sample/channel?

3. Calculate the *probability density function* and *entropy* of the audio signal *bell1.wav* provided in the CD. What compression factor can be achieved using the entropy coding method?

4. What is the principle of DPCM coding method?

5. The autocorrelation matrix of an audio signal is given by

$$\begin{bmatrix} 1 & 0.9 & 0.8 \\ 0.9 & 1 & 0.9 \\ 0.8 & 0.9 & 1 \end{bmatrix}$$

Calculate the optimal first and second order predictors. Can you calculate the third order predictor with the given information?

6. Consider a DPCM coder with a first order predictor $s_n = 0.97 * s_{n-1}$. Assume that the signal has a dynamic range [-128,127]. Encode the audio signal *bell1.wav*. Use the quantization step-sizes {1, 2, 3,4, 5, 6}, and plot the rate-distortion performance.

7. Explain how the psychoacoustics properties of the human ear can be exploited for audio compression.

8. How is the bit-allocation performed in a perceptual subband coder? How is the masking threshold determined for an audio signal?

9. In audio coding, the audio signal is generally divided into blocks of audio samples., which are then coded individually. What are the advantages of this blocking method? Compare the performance of long and short blocks.

10. Briefly review the different audio compression standards. Which standards are targeted for telephony applications? Which standards have been developed for general purpose high quality audio storage and transmission?

11. Explain the MPEG-1 audio compression standard. Why were three coding layers developed? Compare the complexity and performance of the different layers.

12. How are the audio data samples grouped in different MPEG-1 layers?

13. Explain briefly the MP3 audio compression algorithm. How does it achieve superior performance over MPEG-1 layers 1 and 2?

14. Draw a schematic of MPEG Advance Audio Coding method. How does it different from the MPEG-1 compression methods?

15. Explain the encoding and decoding principles of the AC-3 codec. How is it different from MPEG-1 codec?

16. What is advantage of encoding mantissa and exponents separately in AC-3 codec?

Chapter 8

Digital Image Compression Techniques

The term *image compression* refers to the process of reducing the amount of data needed to represent an image with an acceptable subjective quality. This is generally achieved by reducing various redundancies present in an image. In addition, the properties of the human visual system can be exploited to further increase the compression ratio.

In this Chapter, the basic concept of information theory as applied to image compression will be presented first. Several important parameters such as entropy, rate and distortion measure are then defined. This is followed by the lossless and lossy image compression techniques. An overview of the JPEG standard and the newly established JPEG2000 standard are presented at the end.

8.1 PRINCIPLES OF IMAGE COMPRESSION

Similar to audio data, generally image data is also highly redundant. High compression ratio can therefore be achieved by removing the redundancies. As in audio, images generally have statistical redundancy that can be exploited. We note that a still image has two spatial coordinates instead of temporal coordinate as in audio. Therefore, the images have spatial redundancy instead of temporal redundancy. The 2-D structure of an image can be exploited to achieve a superior compression. Lastly, properties of human visual system can be exploited to achieve a higher compression.

Statistical redundancy: Refers to the non-uniform probabilities of the occurrences of the different pixel values. The pixel values that occurr more frequently should be given a larger number of bits than the less probable values.

Spatial redundancy: Refers to the correlation between neighboring pixels in an image. The spatial redundancy is typically removed by employing compression techniques such as predictive coding, and transform coding.

Structural redundancy: Note that the image is originally a projection of 3-D objects onto a 2-D plane. Therefore, if the image is encoded using structural image models that take into account the 3-D properties of the scene, a high compression ratio can be achieved. For example, a segmentation coding approach that considers an image as an assembly of many regions and encodes the contour and texture of each region separately, can efficiently exploit the *structural* redundancy in an image/video sequence.

Psychovisual redundancy: The human visual system (HVS) properties can be exploited to achieve superior compression. The typical HVS properties that are used in image compression are i) greater sensitivity to distortion in smooth areas compared to areas with sharp changes (*i.e.*, areas with higher spatial frequencies), ii) greater sensitivity to distortion in dark areas in images, and iii) greater sensitivity to signal changes in the luminance component compared to the chrominance component in color images.

Although the image compression techniques are based on removing different types of redundancies, the compression techniques are generally classified into two categories – i) lossless technique, and ii) lossy technique. In lossless technique, the reconstructed image is identical to the original image, and hence no distortion is introduced. Typical applications of lossless techniques are medical imaging where distortion is unacceptable, and satellite images where the images might be too important to introduce any noise. However, the lossless techniques do not provide a high compression ratio (less than 3:1). On the other hand, lossy compression techniques generally provide a high compression ratio. Furthermore, these techniques introduce noise in the images that might be acceptable for many applications, including the World Wide Web.

In this Chapter, two lossless techniques are presented: i) entropy coding, and ii) run-length coding. The entropy coding techniques remove the statistical statistical redundancy present in an image, whereas the run-length coding technique removes the spatial redundancy. A few highly efficient lossy compression techniques are then presented. The DPCM and transform coding are primarily based on the removal of the spatial redundancy. However, the lossless techniques and HVS properties can be jointly exploited to achieve a superior compression.

8.2 LOW COMPLEXITY COMPRESSION TECHNIQUES

In this section, three low complexity compression techniques, namely entropy coding, run-length, and predictive coding [1], are presented.

8.2.1 Entropy Coding

It was demonstrated in Chapters 6 and 7 that the non-uniform probability density function (*pdf*) of text and audio symbols can be exploited to achieve compression. In the case of images, the pixel values also do not have uniform *pdf*. Hence, entropy coding can also be employed to compress two-dimensional images. Given an image source, the entropy of a generated image can be calculated. A suitable encoder can then be designed to achieve a bit-rate close to the entropy. As in the case of audio, Huffman and arithmetic coding techniques can help to achieve image compression.

Figure 8.1 shows the *pdf* of the gray levels of individual pixels corresponding to the Lena image. The entropy corresponding to this *pdf* is 7.45 bits/pixel. Note that the entropy is very close to the PCM rate (8 bits), and the compression ratio is negligible. This is true for most natural images, and hence entropy coding is not generally applied directly to encode the pixels. It will be shown later in this section that significant compression can be achieved by entropy coding the transform coefficients of an image.

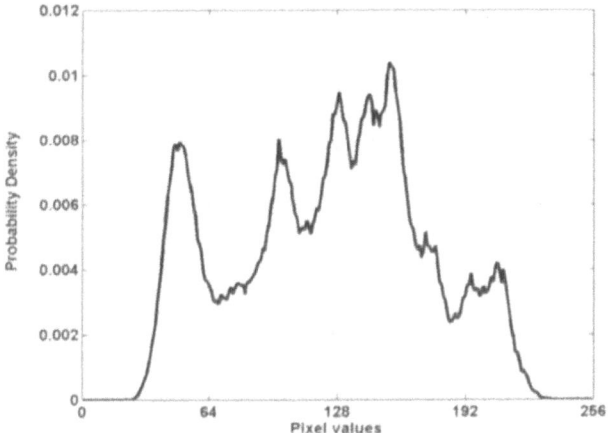

Figure 8.1. The probability density function of the Lena image. The entropy is 7.45 bits/pixel.

8.2.2 Run-length Coding

Run-length coding (RLC) is also a lossless coding technique, and provides a good coding performance for images with identical consecutive symbols. In RLC, a *run* is a sequence of consecutive pixels with identical values (generally zeros or ones) along a given direction (*e.g.*, horizontal or vertical). Instead of coding the individual pixels, the pixels are encoded one run at a time.

In 1-D run-length coding, each scan line is encoded independently, and consists of black runs (*i.e.*, consecutive zeros) and white runs (consecutive ones). In order to make the coding simpler, the first run in a scan line is always considered to be a white run. In case the first pixel is a black pixel, the first run-length will be zero. Each scan line is terminated with a special end-of-line (EOL) symbol.

■ Example 8.1

Determine the run-length code of a scanned fax image whose two scan lines are shown below (assume that 1 and 0 correspond to white, and black pixels, respectively).

Fax image=[11111111111100000000000000000000000011111111111111111

0000000000000011111111111111111111110000000000000000]

In the first scan line, there are 11 ones, followed by 22 zeros, followed by 17 ones. The line should be terminated by the EOL symbol. The second scan line starts with black pixel. As a result, the run-length code of the second line should start with a white run of zero length. This should be followed by 14 zeros, followed by 20 ones, followed 16 zeros. The line should again be terminated by the EOL symbol. The overall run-length code would look as follows:

$$\text{RLC Code} = [\ldots,11,22,17,\text{EOL},0,14,20,16,\text{EOL},\ldots] \tag{8.1}$$

■

Eq. (8.1) shows that the RLC code can provide a compact representation of long runs.

The number of bits required to encode the RLC will depend on how the RLC symbols are encoded. Typically, these symbols are encoded using an entropy coder (Huffman or arithmetic coder). Assume that a given fax image has L pixels in a scan line. Then the run-lengths l can have a value between 0 and L (both inclusive). The average run-lengths of black pixels ($\bar{l}_b$) and white pixels ($\bar{l}_w$) can be expressed as

$$\bar{l}_b = \sum_{l=0}^{L} l.p_b(l) \tag{8.2}$$

and

$$\bar{l}_w = \sum_{l=0}^{L} l.p_w(l) \tag{8.3}$$

where $p_b(l)$ and $p_w(l)$ are the probabilities of the occurrence of black run-length l and white run-length l, respectively. The entropies of the black (h_b) and white (h_w) runs can be calculated as follows:

$$h_b = -\sum_{l=0}^{L} p_b(l)\log_2 p_b(l) \tag{8.4}$$

$$h_w = -\sum_{l=0}^{L} p_w(l) \log_2 p_w(l) \tag{8.5}$$

The compression ratio that can be achieved using RLC is on the order of

$$\eta = \frac{\bar{l}_b + \bar{l}_w}{h_b + h_w}. \tag{8.6}$$

It is observed in Eq. (8.6) that longer runs (*i.e.*, larger numerator) provide a higher compression ratio, which is expected. Eq. (8.6) provides an upper limit of the compression ratio. Generally, a compression ratios achieved in practice is 20-30% lower.

The RLC was developed in the 1950s and has become, along with its two-dimensional extensions, the standard approach for facsimile (FAX) coding. The RLC cannot be applied to images with high details, as the efficiency will be very low. However, significant compression may be achieved by first splitting the images into a set of bit planes which are then individually run-length coded.

RLC is also used to provide high compression in transform coders. Most of the high frequency coefficients in a transform coder become zero after quantization, and long runs of zeros are produced. Run-length coding can then be used very efficiently along with a VLC. Run-length coding is generally extended to two dimensions by defining a *connected area* to be a contiguous group of pixels having identical values. To compress an image using two-dimensional RLC, only the values that specify the connected area and its intensity are stored/transmitted.

The lossless compression techniques generally result in a low compression ratio (typically 2 to 3). Therefore, they are not employed when a high compression ratio is required. A high compression ratio is generally achieved when some loss of information can be tolerated. Here, the objective is to reduce the bit rate subject to some constraints on the image quality. Some of the most popular lossy compression techniques are predictive coding, transform coding, wavelet/subband, vector quantization, and fractal coding. In the next few sections, we briefly discuss each of these coding techniques.

8.2.3 Predictive Coding

Predictive coding exploits the redundancy related to the predictability and smoothness in the data. For example, an image with a constant gray level can be fully predicted from the gray level value of its first pixel. In images with multiple gray levels, the gray level of an image pixel can be predicted from the values of its neighboring pixels. The *differential pulse code modulation (DPCM)* scheme discussed in Chapter 7 can be extended to two-dimension for encoding images. The following are some examples of typical predictors:

$$\hat{s}_n = 0.97 s_{n-1}$$ 1st order, 1-D predictor   (8.7)

$$\hat{s}_{m,n} = 0.48 s_{m,n-1} + 0.48 s_{m-1,n}$$ 2nd order, 2-D predictor   (8.8)

$$\hat{s}_{m,n} = 0.8 s_{m,n-1} - 0.62 s_{m-1,n-1} + 0.8 s_{m-1,n}$$ 3rd order, 2-D predictor   (8.9)

■ **Example 8.2**

Using the predictor given in Eq. (8.9), calculate the predicted error output of the following 4x4 image. Assume no quantization of the error signal.

$$\begin{bmatrix} 20 & 21 & 22 & 21 \\ 18 & 19 & 20 & 19 \\ 19 & 15 & 14 & 16 \\ 17 & 16 & 15 & 13 \end{bmatrix}$$

In order to calculate the DPCM output for the first row and the first column, the 1st order predictor given in Eq. (8.7) is used. For the other rows and columns, the 3rd order predictor given in Eq. (8.9) is used. The predicted pixel values are given below.

$$\begin{bmatrix} 20 & 19.4 & 20.37 & 21.34 \\ 19.4 & 18.8 & 19.78 & 19.16 \\ 17.46 & 19.24 & 16.22 & 14.00 \\ 18.43 & 13.82 & 14.70 & 16.12 \end{bmatrix}$$

The DPCM output can be calculated by subtracting the predicted output from the original pixel values, and are given below.

$$\begin{bmatrix} X & 1.6 & 1.63 & -0.34 \\ -1.4 & 0.20 & 0.22 & -0.16 \\ 1.54 & -4.24 & -2.22 & 2.00 \\ -1.43 & 2.18 & 0.30 & -3.12 \end{bmatrix}$$

■

■ **Example 8.3**

The Lena image is encoded using the DPCM predictor provided in Eq. (8.9). The prediction error sequence is quantized with step-sizes {10, 15, 20, 25, 30}. The entropy of the quantized coefficients are used to calculate the overall bit-rate. Figure 8.2(a) shows the bit-rate versus PSNR of the reconstructed signal. Typical prediction error coefficients are shown in Fig. 8.2(b). It is observed that the prediction performance is very good in most regions, except near the edges. ■

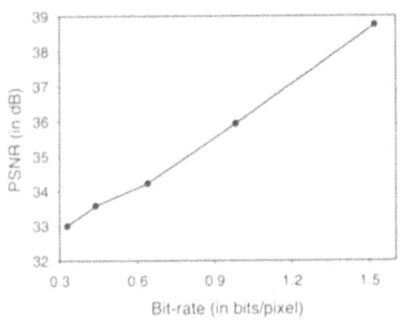
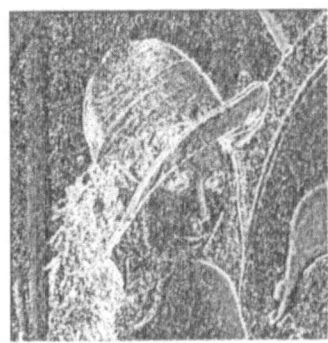

Figure 8.2. DPCM performance. a) Bit-rate versus PSNR, b) Prediction Error (multiplied by 20 for better viewing). White pixels represent regions where prediction is not very successful. Note that high prediction error prediction error (*i.e.*, white pixels) occurs mostly at the edges.

8.3 TRANSFORM CODING

Chapter 5 demonstrated how unitary transforms such as DFT, DCT and DWT decorrelate the image data, and compact most energy in a few coefficients. The transform coding takes advantage of this energy compaction property of the unitary transform to achieve a high compression [1]. In transform coding, the image data is first transformed from spatial to frequency domain by a unitary transform. The transform coefficients are then quantized and encoded. A typical transform coding scheme is shown in Fig. 8.3.

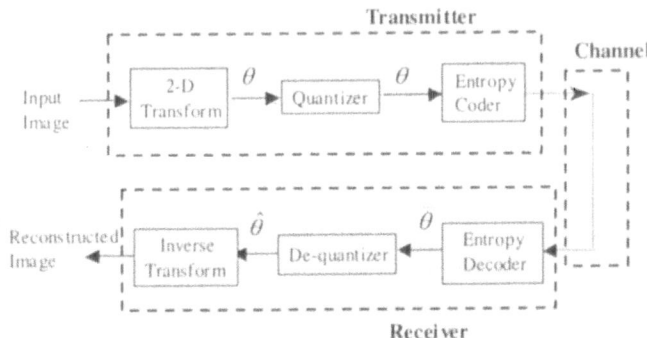

Figure 8.3. Block diagram of a transform coding scheme. θ : Transform coefficient matrix, $\bar{\theta}$: Quantized transform coefficient matrix, $\hat{\theta}$: Reconstructed transform coefficient matrix.

In order to obtain significant compression, the quantization and bit allocation procedure in a transform coding technique generally requires optimization. One of the popular criteria for optimization is to minimize the mean square error (MSE) of the reconstructed image, which is identical to the MSE of the (unitary or orthonormal) transform coefficients. Depending on the statistical characteristics of the transform coefficients, an optimum non-uniform quantizer can be used. However, designing such a quantizer might be difficult since it is data dependent. In most cases, a fixed non-uniform quantizer is used for quantizing the transform coefficients. The quantized coefficients are then encoded using entropy coding (e.g., Huffman or arithmetic coding). The transform coefficients are assigned bits depending on their contribution to error variance in the spatial domain.

8.3.1 Unitary Transform

The non-unitary transforms can also have a very good energy compaction capability. However, unitary (or orthonormal) transforms have two important properties (see section 5.6) that are very useful in image coding applications. First, the total energy in the frequency domain is equal to that of the spatial domain. Second, the mean square reconstruction error is equal to the mean square quantization error.

These two properties are very helpful for designing an MSE quantizer. If the desirable quality of the encoded image is specified using an objective criterion (*e.g.* in terms of SNR), the required MSE can be calculated. The transform coefficients are then quantized in such a way that the actual MSE is close to the desirable MSE. Since the quantized coefficients are encoded losslessly, the reconstructed image will satisfy the target SNR.

Optimum Transform

Since there is a large number of image transforms, it is important to find the optimum transform that provides the maximum compression performance. Naturally, the transform that decorrelates the input image data completely will provide the best compression performance. Because there is no correlation, the autocorrelation matrix of the transform coefficients will be diagonal. Second, the optimum transform should repack the energy in the first few coefficients. If we calculate the energy of the first L (where L is arbitrary) transform coefficients resulting from various transforms, the energy provided by the optimum transform should be the maximum.

The unitary transform that satisfies both the criteria is the Karhunen-Loeve transform (KLT). However, the KLT is image dependent, and has a high computational complexity. Therefore, in practice, image independent sub-optimal transforms such as discrete Fourier, sine, and cosine transforms are used because they have a lower computational complexity compared to the KLT. Among the sub-optimal transforms, the rate distortion performance of discrete cosine transform (DCT) is closest to the KLT for most natural images. It can be demonstrated that for a first order Markov source model, the DCT basis functions become identical to the KLT basis functions as the adjacent pixel correlation coefficient approaches unity. Natural images generally exhibit high pixel-to-pixel correlation, and therefore DCT provides a compression performance virtually indistinguishable from KLT. In addition, DCT has a fast implementation like DFT, with a computational complexity of $O(N \log N)$ for N-point transform. Unlike DFT, the DCT avoids generation of spurious spectral components at the boundaries resulting in higher compression efficiency. Hence, DCT has been adopted as the transform kernel in image and video coding standards, such as JPEG, MPEG and H.261.

It was mentioned in Chapter 5 that the DCT of an Nx1 sequence is related to the DFT of the 2Nx1 even symmetric sequence (see section 3, Chapter 5). Because of this even symmetry, the reconstructed signal from quantized DCT coefficients better preserves the edge.

■ **Example 8.4**

Consider an 8-point signal [0 2 4 6 8 10 12 14]. Calculate the DFT and the DCT of the signal. In order to compress the signal, ignore the smallest three coefficients (out of 8) in both Fourier and DCT domains, and reconstruct the signal. Compare the results.

The DFT and DCT coefficients of the signal are shown in Fig. 8.4(a). Note that the DFT coefficients are complex, and only the absolute values are shown. The three smallest coefficients are made zero, and the inverse transforms are calculated. When the reconstructed signal is rounded off to the nearest integers, the DCT coefficients provide error free reconstruction. However, the DFT provides substantial error. The reconstructed signals are shown in Fig. 8.4(b) where the reconstruction error is particularly noticeable at the edges. ■

8.3.2 Block Transform

Fourier-based transforms (*e.g.* DCT and DFT) are efficient in exploiting the low frequency nature of an image. However, a major disadvantage of these transforms is that the basis functions are very long. If a transform coefficient is quantized, the effect is visible throughout the image. This does not create much problem for the low frequency coefficients that are coded with higher precision. However, the high frequency coefficients are coarsely quantized, and hence the reconstructed quality of the image at the edges will have poor quality. A sharp edge in an image is represented by many transform coefficients (that cancel each other outside the edge area) that must be preserved intact and in the same relationship to one another to achieve good fidelity of the reconstructed image. Second, an image is generally a nonstationary signal where different parts of an image have different statistical properties. If the transform is calculated over the entire image, this nonstationarity will be lost, resulting in a poor compression performance.

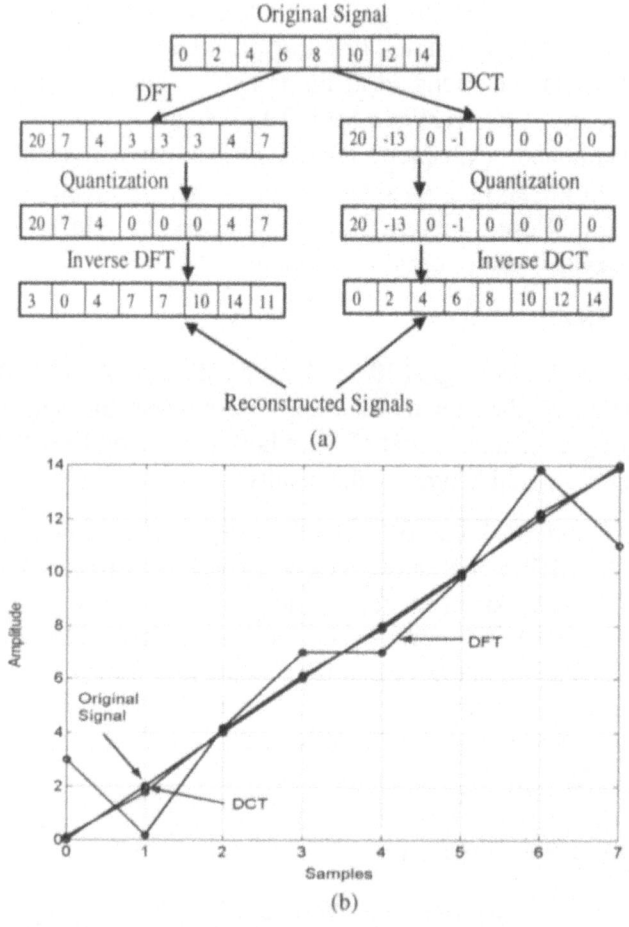

Figure 8.4. Edge preservation performance of the DFT and DCT. a) The DFT and DCT coefficients at various stages, b) the reconstructed signals. Note that the amplitude of the DFT coefficients are shown in Fig. (a).

Typical implementations of the Fourier-based transform coding involves first dividing the original image into 8x8 or 16x16 subblocks (more details will be provided in section 9.7). Each block is then transformed, quantized, and encoded separately. Here, the effect of the coefficient quantization is limited to the corresponding block. Another advantage of block coding is the reduced overall computational complexity. This is explained using the following example.

■ **Example 8.5**

Consider a 512x512 image. Calculate the complexity of a 2-D DFT calculation using the radix-2 FFT method. Divide the image into blocks of 8x8. Calculate the complexity of the 2-D DFT calculation for all blocks. Compare the two complexities.

The complexity of the $N \times N$ DFT calculation using separable approach is $2N \times (N/2) \log_2 N$ or $N^2 \log_2 N$ butterfly operation. Therefore, the direct 512×512 DFT calculation will have a complexity of 2.4×10^6 butterfly operations.

If the image is divided into blocks of 8×8, there will be 4096 blocks. The complexity of 2-D DFT for each block is $8^2 \log_2 8$ or 192 operations. Consequently, the overall complexity will be 4096*192 or 0.79×10^6 butterfly operations. As a result, the blocking method reduces the complexity to one-third. ■

However, blocking of data introduces a number of undesirable side effects. Most objectionable is the blocking effect that shows a quilted appearance of the image. In addition, Gibbs phenomenon of the spectral method causes a loss of contrast in the image when high frequency coefficients have quantization errors. The blocking method also imposes an upper limit to the actual compression ratio achieved, since one must save a high resolution DC term and the coefficients of at least the lowest frequencies for each block.

8.3.3 Wavelet Coding

Recently, wavelets have become very popular in image processing, specifically in coding applications [2, 3] for several reasons. First, wavelets are efficient in representing nonstationary signals because of the adaptive time-frequency window. Second, they have high decorrelation and energy compaction efficiency. Third, blocking artifacts and mosquito noise (present in a block transform coder) are reduced in a wavelet-based image coder. Finally, the wavelet basis functions match the human visual system characteristics, resulting in a superior image representation.

A simple image coding scheme can be implemented by using a wavelet transform in the general transform coding schematic as shown in Fig. 8.3. The wavelet transform may be unitary (orthogonal) or non-unitary. If the transform is orthogonal, the properties of the unitary transform is exploited to achieve a superior compression performance. The wavelet decomposition may be dyadic (only the lowest scale is decomposed recursively), regular (full decomposition) or irregular. The depth of the tree is generally determined by the size of the image and the number of wavelet filter taps. With each decomposition, the number of rows and columns of the lowest passband is halved. For efficient decomposition, the number of rows and columns of the band to be decomposed should not be less than the number of filter taps. In practice, the depth of the tree ranges from 3 to 5.

After each band is quantized with its corresponding quantization step size, the DWT coefficients are encoded using entropy coding. Since the statistics of different bands vary, the bands are generally encoded independently. The remaining nonstationarity within a band can be easily compensated by an adaptive model. Since adaptive coding is a memory process, the order in which the coefficients are fed into the coder is important. The higher the local stationarity of coefficients, the better the adaptation. Note that various types of scanning, such as horizontal, zigzag, and Peano-Hilbert scanning, are employed in practice.

Figure 8.5 shows the probability density function of the HH band (the highpass diagonal band of 1st stage decomposition) of the wavelet coefficients of the Lena image (see Fig. 8.6(b)). The entropy corresponding to this band is 3.67 bits/pixel, resulting in a compression ratio of 2.2:1 (with respecto the 8bits/pixel PCM rate). If the coefficients are uniformly quantized with a step-size of 8, the entropy decreases further to 0.86 bits/pixel, resulting in a compression ratio of 9.3:1. The core compression performance in a transform coding scheme is achieved by reducing the overall entropy of the transform coefficients by quantizing the highpass coefficients.

Although the schematic shown in Fig. 8.3 can provide good coding performance, high performance wavelet codecs, such as a zero-tree and a SPHIT coder, employ a different coding scheme. Later in this Chapter, the upcoming JPEG2000 standard image coding technique, which is based on DWT, will be presented.

8.3.4 Comparison of DCT and Wavelets

The DCT and DWT are the two most important transforms in image coding. Although the block DCT and wavelet coding may look different, there are some similarities. It was shown in Chapter 5 that wavelets provide both spatial and frequency (or scale) information. The following example will demonstrate that DCT also provides similar information.

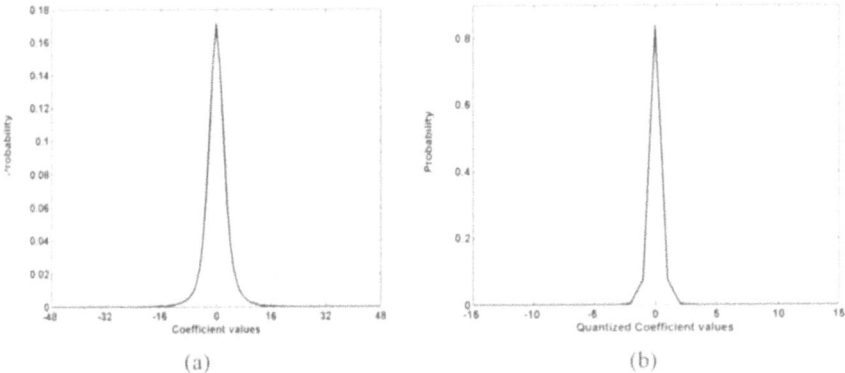

(a)   (b)

Figure 8.5. Probability density function of wavelet coefficients. a) coefficients rounded to nearest integer (entropy=3.67 bits/pixel), b) coefficients rounded to nearest integer after quantization by 8 (entropy=0.86 bits/pixel).

■ Example 8.6

Consider the 512x512 Lena image. Divide the image into non-overlapping blocks of 8x8. Calculate the DCT of each block, and calculate the mean energy of the DC, and 63 AC coefficients. In addition, decompose the image for three-stages using a Daub-4 wavelet. Calculate the mean energy of the lowpass and the nine highpass bands. Compare the two sets of energies.

The Lena image is divided into 8x8 blocks. There are 4096 such blocks. The DC coefficients from each block are extracted and arranged in 2-D according to the respective block's relative position. Similarly, all 4096 AC(0,1) coefficients are extracted and represented in 2-D, while all other AC coefficients are extracted. The 64 2-D coefficient blocks are arranged as shown in Fig. 8.6(a), and the 2-D DWT coefficients are shown in Fig. 8.6(b). It is observed that both DCT and DWT coefficients provide spatial (i.e., the image structure) as well as frequency (i.e., the edge information) information. Note that the spatial information provided by the DCT is due to the block DCT (if one calculates 512x512 DCT, the spatial information will not be available). Figs. 8.6(c) and 8.6(d) show the first four bands of the DCT and DWT coefficients. It is observed that different bands provide lowpass information, and horizontal, vertical and diagonal edges.

The main difference between the DCT and DWT coefficients lies in the highpass bands. The highpass DCT bands provide higher frequency resolution, but lower spatial resolution. As a result, there are more frequency bands, but it is difficult to recognize the spatial information (can you identify the Lena image by looking at a high frequency band?). On the other hand, the wavelet subbands provide higher spatial resolution, and lower frequency resolution. As a result, the number of subbands is few, but the spatial resolution is superior – one can recognize the Lena image looking at a highpass band.

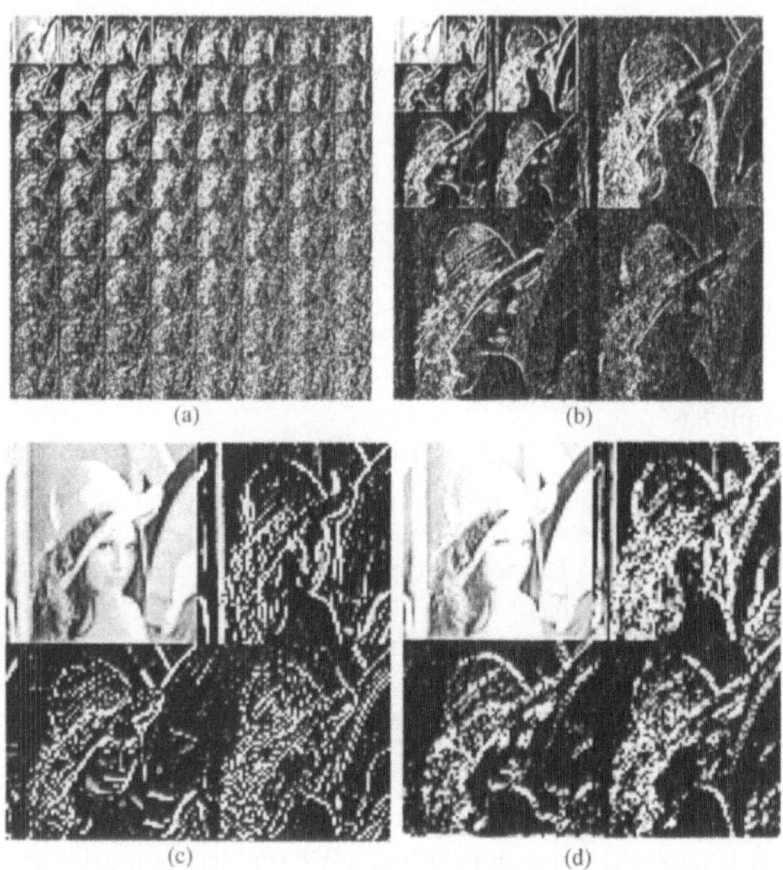

Figure 8.6. Comparison of DCT and DWT. a) The DCT coefficients rearranged into bands of equal frequencies, b) DWT coefficients, c) first 4 DCT bands, and d) first four wavelet bands.

The energy compaction performance of the DCT and the DWT is compared in Fig. 8.7. The average root mean square energy (RMSE) of coefficients of different DCT bands is shown in Fig. 8.7(a), while the RMSE of the DWT coefficients are shown in Fig. 8.7(b). It is observed that both transforms provide comparable energy compaction performance. ■

8.4 OTHER CODING TECHNIQUES

In addition to the image coding methods discussed above, there are several advanced coding methods such as vector quantization and fractal-based methods. A brief overview of these methods is presented below.

8.4.1 Vector Quantization

A fundamental result of Shannon's rate-distortion theory is that better performance can always be achieved by coding vectors instead of scalars [4]. A vector quantizer (VQ) can be defined as a mapping Q of K-dimensional Euclidean space R^K into a finite subset Y of R^K. In other words:

$$Q : R^K \rightarrow Y \qquad \text{where } Y = \left(x_i'; i = 1,2,.......N\right)$$

where Y is the set of reproduction vectors, and is called a VQ codebook or VQ table, and N is the number of vectors in Y. The VQ is similar to the dictionary-based coding (discussed in Chapter 6) in spirit. They both achieve compression by sending indices rather than the actual symbols. The dictionary in VQ is called the codebook.

The principle of vector quantization is explained in Fig. 8.8. Assume that both the encoder and decoder have a codebook with N codewords (or code vectors), with each codewords having a size of K pixels. The input image is divided into blocks of K pixels, with each block compared with all codewords in the codebook. The address or index of the closest matched codeword is then transmitted instead of the data vector itself.

1055	86	40	22	15	10	7	5
53	37	25	17	11	8	6	4
21	21	19	13	9	7	5	4
12	12	11	9	7	5	4	3
7	7	7	7	5	4	3	3
5	5	5	4	4	3	3	3
3	3	3	3	3	3	3	3
3	3	3	3	3	3	3	2

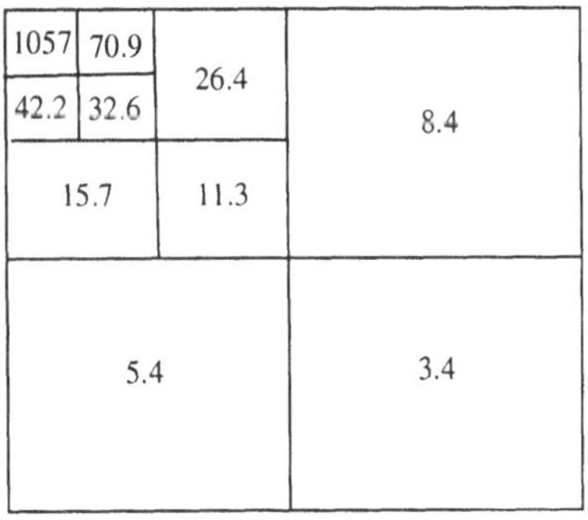

1057	70.9	26.4	8.4
42.2	32.6		
15.7	11.3		
5.4		3.4	

Figure 8.7. Energy compaction of DCT and DWT. a) average root mean square energy (RMSE) of DCT coefficients, and b) average RMSE of DWT coefficients.

At the decoder, the index is mapped back to the codeword, and the codeword is used to represent the original data vector. If an image block has K pixels, and each pixel is represented by p bits, theoretically $\left(2^p\right)^K$ combinations are possible. Hence, for lossless compression, a codebook with $\left(2^p\right)^K$ codewords is required. In practice, however, there are only a limited number of combinations that occur most often, which reduces the size of the codebook considerably. This is the basis for achieving compression in vector quantization.

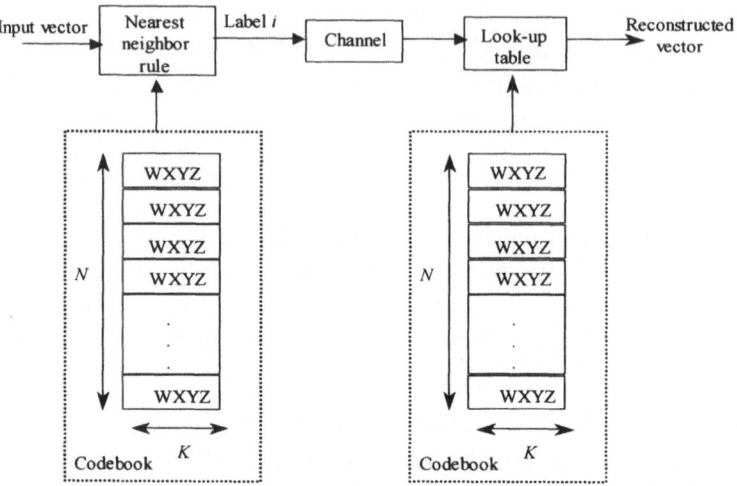

Figure 8.8. Simplified schematic of the vector quantization technique.

Although there is similarity between the VQ and dictionary methods, there are also important differences. The dictionary-based coding is a lossless compression method whereby the dictionary theoretically contains all possible codewords. In VQ, however, the dictionary (or codebook) is of finite size, and an index is generated by a close (not necessarily exact) match between the incoming symbol and the codewords (or symbols) in the dictionary.

The major drawback of vector quantization is that the quantization process is highly image dependent and its computational complexity grows exponentially with vector dimension. Also, there is a problem in designing a good codebook that is representative of all the possible occurrences of pixel combinations in a block.

8.4.2 Fractal Image Compression

In simplest terms, a fractal is an image of a texture or shape, expressed as one or more mathematical formulas. In terms of fractal geometry, a fractal is a geometric form whose irregular details recur at different scales and angles that can be described by fractal transformations (*e.g.* an affine transformation). Fractal transformations can be used to describe most real world pictures.

Fractal image compression [5] is the inverse of fractal image generation. Instead of generating an image from a given formula, fractal image compression technique searches for sets of fractals that describe and represent the digital image. Once the appropriate sets of fractals are determined, they are reduced to very compact fractal transform codes or formulas. The codes are rules for reproducing the various sets of fractals, which in turn regenerate the entire image. Since fractal transform codes require a very small amount of data to be expressed and stored as formulas, fractal compression results in very high compression ratios.

The major drawback of fractal coding is that it is image dependent. For each image, a distinct set of rules must be specified. Obviously, it is easier to specify a set of images showing a repeated pattern than it is for a picture with a number of distinct features. In addition to being image specific, fractal coding is also a computationally intensive technique. However, the computations required are iterative and make possible high-speed hardware implementations. Also, fractal coding is highly asymmetric as the computational complexity of the decoder is much less than that of the encoder.

8.5 IMAGE COMPRESSION STANDARDS

The application of images in our daily lives is growing at a rapid rate. It is crucial to establish image coding standards in order to make the compressed images compatible in various applications. Several coding techniques have been presented in this Chapter. The choice of compression technique depends on its computational complexity and coding performance. Predictive coding techniques have lower computational complexity and memory requirements. However, their coding performance is relatively poor resulting in a loss of subjective image quality at high compression ratios. Transform coding generally achieves compression ratios higher than the predictive coding for a given distortion value. It also achieves a better subjective image quality by distributing the reconstruction error over a large image region. Unfortunately, it has higher computational complexity and memory requirements. Wavelet/subband coding uses a filterbank structure and achieves a good coding performance. The subjective quality of the reconstructed images is better than block transform coding. Vector quantization and fractal coding achieves high compression, but their computational complexity is very high. As a result, none of these techniques provide optimum performance in all scenarios. It has been found that a compression scheme combining transform coding, predictive coding and entropy coding provides a high compression ratio at a reasonable cost of implementation.

In the last few decades, several image compression standards have been established by the International Consultative Committee for Telephone and Telegraph (CCITT), and the International Standard Organization (ISO).

Note that the CCITT has since been renamed as the ITU-T (ITU: International Telecommunication Union). The CCITT has developed CCITT G3 [10] and G4 [11] schemes for transmitting fax images. Later, the ISO established another (more efficient) fax coding standard known as JBIG (Joint Bi-level Imaging Group). The first general purpose still image compression standard has been developed by the Joint Photographic Experts Group (JPEG) [6], which was formed by the CCITT and ISO in 1986. The JPEG standard algorithm encodes both gray level and color images. Most recently, the ISO has established the JPEG-2000 standard that provides superior performance compared to the JPEG standard, in addition to several useful functionalities. In the remainder of this Chapter, a brief overview of the JPEG and JPEG-2000 coding schemes is presented.

8.6 THE JPEG IMAGE COMPRESSION STANDARD

The JPEG standard provides a framework for high quality compression and reconstruction of continuous-tone grayscale and color images for a wide range of applications. The standard specifies details of the compression and decompression algorithms for various application environments. It has four modes of operations that generally cover most image compression environments:

(a) Baseline Sequential

(b) Progressive coding

(c) Hierarchical coding

(d) Lossless Compression

The baseline sequential mode provides a simple and efficient algorithm that is adequate for most image coding applications. The other modes are employed for more sophisticated applications.

8.6.1 Baseline Sequential Mode

The baseline sequential mode provides a simple and efficient algorithm that is adequate for most image coding applications. A schematic of the baseline JPEG encoder is shown in Fig. 8.9. The compression is performed in three steps: DCT computation, quantization and variable-length coding. The input and output data precision is generally limited to 8 bits, whereas the quantized DCT values are restricted to 11 bits. The original image is first partitioned into non-overlapping blocks of 8x8 pixels. Figure 8.10(a) shows such a block chosen arbitrarily from the Lena image. In order to decrease the average energy of the image pixels, each pixel is level-shifted by subtracting the quantity 2^{n-1}, where n is the number of bits required to represent each pixel value. For 8 bit images, 128 is subtracted from each pixel (see Fig. 8.10(b)). The 2-D DCT of the block is then calculated (see Fig. 8.10(c)).

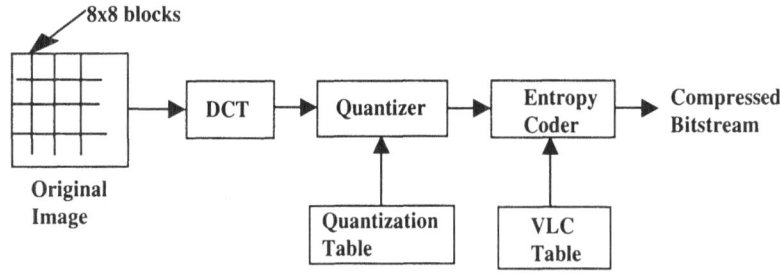

Figure 8.9. Simplified schematic of a Baseline JPEG Encoder.

The DCT coefficients are then quantized to achieve a high compression ratio. The JPEG standard suggests a default visually adapted quantization table (shown in Fig. 8.11) to quantize the DCT coefficients. The quality of the reconstructed image can be controlled by linearly scaling up or down the quantization table. The quantization table with larger step sizes will result in a higher compression; however, the quality of the reconstructed image will degrade.

(a)

104	108	107	101	94	95	98	102
96	100	103	100	96	74	75	73
77	69	70	87	84	64	64	67
71	60	52	59	64	56	54	57
58	53	51	54	52	51	52	52
53	50	53	52	52	58	51	47
48	53	53	51	53	55	51	53
47	48	48	47	55	47	51	48

(b)

-24	-20	-21	-27	-34	-33	-30	-26
-32	-28	-25	-28	-32	-54	-53	-55
-51	-59	-58	-41	-44	-64	-64	-61
-57	-68	-76	-69	-64	-72	-74	-71
-70	-75	-77	-74	-76	-77	-76	-76
-75	-78	-75	-76	-76	-70	-77	-81
-80	-75	-75	-77	-75	-73	-77	-75
-81	-80	-80	-81	-73	-81	-77	-80

(c)

-495	20	-8	0	10	-1	-3	3
135	22	-3	-9	7	1	-3	0
59	1	-1	-10	-9	-3	-1	3
17	-3	9	-3	-14	1	6	-4
-5	-7	14	3	-2	0	-1	0
2	-10	7	3	0	-2	2	-4
-2	-9	-1	3	3	3	1	-2
1	-7	0	-4	2	2	-1	-2

Figure 8.10. The forward DCT of a block. a) The original pixels, b) 128 shifted pixels, and c) DCT coefficients.

The quantization table, shown in Fig. 8.11(a), has been derived following the contrast sensitivity of the human visual system. Imagine a scenario where a 720×576 image is being viewed from a distance of six times the screen width. If the image is encoded by JPEG, the quantization step-sizes corresponding to Fig. 8.11(a) will produce distortion at the threshold of visibility. In other words, if quantization step-sizes mentioned in the default table is used, the reconstructed image will be almost as good as the original. Note the quantization step sizes of a few high frequency DCT coefficients, such as (0,1) and (0,2), are smaller compared to the step-size for the DC (0,0) coefficient. This is because of the frequency sensitivity of the HVS. Coefficient (0,1) corresponds to a cosine wave of ½ cycle with a span of 8 pixels. Therefore, it corresponds to a spatial frequency of 1 cycle/16 pixels, which is equivalent to 5 cycles/degree. The coefficient (0,2) corresponds to roughly 7.5 cycles/degree. As mentioned in Chapter 3, the frequency sensitivity is highest in the region of 5-8 cycles/degree. Consequently, the corresponding quantization steps are smaller. In addition, the frequency sensitivity in the horizontal direction is higher than that in the vertical direction, making the steps smaller in the horizontal direction. Note that when the viewing assumptions are violated, the default table may not provide the best compression. Thus, JPEG also allows user defined quantization tables.

Note that in order to reconstruct the image, the quantized coefficients have to be de-quantized, which is the inverse of the quantization step. The dequantized coefficients (Fig. 8.11(c)) are obtained by multiplying the quantized coefficients (Fig. 8.11(b)) with the quantization table (Fig. 8.11(a)).

16	11	10	16	24	40	51	61
12	12	14	19	26	58	60	55
14	13	16	24	40	57	69	56
14	17	22	29	51	87	80	62
18	22	37	56	68	109	103	77
24	35	55	64	81	104	113	92
49	64	78	87	103	121	120	101
72	92	95	98	112	100	103	99

DC Coefficients / AC Coefficients

-31	2	-1	0	0	0	0	0
11	2	0	0	0	0	0	0
4	0	0	0	0	0	0	0
1	0	0	0	0	0	0	0
0	0	0	0	0	0	0	0
0	0	0	0	0	0	0	0
0	0	0	0	0	0	0	0
0	0	0	0	0	0	0	0

-496	22	-10	0	0	0	0	0
132	24	0	0	0	0	0	0
56	0	0	0	0	0	0	0
14	0	0	0	0	0	0	0
0	0	0	0	0	0	0	0
0	0	0	0	0	0	0	0
0	0	0	0	0	0	0	0
0	0	0	0	0	0	0	0

(a)   (b)   (c)

Figure 8.11. Quantization of DCT coefficients. a) Quantization table, b) the quantized coefficients, and c) The dequantized coefficients.

In order to achieve a good compression ratio, the quantized coefficients shown in Fig. 8.11(b) should be represented losslessly using as few bits as possible. To achieve this, the quantized 2D DCT coefficients are first converted to a 1D sequence before being encoded losslessly. For a typical image block, the significant DCT coefficients (i.e., coefficients with large amplitude) are generally found in the low frequency region that is in the top-left corner of the 8x8 matrix (see Fig. 8.11(b)). A superior compression can be achieved if all non-zero coefficients are encoded first. After the last non-zero coefficient is transmitted, a special end-of-block (EOB) code is sent to indicate that the remainders of the coefficients in the sequence are all zero. A special 45° diagonal zigzag scan (see Fig. 8.12) is employed to generate the 1-D sequence of quantized coefficients.

For encoding, the 64 quantized DCT coefficients are classified into two categories – one DC coefficients, and 63 AC (i.e., high frequency) coefficients (see Fig. 8.11(b)). The DC coefficients of the neighboring blocks have substantial redundancy for most natural images. Therefore, the DC coefficient from each block is DPCM coded. The AC coefficients of a typical image have a nonuniform probability density function, which can be encoded efficiently by entropy coding. In addition, long runs of zero value coefficients are likely to occur before the EOB. Hence, a combination of Huffman and run-length coding is employed to encode the AC coefficients. JPEG also suggests an arithmetic coder, known as the QM coder [6]. The arithmetic coder in JPEG generally provides better performance than the Huffman coder, but at the expense of increasing complexity.

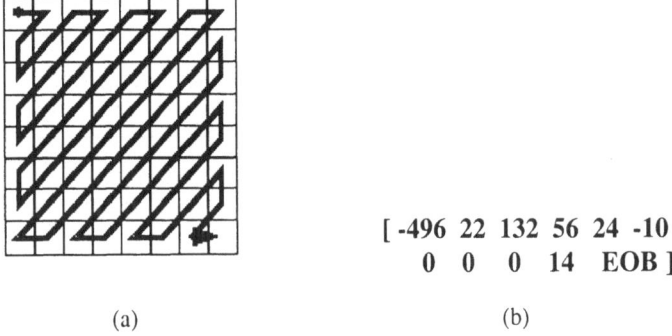

$$[-496 \quad 22 \quad 132 \quad 56 \quad 24 \quad -10$$
$$0 \quad 0 \quad 0 \quad 14 \quad EOB]$$

(a)   (b)

Figure 8.12. Zigzag scan of DCT coefficients. (a) Scanning, (b) scanned coefficients corresponding to Fig. 8.11(b).

The decoding scheme is the inverse of the encoding scheme. The compressed bit-stream is first Huffman and run-length decoded. The DCT coefficients are then dequantized, inverse transformed, and level shifted to obtain the reconstructed pixel values. The de-quantization process is the inverse of the quantization step. The dequantized coefficients (Fig. 8.11(c)) are obtained by multiplying the quantized coefficients (Fig. 8.11(b)) with the quantization table (Fig. 8.11(a)). The inverse transform and level-shifting operations are shown in Fig. 8.13. The inverse DCT coefficients corresponding to Fig. 8.11(c) are shown in Fig. 8.13(a). When these coefficients are level-shifted by 128, the reconstructed pixel values are obtained. Note that the reconstructed coefficients shown in Fig. 8.13(b) are not identical to the original pixel values shown in Fig. 8.10(a). This is due to the quantization of the DCT coefficients. The error coefficients corresponding to this block (i.e., the difference between Fig. 8.10(a) and Fig. 8.13(b)) are shown in Fig. 8.13(c). It is observed that the maximum and minimum values of the error coefficients are 13 and −10, respectively. The signal-to-noise ratio (SNR) for this block corresponding to the default quantization table is 11.94 dB (15.6 in absolute values), and the corresponding PSNR value is 34.28 dB.

■ **Example 8.7**

In this example, we evaluate the performance of a simplified JPEG-like coder. The image is divided into 8x8 blocks, and 128 is subtracted from the pixel values. The DCT coefficients corresponding to the Lena image are quantized using the JPEG default quantization table. In order to achieve different rate-distortion performance, we scale the quantization table with factors {0.8, 1.0, 1.2, 1.8, 2.4}. To simplify the implementation, we assume that all 64 quantized DCT coefficients of a block will be encoded using an ideal entropy coder (although we do not know how to implement such an ideal entropy coder). In other words, we use the entropy to calculate the bit-rate instead of the Huffman and run-length coding as with the standard.

The rate distortion performance is shown in Fig. 8.14. ■

-20	-20	-22	-24	-28	-33	-37	-39
-32	-33	-34	-36	-39	-44	-47	-50
-50	-50	-51	-52	-55	-59	-62	-64
-65	-64	-64	-65	-67	-70	-73	-75
-73	-73	-72	-72	-73	-75	-77	-79
-77	-77	-75	-75	-75	-76	-77	-79
-80	-79	-77	-76	-75	-76	-77	-78
-81	-80	-78	-77	-76	-76	-77	-77

108	108	106	104	100	95	91	89
96	95	94	92	89	84	81	78
78	78	77	76	73	69	66	64
63	64	64	63	61	58	55	53
55	55	56	56	55	53	51	49
51	51	53	53	53	52	51	49
48	49	51	52	53	52	51	50
47	48	50	51	52	52	51	51

-4	0	1	-3	-6	0	7	13
0	5	9	8	7	-10	6	-5
-1	-9	-7	11	11	-5	-2	3
8	-4	-12	-4	3	-2	-1	4
3	-2	-5	-2	-3	-2	1	3
2	-1	0	-1	-1	6	0	-2
0	4	2	-1	0	3	0	3
0	0	-2	-4	3	-5	0	-3

(a)   (b)   (c)

Figure 8.13. The inverse DCT of a block. a) The inverse DCT coefficients, b) Reconstructed pixel values, c) Reconstructed error values.

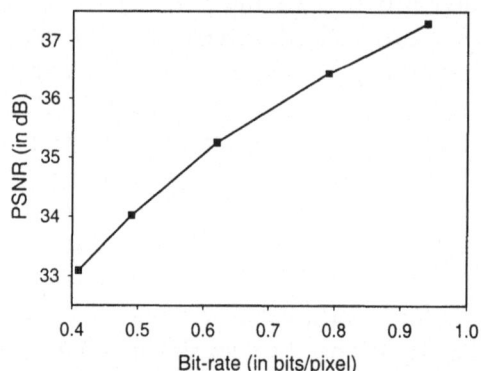

Figure 8.14. Performance of a simplified JPEG-like coder.

So far, we have mainly focused on the coding of one block. However, an image typically has several blocks. These blocks are encoded separately, encoded, and concatenated to form the JPEG bitstream. A typical organization of the JPEG bitstream is shown in Fig. 8.15.

The performance of the JPEG baseline coding algorithm for coding the gray level Lena image (512x512) is shown in Fig. 8.16. It is observed that even at 0.56 bits per pixel (corresponding to a compression ratio of 14:1), the reconstructed image quality is very good. However, at a bit rate lower than 0.2 bits per pixel (compression ratio greater than 40:1), the quality degrades rapidly.

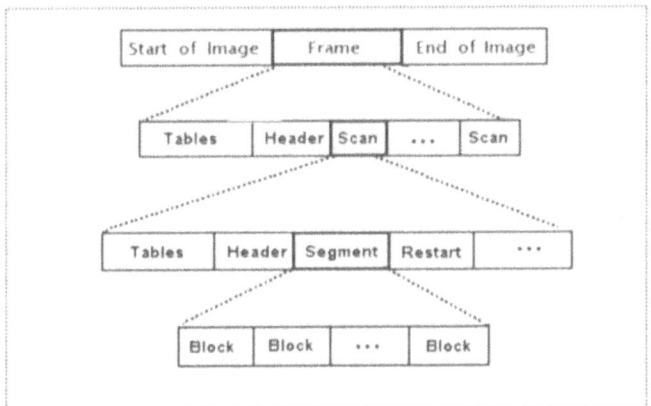

Figure 8.15. JPEG Bit Stream

The coding performance for the color Lena image for different bit-rates (such as 0.95, 0.53, 0.36, and 0.18 bits per pixel) is included in the accompanying CD. It is demonstrated that at 0.53 bpp (compression ratio of 45:1), the reconstructed image quality is reasonably good. Note that a higher compression ratio can be achieved for color image since the sensitivity of the chrominance components is low for the HVS.

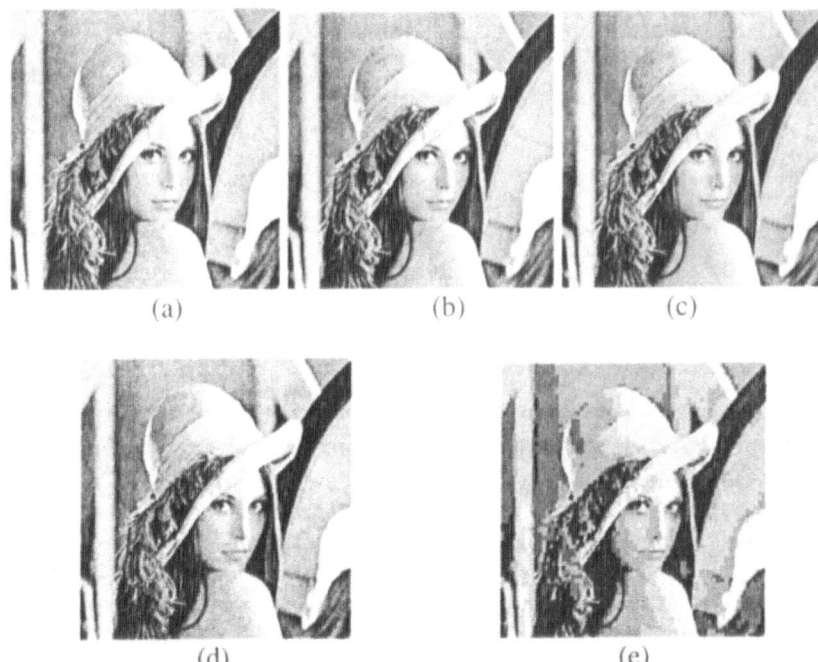

Figure 8.16. Reconstructed images from JPEG Bitstream at different bitrates. (a) 0.90, (b) 0.56, (c) 0.37, (d) 0.25, (e) 0.13

8.6.2 Other JPEG Modes

The baseline sequential mode provides a simple and efficient algorithm that is adequate for most image coding applications. However, to satisfy the requirements of a wider range of applications, three other modes of encoding have also been defined, and are presented below.

In progressive coding, the image is encoded in multiple scans for applications in which transmission time is long and the user prefers to view the image building up in multiple coarse-to-clear passes. In heirarchical coding, the image is coded at multiple resolutions, so that lower resolutions versions may be accessed without first having to decompress the image at its full resolution. In lossless coding, the image is encoded to guarantee the exact recovery of every source image sample value (of course, if the channel is error free!). The progressive and hierarchical modes generally use a modified version of the baseline algorithm. We now present a brief description of these modes.

Progressive Mode:

In the baseline sequential mode, the image is displayed after full decompression. However, when the image is accessed over a low bandwidth network, the slow decompression might be annoying at times, especially when a user is searching for a specific image. Progressive mode offers an alternative method of image coding and reconstruction. Here, the DCT coefficients are encoded in multiple scans using two different methods – spectral selection and successive approximation. In the spectral selection method, the 63 AC coefficients of a block can be partitioned into nonoverlapping segments. These segments are then encoded separately. While decoding, a low resolution image is formed by decoding the DC coefficients from all the blocks. A higher resolution image can then be formed by decoding the first AC segment from all blocks. The resolution of the image can be further increased by decoding the second AC segments corresponding to all the blocks. The finest resolution can be obtained by decoding all segments from all blocks. In other words, the quality of the reconstructed improves progressively. In the successive approximation method, all 63 AC coefficients are encoded in all scans, but the precision of the coefficients increases progressively.

Hierarchical Mode:

This mode provides a progressive coding with increasing spatial resolution between progressive stages. This is also known as pyramidal mode where an image is encoded as a succession of increasingly higher resolution subimages. This mode provides the best performance (among all three lossy modes) at low bit-rate. However, at higher bit-rates, the performance degrades compared to the other modes.

Lossless Mode:

In this mode, the reconstructed image is identical to the original image. Here, the DCT is not used for energy compaction. Instead, a predictive technique similar to the coding of the DC coefficients is employed, and a given pixel is predicted from the neighboring three causal pixels. There are eight modes of prediction; the mode that provides the best prediction is employed to code a pixel.

8.7 THE JPEG-2000 STANDARD

The JPEG standard provides good compression performance, and has been a great success considering that the majority of the images currently available on the WWW are in JPEG format. The major limitation of the JPEG algorithm is that the coding perfromance degrades rapidly at low bitrate (especially below 0.25 *bpp*). Hence, the ISO has finalized a new coding standard known as JPEG-2000 [7]. This standard is targeted for a wide range of applications such as Internet, printing, remote sensing, facsimile, medical, digital library, and E-commerce. In addition to significant improvement in coding performance over the JPEG standard, it will have several other features such as:

- Lossy to lossless compression in a single codestream
- Coding static/dynamic region of interest with a higher quality
- Error resilience coding
- Spatial/quality scalability
- Content-based description

JPEG-2000 employs the discrete wavelet transform (DWT) instead of the discrete cosine transform that is used in many image and video coding standards. The core compression algorithm is primarily based on the Embedded Block Coding with Optimized Truncation (EBCOT) of the bitstream [8].

In JPEG2000, an image is independently partitioned into non-overlapping blocks (or tiles) to reduce the memory requirements and efficiently process the regions of interest in the image. The block schematic of the JPEG2000 encoder is shown in Fig. 8.17. Various operations, such as wavelet transform, scanning, quantization and entropy encoding are performed independently on all tiles of the image. For each tile, the DWT is applied on the source image data, and the wavelet coefficients are scanned and quantized. Entropy coding is then performed to compress the quantized coefficients. A rate-control mechanism is used along with the quantizer to satisfy the bit rate requirements by truncating the coded data generated from each code block. Note that the JPEG2000 standard employs the "block coding" scheme, which is generally not used in high performance wavelet coding schemes, for several reasons. First, local variations in the statistics of the image from the block-to-block can be exploited to achieve superior compression. Second, it will provide support for applications requiring random access to the image. Finally, parallel implementation of the coder will be easier, and the memory requirement will be smaller.

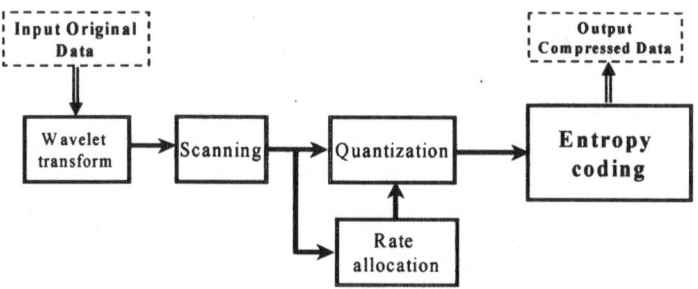

Figure 8.17. Block diagram of the JPEG2000 encoding process.

Discrete Wavelet Transform

In JPEG2000, the 2-D DWT is applied to the image on a tile-by-tile basis. Each stage of decomposition of the lowpass image produces 4 subbands, typically denoted by the "LL", "LH", "HL" and "HH" subbands. The standard supports arbitrary wavelet kernels and decompositions. Different wavelet filters can be used for different directions and resolution levels. The default wavelet filters (see Tables 8.1 and 8.2) include the integer wavelet (5,3) for lossless compression, and Daubechies (9,7) floating point wavelet for lossy compression.

Table 8.1. Le Gall 5/3 analysis and synthesis filter coefficients.

	Analysis Filter Coefficients		Synthesis Filter Coefficients	
i	Low-Pass Filter	High-Pass Filter	Low-Pass Filter	High-Pass Filter
0	6/8	1	1	6/8
+/- 1	2/8	-1/2	1/2	-2/8
+/- 2	-1/8	--	--	-1/8

Optimized Quantization

For lossless compression, the DWT coefficients (DWTC) are coded with their actual values. However, the DWTC must be quantized to achieve high compression ratios. The quantization is tile specific, lossy, and non-reversible. Uniform scalar dead-zone quantization is performed within each subband, with different levels of quantization for each subband in part I of the standard. The lower resolution subbands are generally quantized using smaller step-sizes since the human visual system is more sensitive at lower frequencies. The sign of the DWT coefficients is encoded separately using the context information. The quantization scheme may be implemented in two ways. In the first (explicit) method, quantization step-size for each subband is calculated independently, whereas in the second (implicit) method, quantization steps derived from the LL subband is used to calculate the quantization steps for other subbands. Part II of the standard allows both generalized uniform scalar dead-zone quantization and trellis-coded quantization (TCQ).

Table 8.2. Daubchines 9/7 analysis and synthesis filter coefficients.

Analysis Filter Coefficients		
i	Low-Pass Filter	High-Pass Filter
0	0.6029490182363579	1.115087052456994
+/- 1	0.2668641184428723	-0.5912717631142470
+/- 2	-0.07822326652898785	-0.05754352622849957
+/- 3	-0.01686411844287495	0.09127176311424948
+/- 4	0.02674875741080976	
Synthesis Filter Coefficients		
i	Low-Pass Filter	High-Pass Filter
0	1.115087052456994	0.6029490182363579
+/- 1	0.5912717631142470	-0.2668641184428723
+/- 2	-0.05754352622849957	-0.07822326652898785
+/- 3	-0.09127176311424948	0.01686411844287495
+/- 4	--	0.02674875741080976

Packets and Code-Blocks

Each sub-band in the wavelet domain is partitioned into non-overlapping rectangular blocks. Three spatially consistent rectangles (one from each sub-band) at each resolution level comprise a packet partition location or precinct. Each precinct is then divided into non-overlapping rectangles, called *code-blocks*. A code-block is the smallest independent unit in the JPEG2000 standard. Partitioning the wavelet coefficients into code-blocks facilitates coefficient modeling and coding, and it makes the processing more efficient and flexible in organizing the output compressed bit stream. Here, the DWT coefficients are sliced into bit-planes in units of code-block, and the bit-planes are coded to generate the output bit streams with respect to the required order.

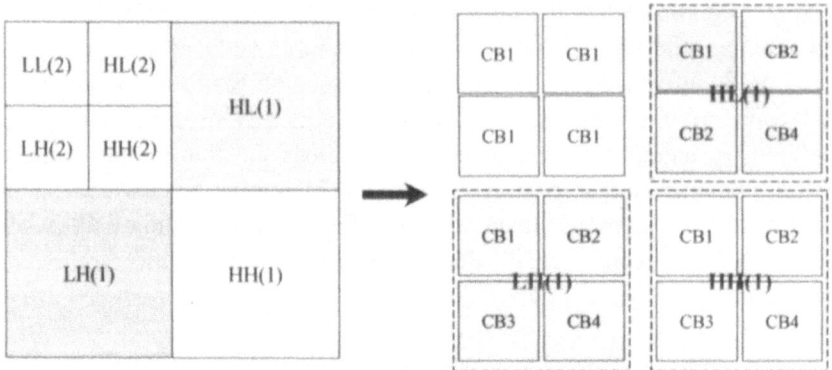

Figure 8.18. Sub-bands and code-blocks

Bit-Plane

A number can be represented in several formats such as decimal and hexadecimal. The most popular format in digital signal processing is the binary representation. The order of the number's bits descends from the most significant bit (MSB) to the least significant bit (LSB). A bit-plane is the collection of bits of a given significance corresponding to a set of decimal numbers. An example is given in Table 8.3. The four decimal numbers (21, 1, 0, 10) are expressed in binary format row-wise such that the bits corresponding to a particular weight are in the same column. The rows represent the original decimal number and the columns represent the bit-planes. For a set of data, the number of bit-planes is determined by the absolute maximum value in the data set.

Table 8.3. Bit-planes in JPEG-2000. "bp" in the table refers to bit-plane

Coefficients	bp5 (MSB)	bp4	bp3	bp2	bp1 (LSB)
21	1	0	1	0	1
1	0	0	0	0	1
0	0	0	0	0	0
10	0	1	0	1	0

Entropy Coding

The entropy coding in JPEG2000 employs a bit-modeling algorithm. Here, the wavelet coefficients from each code-block are represented in the form of a combination of bits that are distributed in different bit-planes. Furthermore, these bits are reassembled to coding passes according to their significance status. The use of bit modeling provides hierarchical representation of wavelet coefficients by ordering the bit-planes of wavelet coefficients from the MSB to the LSB. Hence, the formed bit streams with inherent hierarchy can be stored or transferred at any bit rate without destroying the completeness of the content of the image. The entropy coding process is a fully reversible (*i.e.*, lossless) compression step. The bit-modeling algorithm is applied to all the coefficients contained in each code-block. The compressed bit streams from each code-block in a precinct comprise the body of a packet.

Lossless compression.

Loss-less compression is achieved by using a quantization step size equal to one for all sub-bands (*e.g.* no quantazation is performed), and by keeping all coding passes in the final bit-stream. Moreover, a reversible DWT is used. The default reversible transform is implemented by means of the Le Gall 5-tap/3-tap filter (see Table 8.1).

Lossy compression

In lossy compression, a rate-control mechanism is used, along with the quantization process, to meet a particular target bit-rate or transmission time. Table 8.2 shows a 9-7 taps wavelet that is the default wavelet for the lossy compression. For each code-block, a separate bit-stream is generated without utilizing any information from any of the other code-blocks. The bit-streams have the property that these can be truncated to a variety of discrete lengths, and the distortion incurred when reconstructing from each of these truncated subsets is estimated and denoted by a distortion metric (*e.g.* mean square error). During the encoding process, these lengths and their corresponding distortions are computed and temporarily stored in a compact form along with the compressed bit-stream.

Once the bit-streams corresponding to all code-blocks have been calculated, a post-processing operation is performed. This operation determines the extent to which each code-block's embedded bit-stream should be truncated in order to achieve a given target bit-rate. The final bit-stream is composed from a collection of "layers". The first (i.e., the lowest quality) layer is formed from the optimally truncated code-block bit-streams in the manner described above. Each subsequent layer is formed by optimally truncating the code-block bit-streams to achieve successively higher target bit-rates, and including the additional coded-data form each code-block required to augment the information represented in previous layers. In other words, a layer is a collection of a few consecutive coding passes from each code-block. Some layers may include as little as zero coding passes for any or all code-blocks. Fig. 8.19 shows an example of the code-block contributions to bit-stream layers for five layers with seven code-blocks. Note that all code-blocks need not contribute to every layer and that the number of bytes contributed by code-blocks to any given layer is highly variable.

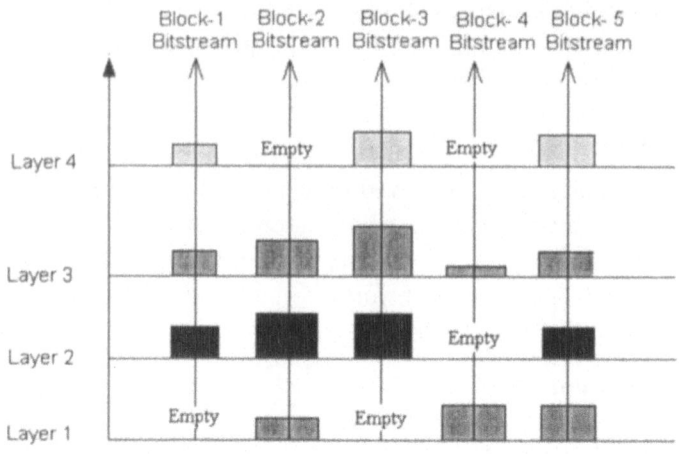

Figure 8.19. Code-block contributions to bit-stream layers.

Enhanced Features

Some enhanced features are also available in JPEG2000 standard. These include region of interest (ROI) coding, error resilience, and manipulation in the compressed domain. In ROI coding, part of an image is stored in a higher quality than the rest of the image. There are two methods to achieve this. In the first method, known as the *mask computation* method, a ROI mask is employed to mark out a region of image that is to be stored with higher quality. During quantization, the wavelet coefficients of this region are quantized less coarsely. In the second method, known as the *Scaling Base and Maxshift* Method, the highest background coefficient value is first found before the DWTC of the ROI are shifted upward. After this shifting, all the ROI coefficients are larger than the largest background coefficient, and can be easily identified.

Error resilience syntax and tools are also included in the standard. The error resilience is generally achieved by the following approaches: data partitioning and resynchronization, error detection and concealment, and Quality of Service (QoS) transmission based on priority. Error resilience can be achieved at two levels: entropy coding level (code block) and packet level (resynchronization markers).

Coding Performance

The coding performance of the JPEG2000 algorithm has been evaluated with JASPER [9] implementation. The performance corresponding to the Lena image is shown in Fig. 8.20. The coding has been performed with six levels of wavelet decomposition, a code-block size of 64x64, a precinct size of 512x512, and one quality layer. It is observed that the reconstructed image quality is good even at a compression ratio of 60:1, and 150:1 for gray level, and color images, respectively.

The JPEG2000 coding performance for color Lena image for different bit-rates (such as 0.95, 0.53, 0.36, and 0.18 bits per pixel) is included in the accompanying CD. It can be shown that even at 0.18 *bpp* (compression ratio of 133:1), the reconstructed image quality is reasonably good.

The performance of JPEG2000 is compared with JPEG standard in Fig. 8.21. It is observed that for a given bit-rate, JPEG2000 provides about 3-4 dB improvements over the JPEG algorithm.

8.8 IMAGE FORMATS

The digital image data is represented in a wide variety of formats [12]. Some of the formats express images losslessly, whereas other formats represents images using a lossy coding method. The following is a brief description of the most commonly used formats.

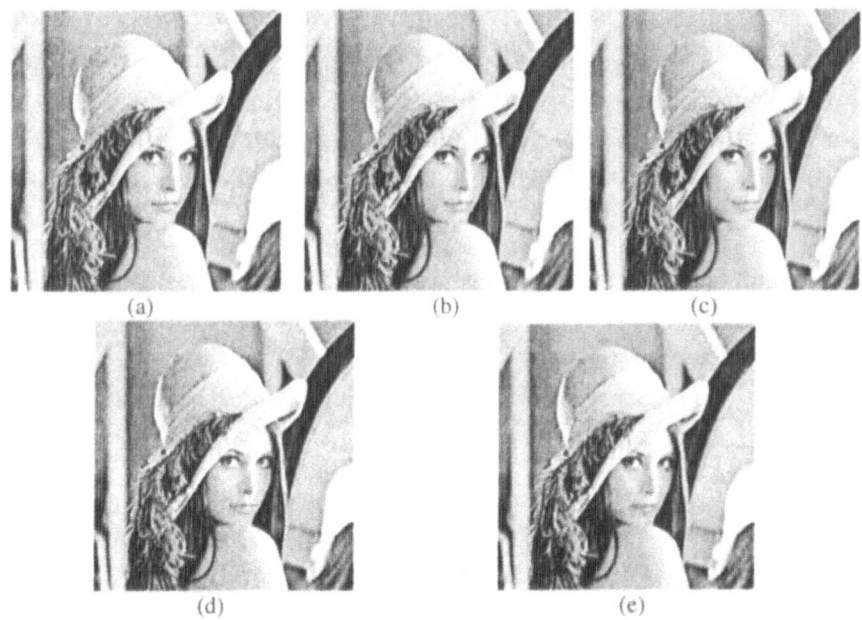

Figure 8.20. Reconstructed images from JPEG2000 Bitstream at different bitrates. (a) 0.90, (b) 0.56, (c) 0.37, (d) 0.25, and (e) 0.13 bits/pixel

The *portable graymap* (PGM) is a simple format for gray level image. The data is stored in the uncompressed form. The pixel values are stored either in ASCII or in binary format. The data is preceded by a few bytes of header information (to store primarily the file type, and the number of rows and columns).

The *portable pixmap* (PPM) is the extension of the PGM format for representing color pixels in uncompressed form. It can store the R, G, and B channels in different orders.

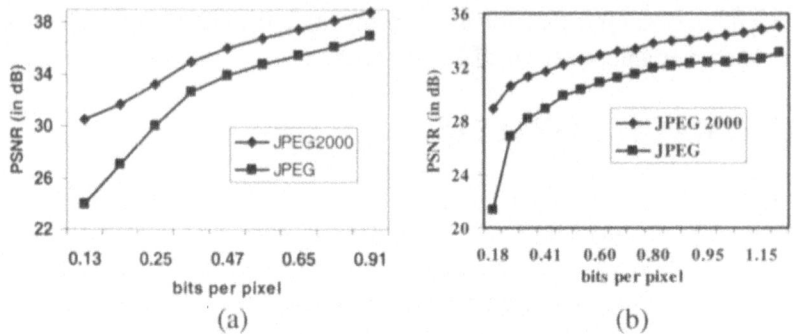

Figure 8.21. Performance comparison of JPEG2000 and JPEG standard. a) gray level, and b) color Lena image.

The GIF is another popular format that uses a variant of Lempel-Ziv-Welch algorithm (see Chapter 6) to achieve compression. It is limited to only 8-bit (256) color images, and hence suitable for images with few distinctive colors. It supports interlacing, and simple animation. It can produce grayscale images using a gray palette.

The *JPEG* uses the JPEG compression algorithm (see section 8.6) to reduce the file size. The users are allowed to set the desired level of quality or compression ratio.

The *Tagged Image File Format* (TIFF) stores many different types of images (*e.g.*, monochrome, grayscale, 8-bit and 24-bit RGB) that are tagged. The images can be stored losslessly, or in a lossy format to reduce the file size.

The PNG format is intended to replace the patented GIF and complex TIFF formats. It supports three main image types: truecolor, grayscale and palette-based ("8-bit"). The JPEG format supports only the first two image types while the GIF truly supports only the third image type. It can save up to 48 bit true color, and 16 bit gray scale. The compression performance is 5-20% better than GIF for 8-bit images.

REFERENCES

1. R. C. Gonzalez and Richard E. Woods, *Digital Image Processing*, Addison Wesley, 1993.

2. M. K. Mandal, S. Panchanathan and T. Aboulnasr, "Choice of wavelets for image compression," *Lecture Notes in Computer Science*, Vol. 1133, pp. 239-249, Springer Verlag, 1996.

3. J. M. Shapiro, "Embedded image coding using zerotrees of wavelet coefficients," *IEEE Trans. on Signal Processing*, Vol. 41, pp. 3455-3462, Dec. 1993.

4. Y. Linde, A. Buzo and R. Gray, "An algorithm for vector quantizer design", *IEEE Trans. on Communications*, Vol. COM-28 (1), pp. 84-95, Jan 1980.

5. M. Nelson and J. –L. Gailly, *The Data Compression Book*, M & T Books, 1995.

6. W. B. Pennebaker and J. L. Mitchell, *JPEG Still Image Compression Standard*, Van Nostrand Reinhold, New York, 1992.

7. ISO/IEC JTC1/SC29/WG1, Document N1646R, *JPEG Part I Final Committee Draft*, Version 1, March 16, 2000.

8. D. Taubman, "High performance scalable image compression with EBCOT," *IEEE Tran. on Image Processing*, Vol. 9, No. 7, pp. 1158-1170, 2000.

9. JasPer: Software Implementation of JPEG2000, Version 1.500.3, http://www.ece.ubc.ca/~mdadams/jasper/.

10. CCITT Recommendation T.4, *Standardization of Group 3 facsimile apparatus for document transmission*.

11. CCITT Recommendation T.6, *Facsimile coding schemes and coding control functions for Group 4 facsimile apparatus*.

12. Graphics File Formats FAQ, http://www.faqs.org/faqs/graphics/fileformats-faq/

QUESTIONS

1. Do you think applying entropy coding directly on images will provide a high compression ratio? Why?

2. You are transmitting an uncompressed image using packets of 20 bits. Each packet contains a start bit, and a stop bit (assume that there is no other head). How many minutes would it take to transmit a 512x512 image with 256 gray levels if the transmission rate is 2400 bits/sec?

3. Justify the use of run-length coding in FAX transmission.

4. Determine the run-lengths for the following scan line. Assume that "0" represents a black pixel, and the scan line starts with a white-run.

    [00000011111111100011111110000000001111111100000000000]

5. Explain the principle of transform domain image compression. For what types of images does the transform domain image compression algorithm provide a high compression ratio?

6. What are the advantages of the transform coding?

7. Explain the advantage of using unitary transform over non-unitary transform for image coding.

8. What is the optimum transform for image compression? Can a fixed transform provide optimal performance for all images?

9. Can it be said that the DCT provides better energy compaction compared to the DFT for all images?

10. The first eight pixels of a scanned image are as follows: $f = [50, 60, 55, 65, 68, 62, 67, 75]$. Calculate the DFT and DCT coefficients of the signal. Retain the 5 largest magnitude coefficients in both cases, and drop the three lowest magnitude coefficients. Reconstruct the signal by calculating inverse

transforms, and determine which transform reconstructs the edge better? Compare the results with those obtained in Example 8.4.

11. Why is block coding generally used in the DCT based image coding?

12. What is transform coding gain? What should be its value for efficient compression?

13. Why is coding a vector of pixels or coefficients more efficient than coding individual coefficients?

14. Compare and contrast the DCT and DWT for image coding.

15. What are the advantages and disadvantages of the vector quantization?

16. When will vector quantization offer no advantage over scalar quantization?

17. What are the advantages and disadvantages of the fractal image coding?

18. Why are the DC coefficients in JPEG coded using a DPCM coding?

19. Explain how the psychovisual properties are exploited in JPEG algorithm.

20. Why was zigzag scanning chosen over horizontal or vertical scanning in JPEG algorithm?

21. What are the advantages of using joint Huffman and run-length coding to encode the AC coefficients in JPEG?

22. How does JPEG algorithm exploit the psychovisual properties?

23. Explain the usefulness of progressive coding.

24. Compare and contrast JPEG and JPEG2000 Standards image coding algorithms.

25. How does JPEG2000 achieve embedded coding?

26. What is EBCOT algorithm?

Chapter 9

Digital Video Compression Techniques

Digital video is the most interesting of all multimedia data. However, there is a price to pay – it requires maximum bandwidth for transmission, and a large space for storage. For example, if you want to transmit the digital television color video (with frame size 720x480) through a network in real time at a rate of 60 frame/sec, you would require about 497 million (=720*480*60*24) bits/sec of bandwidth. This is too high a bitrate for consumer applications. Therefore, efficient video compression techniques are essential in order to make digital video applications feasible. In this Chapter, the fundamentals of video compression technique are first presented. They are followed by brief overview of a few selected general purpose video coding standards.

9.1 PRINCIPLES OF VIDEO COMPRESSION

A video can be considered a three-dimensional light intensity function $i(x, y, t)$, where the amplitude of the function at any spatial coordinate (x, y) provides the intensity (brightness) of the image at that coordinate at a particular time t. Like images, a video can be monochrome or color. A monochrome video is represented in terms of the instantaneous luminance of the light field $i(x, y, t)$, while color video is represented in terms of a set of tristimulus values that are linearly proportional to the amounts of red ($R(x, y, t)$), green ($G(x, y, t)$), and blue ($B(x, y, t)$) light. For digital video all three parameters x, y, and t are discrete.

Since a video is a sequence of image frames ordered in time, the image coding techniques can be directly applied to individual video frames to achieve compression. As mentioned in Chapter 8, the image compression techniques generally exploit the statistical, spatial, structural and psycho-visual redundancies. In addition, a video has temporal and knowledge redundancies that can be exploited to achieve superior performance.

Temporal redundancy: Refers to the correlation among the neighboring image frames in a video. This inter-frame redundancy is typically removed by employing motion estimation and compensation techniques.

Knowledge redundancy: When a video has limited scope, and a common knowledge can be associated with it, efficient compression can be achieved. Consider a videophone call where two people are talking to each other. The pictures of the people do not change during the call. However, there might be little body movements, and some facial expressions (*e.g.*, a smile, crying). At the beginning of the call, if the images of the people in the teleconferencing are sent to the opposite sides, and only the body movements are sent to the other side, a very high compression ratio may be achieved. This is the principle of model-based video coding.

Psychovisual redundancy: There are several forms of psychovisual redundancies that can be exploited to achieve video compression. First, our visual system is not very sensitive to the distortion in chrominance components. Hence, the chrominance components can be quantized more heavily than the luminance components. Second, we have seen in our daily lives that when a video cassette is played, good quality video can be seen. However, when we pause the cassette, the still image does not look as good. This is because our eyes are less sensitive to faster moving objects in a scene. This psychovisual property can be exploited by quantizing more heavily in the region of large motion.

In this Chapter, only the color and temporal redundancy reduction techniques will be discussed. Techniques to exploit other forms of redundancy are outside the scope of this book.

9.2 DIGITAL VIDEO AND COLOR REDUNDANCY

It was mentioned in Chapter 2 that video cameras typically use the $\{R,G,B\}$ color space to represent the color video. Assume that the video camera produces gamma corrected $\{R_N,G_N,B_N\}$ signals that are normalized to Reference white. The luminance and chrominance components of the analog video are then generated by a linear combination of $\{R_N,G_N,B_N\}$. The $\{R,G,B\}$ color space, however, is not very efficient from the compression standpoint. For video transmission, the $\{R,G,B\}$ channels are converted to a luminance (Y) and two chrominance channels.

Digitization of Analog Video

For digital video storage and processing, the analog video has to be converted to digital form. Assume that each component of $\{R_N,G_N,B_N\}$ has a dynamic range of 0-1 volt. The luminance component Y $(= 0.299R_N +$

$0.587G_N + 0.114B_N$) will then have a dynamic range of $[0,1]$ volts. However, the color difference signal $B_N - Y$ and $R_N - Y$ will have a dynamic range of $[-0.886, 0.886]$ and $[-0.701, 0.701]$, respectively. The digitized luminance and chrominance signals are then obtained as [1]:

$$Y_d = 219Y + 16 \qquad\qquad Y_d = [16,235]$$

$$C_B = \frac{112(B-Y)}{0.886} + 128 \qquad C_B = [16,240]$$

$$C_R = \frac{112(R-Y)}{0.701} + 128 \qquad C_R = [16,240]$$

Note that each of the Y_d, C_B and C_R components have 8 bit resolution, and could have the full dynamic range $[0,255]$. However, the levels below 16 and above 235/240 are not used in order to provide working margins for various operations such as coding and filtering.

Color Subsampling

After the digitization we obtain three color channels: Y_d, C_B and C_R. Since our visual system has a low sensitivity for the chroma components (i.e. C_B and C_R), these components are subsampled to reduce the effective number of chroma pixels [1]. Subsampling factors that are typically used are denoted by the following conventions (see Fig. 9.1):

   4:4:4 : No chroma subsamp., each pixel has Y, Cr and Cb values.
   4:2:2 : Horizontally subsample Cr, Cb signals by a factor of 2.
   4:1:1 : Horizontally subsampled by a factor of 4.
   4:2:0 : Subsampling by a factor of 2 in both the hor. and vert. directions.

■ Example 9.1

Determine the reduction in bitrate due to 4:2:2 and 4:2:0 color subsampling.

Assume that there are N color pixels in the video. When there is no subsampling, the video will have a size of $3N$ bytes (assuming 8 bit resolution of each color channel). When 4:2:2 color subsampling is used, there will be N Y samples, $N/2$ Cr and $N/2$ Cb samples. Therefore, the 4:2:2 video will have a size of $(N+N/2+N/2)$ or $2N$ bytes. When 4:2:0 subsampling is used, there will be N Y, $N/4$ Cr, and $N/4$ Cb samples. As a result, the size of the 4:2:0 video will be $(N+N/4+N/4)$ or $1.5N$ bytes. Therefore, a 4:2:2 subsampling provides a 33% reduction in bitrate while 4:2:0 provides a 50% reduction in bitrate. ■

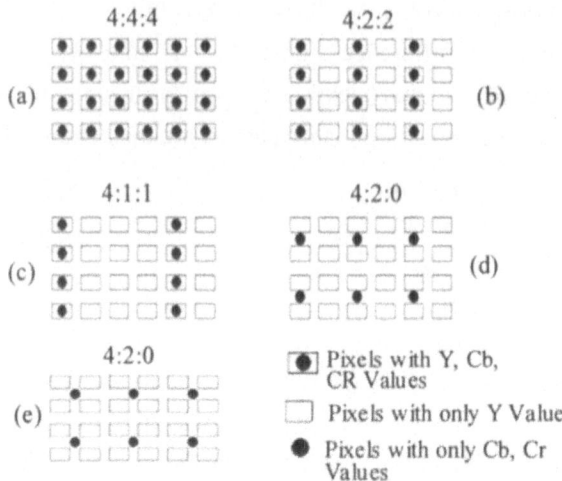

Figure 9.1. Chroma Subsampling. The 4:2:0 subsampling is widely used in video applications, especially at lower bit-rate. Fig. (d) shows the 4:2:0 format adopted in MPEG standard, whereas Fig. (e) shows the format adopted in H.261/H.263 standards.

Digital Video Formats

There are three different analog TV systems used worldwide, and it is difficult to exchange analog video from one standard to another. Although it is possible to digitize video in different TV standards independently, it is important to establish a digital video format that will be more compatible with existing TV standards. International Telecommunication Union (ITU-R), formerly known as CCIR (International Consultative Committee for Radio), has established such a video format that is known as CCIR 601. The standard defines two formats that can be exchanged easily. The first format is for the NTSC video and referred to as 525/60, while the second format is for the PAL system, and is referred to as 625/50.

The parameters of the CCIR 601 standards are shown in Table 9.1. Note that the CCIR-601 employs 4:2:2 color subsampling, and a field rate of 59.94 (for NTSC system) or 50 (for PAL system) fields/sec. For NTSC digital video, the raw data rate is about 331.8×10^6 ((=720*480*60*16) bits/sec. Since this rate is too high for most applications, CCITT (International Consultative Committee for Telephone and Telegraph) Specialist Group (SGXV) proposed a new digital video format called the Common Intermediate Format (CIF). The parameters of the CIF and QCIF (Quarter CIF) format are also shown in Table 9.1. Note that both CIF and QCIF formats are progressive. The CIF format requires approximately 36.46

Mbit/s (=352*288*1.5*29.97*8 Bits/s), and QCIF requires about 9.1 Mbits/s bitrate.

720×480

Table 9.1. Digital Video Data Formats

Parameters	CCIR-601 (4:2:2)		CIF/QCIF (4:2:0)	
	525-line/60 Hz NTSC	625-line/50 Hz PAL/SECAM	CIF	QCIF
Luminance, Y	720×480	720×576	352×288	176×144
Chroma (U, V)	360×480	360×576	176×144	88×72
Field/Frame rate	59.94	50	29.97	29.97
Interlacing	2:1	2:1	1:1	1:1

9.3 TEMPORAL REDUNDANCY REDUCTION

A typical video has a frame rate of more than 25 frames/s, but the video scenes do not change as fast. As a result, most neighboring video frames look very similar. An example of a video sequence, at 10 frames/sec, is shown in Fig. 9.2 where we see that the two neighboring frames will be significantly different only if there is a sudden scene change. For example, a scene change occurs at image-34 (third image at fifth row) in Fig. 9.2. However, sudden scene change does not occur frequently (at most once in a few seconds of video). This temporal redundancy in video can be exploited to achieve a superior video compression through predictive coding whereby a current frame is predicted from a past frame. However, even when two neighboring frames look similar, there might be minor differences that are essentially due to object or camera motion. Hence, better prediction can be achieved if the prediction is performed by compensating motions of different objects present in a scene.

In motion estimation technique, the objects in the current frame are first displaced to their estimated positions in the previous frames, and then the subtraction of two frames is performed. This produces a *difference* frame with much less information compared to simple inter frame difference. For motion compensated frame difference, the only information required to be transmitted is the motion vector's values for each object, but a substantial amount of information is needed if a simple frame difference is transmitted. It has been found that coding of motion compensated frames generally results in a significant reduction (more than 30%) in bit-rate compared to coding simple frame differences despite the overhead of the motion vectors.

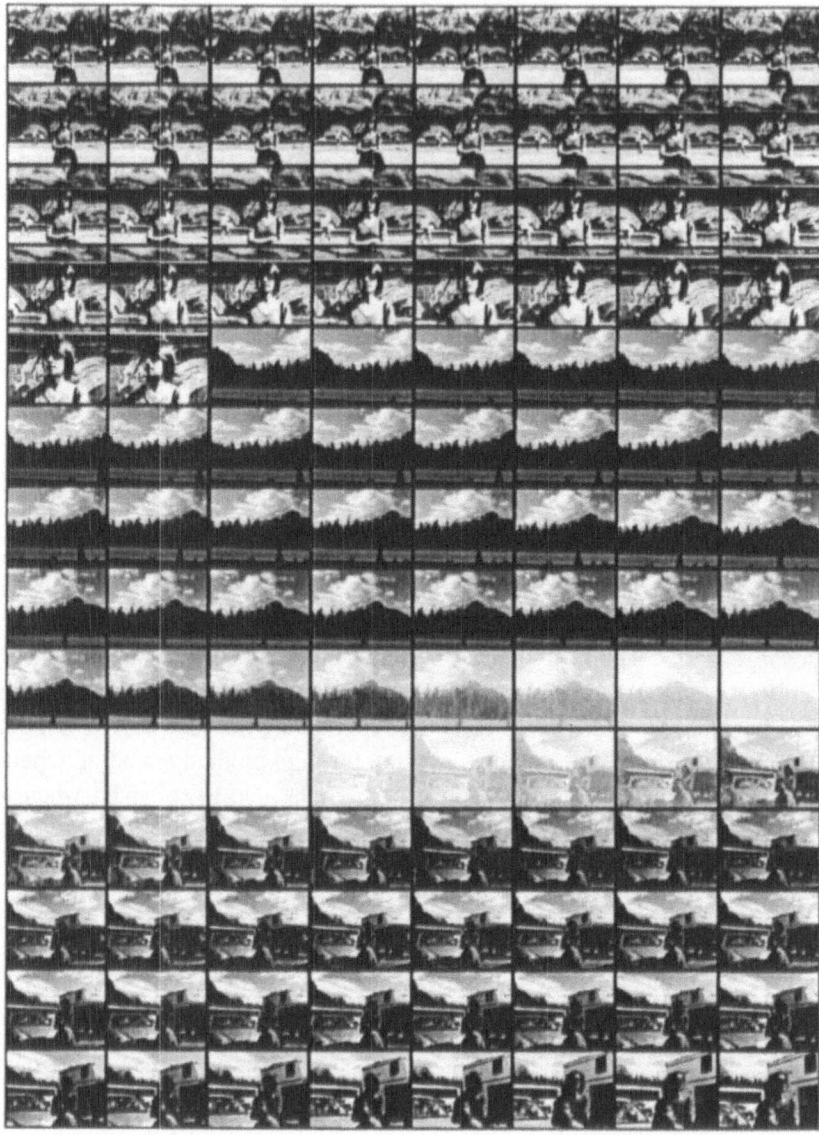

Figure 9.2. An example video sequence at 10 frames/sec. It can be observed that the scenes mostly change gradually. However, in a few instances, there is a sudden change (for example 5th row, 3rd image).

Motion estimation (ME) techniques can be broadly classified into two categories: block-matching and pel-recursive techniques [2]. In block matching techniques, an image is divided into a number of small blocks with the assumption that the pixels within a block belong to a rigid body, and thus have the same motion activity. In the pel-recursive technique, a recursive technique is employed to estimate the motion vector of individual pixels. The pel-recursive techniques provide a fine-grained motion field and thus provide a superior motion estimation, but at the cost of a larger computational complexity. Hence, in most video coding applications, block-based methods are employed. A few selected block-based methods are presented in the next section.

9.4 BLOCK-BASED MOTION ESTIMATION

Consider a video pixel $i(x, y; k)$ where (x, y) denotes the spatial coordinate and k denotes the time. The goal in motion estimation is to find a mapping $d(x, y; k)$ (i.e., a motion vector) that would help reconstruct $i(x, y; k)$ from $i(x, y; k \pm p)$ where p is a small integer. Note that $i(x, y; k - p)$ represents a pixel at the same position in the pth previous frame, while $i(x, y; k + p)$ represents a pixel at the same position in the pth future frame.

The block-based motion estimation is based on the assumption that an image is composed of rigid objects that are moving either horizontally or vertically. In other words, it is possible to find vector $d(x, y; k)$ such that the following relationship is satisfied.

$$i(x, y; k) = i((x, y) - d(x, y; k), k - p) \qquad (9.1)$$

If there is rotation or zoom, the motion estimation will fail. In addition, it is generally assumed that the motion is homogeneous in time, i.e., the objects move at a constant rate in the image plane. If this is true, the following relationship will also be satisfied:

$$i(x, y; k) = i((x, y) + d(x, y; k), k + p) \qquad (9.2)$$

In a block-based scheme, these assumptions are expected to be valid for all points within block b using the same displacement vector d_b. These assumptions are generally satisfied when the blocks are much smaller than the objects, and temporal sampling is sufficiently dense.

In a block-matching motion estimation scheme, each frame is divided into non-overlapping rectangular blocks of size $K \times L$. Each block in the present frame is then matched to a particular block in the previous frame(s) to find the horizontal and vertical displacements of that block. This is illustrated in Fig. 9.3, where the maximum allowed displacement in the vertical and horizontal directions are, respectively, Δ_u and Δ_v. The most frequently used block-matching criteria are *the mean absolute difference* (MAD) and *mean squared error* (MSE). The optimum motion vector (having two components $\hat{u}$ and $\hat{v}$) can be expressed using the MAD and MSE criteria respectively as

$$(\hat{u}, \hat{v}) = \underset{\substack{(u,v) \in Z^2 \\ |u| \leq \Delta_u, |v| \leq \Delta_v}}{\arg\min} \sum_{x=0}^{K-1} \sum_{y=0}^{L-1} |i(x, y; k) - i(x - u, y - v; k - 1)| \quad (9.3)$$

where Z is the set of all integer numbers, which signifies that the motion vectors have one-pixel accuracy. The $(k-1)$ factor is used with the assumption that the motion estimation is being performed with the immediate previous frame. The double sum essentially gives us the total absolute difference between the current block and a candidate block in the previous frame. The "arg min" statement finds out the motion vector $(\hat{u}, \hat{v})$ of the block that produces the minimum absolute difference. The next example illustrates the motion estimation procedure for a small block.

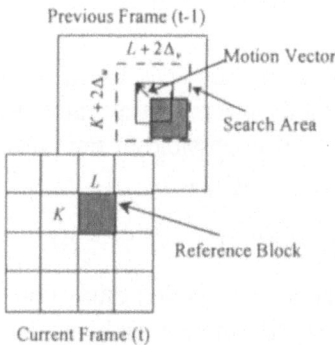

Figure 9.3. Block matching motion estimation process.

■ Example 9.2

Assume that we want to predict the 2x2 block shown in Fig. 9.4 (b) from the reference frame shown in Fig. 9.4(a). Calculate the best motion vector with respect to the MAD criteria, and the corresponding prediction error.

The current block [40, 41; 41, 43] is matched with all (25 in this case) possible candidate blocks. The total absolute difference corresponding to each candidate block is shown in Fig. 9.4(c). It is observed that the minimum absolute difference is 1, and the best-matched block is two pixels above the current block. Hence, the motion vector is (0,-2), meaning that there is no horizontal motion, and a vertical motion of 2 pixels downwards. The prediction difference [0 0 –1 0] is obtained by subtracting the predicted block from the current block. ■

Note that Eq. (9.3) employs the minimum absolute error as the matching criterion. However, sometimes the matching is performed using the minimum mean square error (MSE) criterion with this equation:

$$(\hat{u}, \hat{v}) = \underset{\substack{(u,v) \in Z^2 \\ |u| \le \Delta_u, |v| \le \Delta_v}}{\arg \min} \sum_{x=0}^{K-1} \sum_{y=0}^{L-1} |i(x, y; k) - i(x - u, y - v; k - 1)|^2 \qquad (9.4)$$

The MSE criterion has the advantage that it provides better SNR of the predicted frame. However, there is little difference in the subjective quality of the predicted frame, and it has a higher computational complexity. Hence, the MAD criterion is used more frequently.

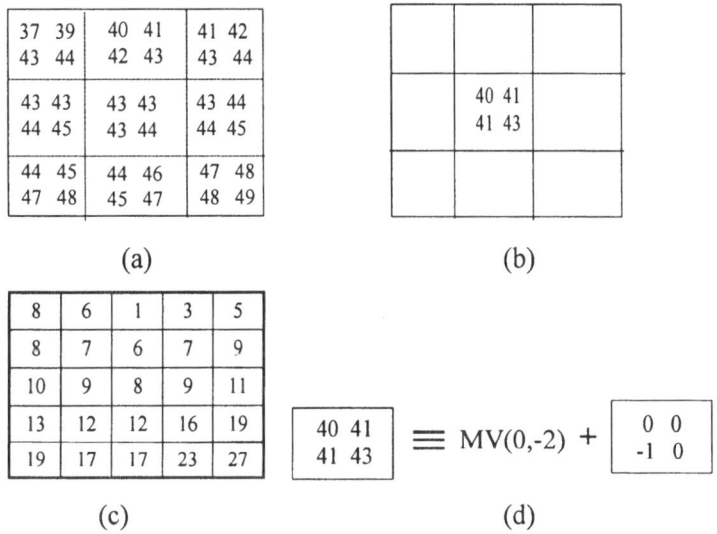

Figure 9.4. Illustration of Motion Estimation. (a) previous reference frame, b) the block from the current frame to be predicted, c) motion prediction error for different search grid points, d) motion prediction with the best motion vector (0,-2), and the corresponding error block.

Eq. (9.3) estimates the motion vector with integer pixel accuracy. However, more accurate ME is possible by estimating motion vectors at a fraction-pixel accuracy, especially with half-pixel accuracy. This is generally done by interpolating the image blocks. Although the fractional pixel ME provides superior motion estimation, it has a higher complexity.

In order to obtain the minimum prediction error, the argument in Eqs. (9.3) and (9.4) should be evaluated for all possible candidate blocks, which is known as the full search algorithm (FSA). A few other motion estimation techniques will be presented shortly.

Evaluation Criteria for Motion Estimation Techniques

A motion estimation algorithm is evaluated using two factors: the motion compensation efficiency, and the computational complexity. The motion compensation efficiency can be measured in terms of the average prediction gain which, for an image block, can be defined as follows:

$$G_{MP} = \frac{\text{Energy of the original image block}}{\text{Motion compensated residual energy}} \qquad (9.5)$$

If the motion compensation is adequate, the residual energy will be small resulting in a high prediction gain G_{MP}. The computational complexity of a motion estimation algorithm should be low in order to facilitate real time implementation. The computational complexity is proportional to the number of points tested by an algorithm for a given search area. However, for real-time hardware implementation, the number of required sequential steps can also become important since the ME of individual blocks of a given step can be evaluated in parallel.

■ Example 9.3

Calculate the motion vectors corresponding to the frame (size: 240x352) shown in Fig. 9.5(b) with respect to the reference frame (these frames have been taken from the well-known *football* sequence for evaluating motion compensation performance) shown in Fig. 9.5(a). Assume a block size of 16x16, and a search window of [-16,16] in both horizontal and vertical directions. Calculate the motion prediction gain, and estimate the motion estimation complexity.

The football sequence is a high motion sequence. If the current frame is predicted from the previous frame without motion estimation (i.e., motion vectors are considered to be zeros), the frame difference energy will be significant. Figure 9.5(c) shows the frame difference signal. The prediction (as defined in Eq. (9.5)) gain is 17.86.

Motion prediction is performed using search windows of [-7, 7] and [-16, 16]. The performance is shown in Figs. 9.5(d) and 9.5(e). It is observed that the search window [-7,7] provides a frame difference that is significantly lower than Fig. 9.5(c). The prediction gain in this case is 55.64. The search window [-16,16] further reduces the error energy (as shown in Fig. 9.5(e), providing a prediction gain of 60.11.

The motion vectors are shown in Fig. 9.6(a). It is observed that most motion vectors are zero (in both horizontal and vertical directions). The histogram of the motion vectors is shown in Fig. 9.6(b). It is observed that 177 motion vectors out of total 330 vectors have zero value. There are 306 motion vectors within search range [-7,7]. Since the [-7,7] search window captures most of the motion vectors, it provides a prediction gain (=55.64) close to that of the [-16,16] search window (=60.11).

(a)   (b)   (c)

(d)   (e)

Figure 9.5. Motion compensation performance. a) Reference *football* frame, b) *football* frame to be predicted, c) frame difference signal with no motion compensation (G_{MP}=17.86), d) frame difference signal with full search motion estimation with search window ($\pm 7, \pm 7$) (G_{MP}=55.64), e) frame difference signal with FSA with search window ($\pm 16, \pm 16$) (G_{MP}=60.11). The DFDs are amplified 5 times for better print quality.

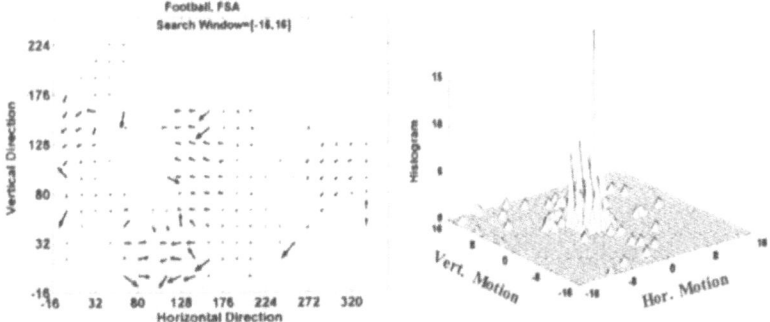

Figure 9.6. Motion vectors corresponding to Fig. 9.5. a) Full search motion vectors (search area =16) of the entire frame, b) histogram of the motion vectors. Each grid-point corresponds to a 16x16 block. A downward motion vectors means that the current block is best predicted from an upper block in the previous frame. A right arrow means that the current block is best predicted from a left-block in the previous frame. The histogram has a peak value of 177 at the center (*i.e.*, zero motion), which has been clipped to 15 in the figure.

Since the frame size is 240x352, and the motion block-size is 16x16, there are 330 blocks in the current video frame. For each block, there are 1089 (=33x33) candidate blocks since the search range is [-16,16]. The total absolute difference calculation for a 16x16 block requires 512 operations (256 subtraction + 256 additions) ignoring the absolute value calculation. For 1089 candidate blocks, the total number of operations will be 557568 (=1089x512) operations. For the entire frame, the approximate number of arithmetic operation will be 184 millions (=557568x330) operations. Note that the blocks at the outer edge of the frame will not have 1089 candidate blocks for motion prediction. Therefore, the actual complexity will be little less than 184 millions.

The *football* frames and the MATLAB code for this example are provided in the accompanying CD. ■

9.4.1 Fast Motion Estimation Algorithms

It was observed in Example 9.3 that the complexity of motion estimation for a video frame (with size 240x352) can be more than 100 MOP. In general, the complexity of the FSA is very high, on the order of $8\Delta_u\Delta_v$ arithmetic operations/pixel. If we consider a frame rate of 30, to achieve a real-time operation the motion estimator should have at least a few GOPS computational capability. It means that a motion estimator that uses FSA will have high implementation costs.

Several techniques have been proposed to reduce the complexity of ME algorithms. Most of these techniques are based on the assumption that the matching criterion (*i.e.*, the prediction error) increases monotonically as the search moves away from the direction of minimum distortion. These algorithms are faster compared to FSA; however, they may converge to a local optimum that corresponds to an inaccurate prediction of the motion vectors. A few selected block-matching techniques are now presented.

Three-Steps Search Algorithm

In this algorithm, the motion vectors are searched in multiple steps. In the first step, the cost function is calculated at the center and eight surrounding locations that form a 3x3 grid. The location that produces the smallest cost function becomes the center point for the next search step, and the search range is reduced by one-half.

An example of motion vector calculation is shown in Fig. 9.7 for a search range [-7,7]. In the first step, the motion vector is calculated at nine points: (0,0), (-4,-4), (-4,0), (-4,4), (0,4), (4,4), (4,0), (4,-4), and (0,-4). Assume that grid (4,4) has produced the minimum cost function. This grid point will be used for the motion vector search at the next step. In the second step, the motion vector is calculated at eight points: (2,2), (2,4), (2,6), (4,6), (6,6), (6,4), (6,2), and (4,2). Note that the distance between the search points is now 2 (compared to 4 in the previous step). Assume further that the grid point (4,6) provides the minimum cost function at the second step. In the third step, the cost function of eight points surrounding the grid point (4,6) will be calculated. The grid point that will provide the minimum cost function will be selected as the best motion vector. In the example, the lowest cost function is provided by grid (3,6), and hence the motion vector would be (6,3) (*i.e.*, the horizontal motion is +6, and the vertical motion is +3).

This algorithm reduces the number of calculations from $(2p+1)^2$ required by the full search (when $\Delta_u = \Delta_v = p$) to the order of $1 + 8\log_2 p$.

2-D Logarithmic Search Algorithm

The 2-D logarithmic search is similar to the three-step search. However, there are some important differences. In each step of this algorithm, the search is performed only at the five locations that include a middle point and four points in the two main directions: horizontal and vertical. The location that provides the minimum cost function is considered the center of the five locations used in the next step. If the optimum is at the center of the five locations, the search area is decreased by one-half, otherwise the search area remains identical to that of the previous step. The procedure continues in recursive manner until the distance between the search points is reduced to 3x3. In the final step, all the nine locations are searched and the position of the minimum distortion is selected as the horizontal and vertical components of the motion vectors.

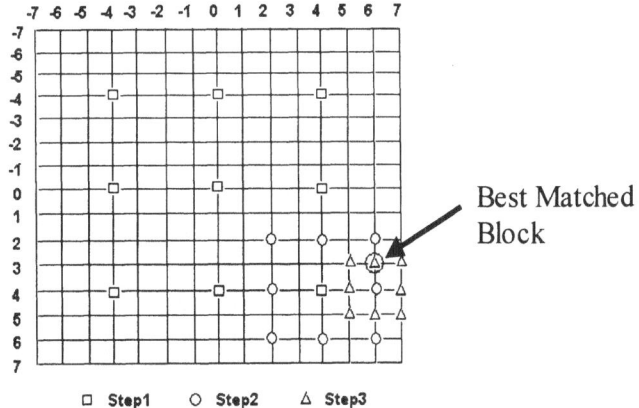

Figure 9.7. Motion Estimation using *three-steps search* algorithm.

An example of a motion vector calculation using a 2-D logarithmic search is shown in Fig. 9.8. In the first step, the motion vector is calculated at five points: (0,0), (-4,0), (0,4), (4,0), and (0,-4). Assume that grid (0,4) has the minimum cost function. Since this is an outer point, the cost function corresponding to three grid points (-4,4), and (4,4) with identical search range will be calculated. Assume that in this sub-step, grid (4,4) has produced the minimum cost function.

In the second step, the cost function for grids (2,4), (4,6), (6,4), and (4,2) will be calculated and be compared with the cost function for grid (4,4). Assume that the grid (4,6) produces the minimum cost function. Since it is an outer grid point, cost functions of the two other grids (2,6) and (6,6) are calculated and compared with the cost function for grid (4,6). Assume that (4,6) provides the minimum cost function of the four points. In the final step, the cost function corresponding to eight-points surrounding grid (4,6) is calculated, and the position of the minimum cost function is considered as the motion vector. In Fig. 9.8, the minimum cost function is shown to be located at (6,3).

This algorithm reduces the number of calculations from $(2p+1)^2$ required by the full search (when $\Delta_u = \Delta_v = p$) to about $2 + 7\log_2 p$.

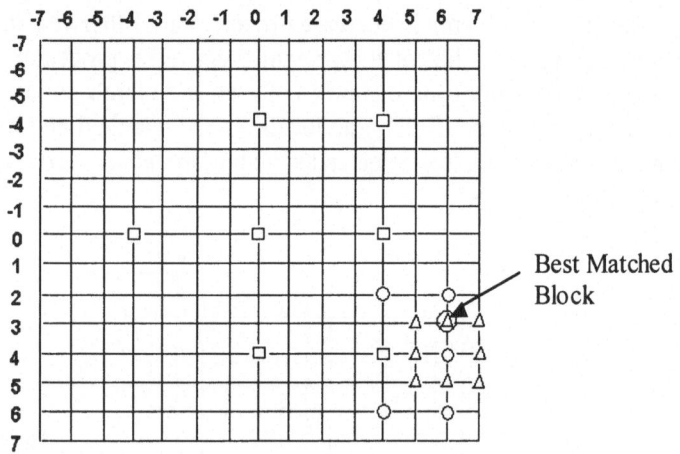

Figure 9.8. Motion Estimation using *2-D logarithmic* search.

Conjugate Direction Search

In this algorithm, the search for the optimal motion vectors is done in two steps. In the first step, the minimum cost function is estimated in one direction. In the second step, the search is carried out in the other direction starting at the grid point found in the first step.

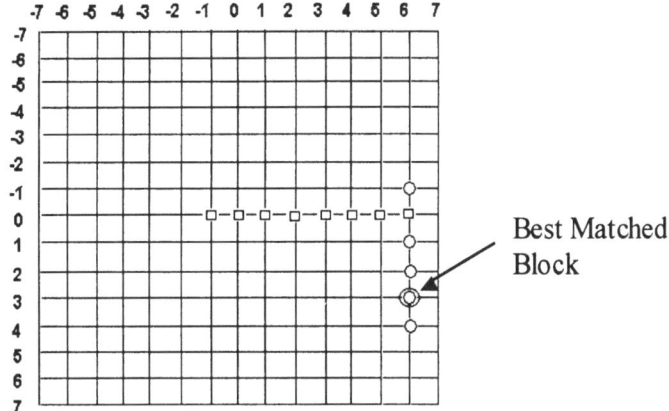

Figure 9.9. Motion Estimation using conjugate direction search.

An example is shown in Fig. 9.9 where the horizontal component of the optimal motion vector is found first. In each step, the cost function of three consecutive grid points are calculated. First, the cost functions for grids (0,-1), (0,0) and (0,1) are calculated. Assume that grid (0,1) provides the minimum cost. Since the direction of the minimum cost function is towards the right, the cost function of (0,2) is calculated next. In each subsequent step, the cost function of the next right grid point is calculated. Assume that the cost function monotonically decreases until grid (0,8), i.e., cost(0,2)> cost(0,3)>cost(0,4)>cost(0,5)>cost(0,6)<cost(0,7). Because cost(0,6) is smaller than cost(0,7), the horizontal coordinate of the motion vector is designated as 6.

In the next step, the best motion vector is searched in the vertical direction. First, the cost functions of grids (-1,6) and (1,6) are calculated. If cost(-1,6)> cost(0,6)<cost(1,6), grid (0,6) will be considered as the overall minimum cost function. Assume that in this particular case, cost(1,6)<cost(0,6)<cost(-1,6). Hence, the search for the optimal vertical component will be carried in the downward direction until a plateau has reached. In this example, it is assumed that cost(1,6)> cost(2,6)>cost(3,6)<cost(4,6)>. Therefore, the grid(3,6) is considered to be the grid with the smallest cost function, resulting in a motion vector of (6,3).

The maximum number of searches using this technique is ($2p + 3$).

■ Example 9.4

In this example, the performance of the three-steps and conjugate directions methods is evaluated using the two football frames considered in Example 9.3. The three-steps and conjugate directions methods provide prediction gains of 47.6 and 36.7, respectively (whereas the full search algorithm provides a gain of 55.64). Figs. 9.10(a) and 9.10(b) show the motion predicted error frames corresponding to the three-steps and conjugate directions methods. Note that the three-steps search method provides a gain close to the FSA search method, at a substantially reduced complexity (you can verify it by executing the MATLAB code, and comparing the runtimes). The conjugate direction method reduces the complexity further. However, the prediction gain also drops. The motion vectors corresponding to the two search methods are shown in Fig. 9.11. ∎

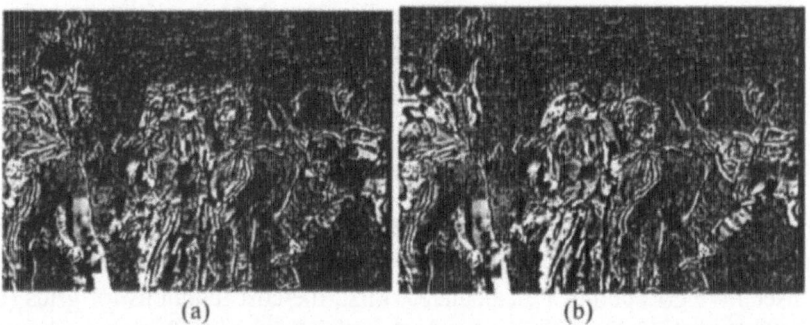

(a)   (b)

Figure 9.10. Motion compensation performance. a) frame difference signal corresponding to the three-steps search with search window $(\pm 7, \pm 7)$ $(G_{MP}=47.6)$, b) frame difference signal corresponding to the conjugate direction method with search window $(\pm 7, \pm 7)$ $(G_{MP}=36.7)$.

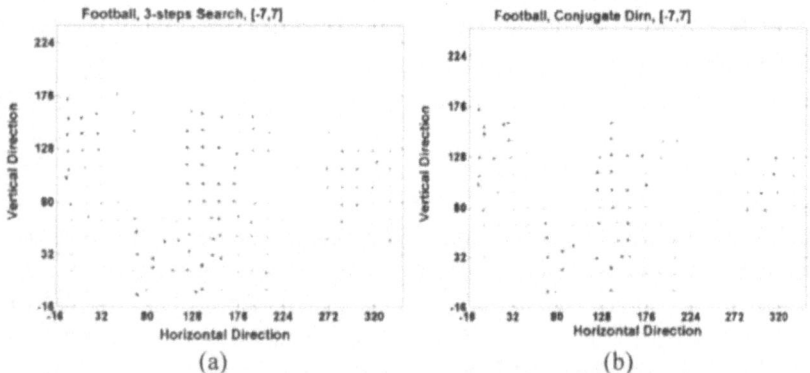

(a)   (b)

Figure 9.11. Motion vectors corresponding to Fig. 9.10. a) Motion vectors calculated with three-steps search, b) motion vectors calculated with conjugate directions search. Each grid-point corresponds to a 16x16 block. A downward motion vectors means that the current block is best predicted from an upper block in the previous frame. A right arrow means that the current block is best predicted from a left-block in the previous frame.

■ Example 9.5

The *football* is a fast moving video sequence, and hence the motion prediction gain is only in the range of 50. In this example, we consider the slow moving sequence *Claire*, which is a typical teleconferencing sequence. Figures 9.12(a) and 9.12(b) shows the reference frame and to-be-predicted frame, respectively. A straightforward frame difference (*i.e.*, with no motion prediction) produces a prediction gain of 360, while the FSA produces a prediction gain of 1580. The corresponding error frames are shown in Fig. 9.12(c)-(d). It is observed that a high prediction gain is achieved in this case. As a result, the overall bit-rate in this case will be lower compared to the football sequence (at similar quality level). ■

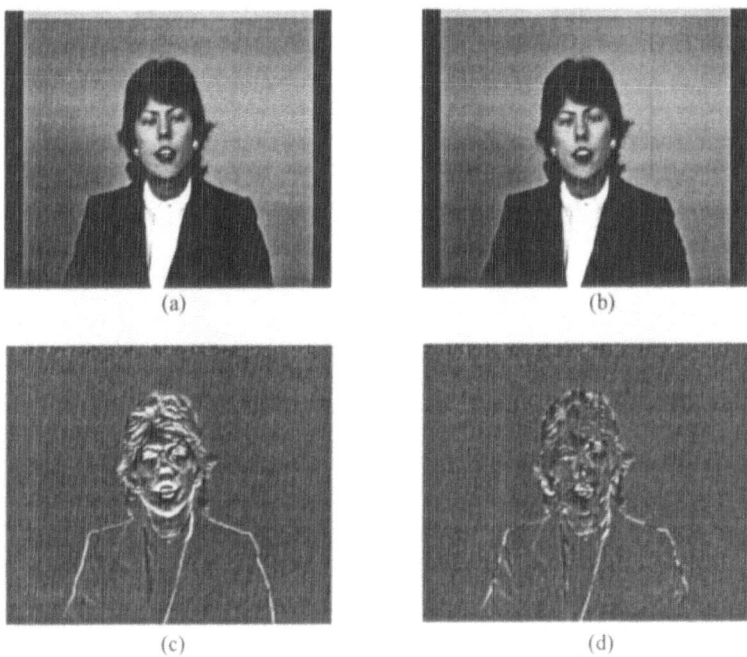

Figure 9.12. Motion compensation performance. a) Reference Claire frame, b) Claire frame to be predicted, c) frame difference signal with no motion compensation (G_{MP}=360), and d) frame difference signal with full search motion estimation with search window ($\pm7,\pm7$) (G_{MP}=1580)

Limitation of Fast Search Algorithm

Most fast search algorithms (especially the three algorithms discussed here), assume a monotonic error surface where there is only one minima (as shown in Fig. 9.13). If this assumption is valid, all techniques will eventually reach the global minima. However, the error surface is not strictly monotonic in most cases. Figure 9.13(b) shows the error surface corresponding to a block in the football sequence. It is observed that there are several local minima, and the fast search algorithm may be trapped in one of these local minima, and thus will provide only sub-optimal performance. In order to avoid local minima, the initial search points should be spread out as far as possible. Among the three algorithms discussed, the conjugate direction search algorithm is most likely to be affected by the local minima's.

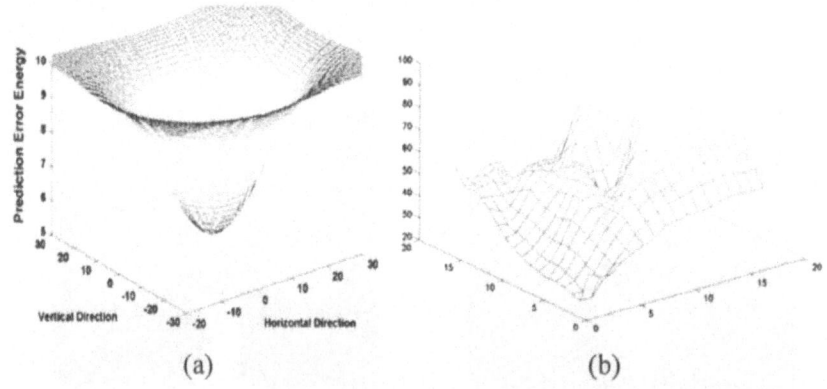

(a)   (b)

Figure 9.13. Error surface. a) Ideal monotonic error surface with one (global) minima, b) an irregular error surface with several local minima's.

Bidirectional Motion Estimation

So far, motion estimation has been considered to be forward predictive. The current frame is predicted from the past frame to satisfy the causality criterion. However, in many instances, the forward prediction fails. Consider the 3rd frame in row-5 of the video shown in Fig. 9.2. Since there is a scene change, we cannot predict this frame from the previous frame. A better performance can be obtained if we use a future frame as the reference. However, the prediction will not be causal, and to implement it, we need to delay the motion estimation by a few frames. It will be shown in the next section that some video coding standards use bi-directional ME where ME is performed simultaneously using past and future reference frame. In addition to scene changes, this technique is also useful when an object in the current frame is occluded in the past frame, but completely visible in the future frame (see Fig. 9.14).

Motion Estimation of the Chroma Components

There are three color components in a digital color video. The motion estimation techniques can be applied individually to each color component. In other words, the Y, Cb, and Cr components of the current frame can be motion predicted from the Y, Cb, and Cr components of the reference frame. However, it has been found that there is a strong correlation among the motion vectors of the three components (when an object moves, it should be reflected equally in all three components). Hence, motion vector is generally not calculated for Cb and Cr components in order to reduce the computational complexity. Instead, a motion vector corresponding to a block in Y component is employed for motion compensation of the corresponding block in Cb and Cr components. However, a suitable scaling is required since the Cb and Cr components may have a size different than the Y component.

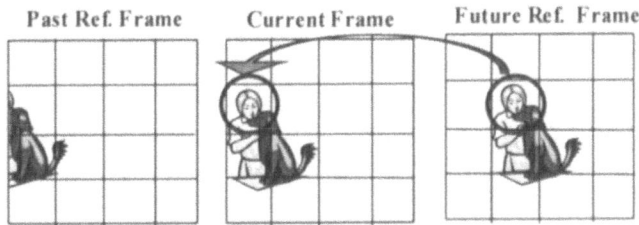

Past Ref. Frame   Current Frame   Future Ref. Frame

Figure 9.14. Example of prediction from a future frame. The head block cannot be predicted from the previous reference frame. The future reference frame can be used here to predict the block.

9.5 VIDEO COMPRESSION STANDARDS

Several video compression standards [3] for different applications have been developed in the last 10 years. H.261 [4] and H.263 [5] have been established for video-telephony and video conferencing applications using ISDN (Integrated Services Digital Network). These standards have been developed primarily by ITU-T (formerly CCITT).

A few general-purpose video coding standards have been developed by the Motion Pictures Expert Group (MPEG) [6] of the International Standards Organization (ISO). These are known as MPEG-1 [7], MPEG-2 [8], and MPEG-4 [9, 10] standards. MPEG-1 specifies a coded representation that can be used for compressing video sequences up to a maximum bit rate of 1.5 Mbit/s. It was developed in response to the growing need for a common format for representing compressed video on various digital storage media such as CDs, DATs, Winchester disks and optical drives. MPEG-2 has been developed for a target rate of up to 50 Mbits/sec, and is intended for applications requiring high quality digital video and audio. MPEG-2 video builds upon the MPEG-1 standard by supporting interlaced video formats and a number of advanced features including those supporting HDTV.

The following is a brief introduction to a few select coding standards.

9.5.1 Motion-JPEG

The Motion JPEG is the simplest nonstandard video codec. It uses the JPEG still image coding standard to encode and decode each frame individually. Since there is no motion estimation, the complexity of the coding algorithm is very small. Unfortunately, the performance of this codec is not very good since it does not exploit the temporal correlation among the video frames. As a result, many early video coding applications used this standard.

9.5.2 The MPEG-1 Video Compression Standard

The MPEG-1 video coding standard employs efficient algorithms to compress fully synchronized audio and video data at a maximum bit rate of 1.5 Mbit/s. The primary objectives of this standard are to facilitate the storage of videos in a CD-ROM and their possible transmission through a variety of digital media. Fig. 9.15 shows the schematic of the MPEG-1 encoder. Block-based motion compensation is employed to remove the temporal redundancy. The residual spatial correlation in the predicted error frames is further reduced by employing block DCT coding similar to JPEG.

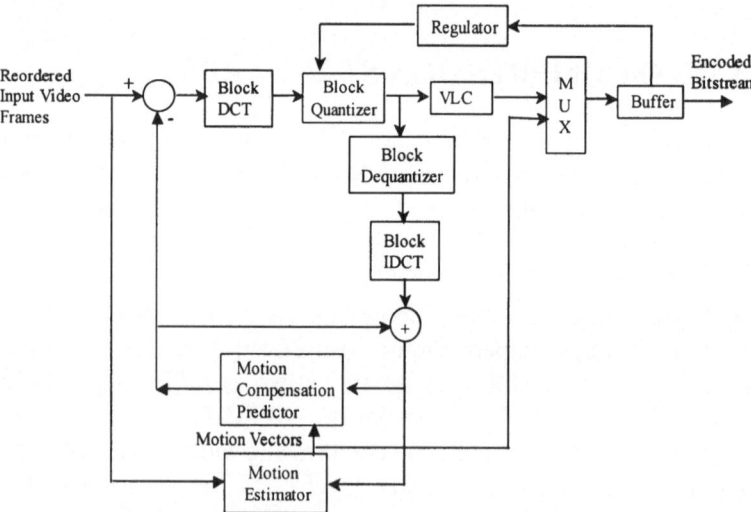

Figure 9.15. Simplified lock diagram of the MPEG-1 video encoder. VLC: variable length coding, MUX: multiplexer.

Because of the conflicting requirements of random access and high compression ratio, the MPEG standard suggests that the video frames be divided into three categories: I, P and B frames. I- (Intra coded) frames are encoded without reference to other frames. They provide access points to the coded sequence where decoding can be performed immediately, but are coded with moderate compression ratio. P- (predictive coded) frames are coded more efficiently using motion compensated prediction from a past I-frame, or another P-frame, and are generally used as a reference for further prediction. B- (bi-directionally predictive coded) frames provide the highest degree of compression, but require both past and future reference frames for motion compensation. B-frames are never used as references for prediction. Fig. 9.16 illustrates the relationship among the three different frame types in a group of pictures or frames (GOP). Note that a GOP begins with an I frame, and ends at the last picture before the next I frame. The GOP length is flexible, but 12-15 pictures is a typical value. Note that the GOP in Fig. 9.16 has a length of 9 pictures.

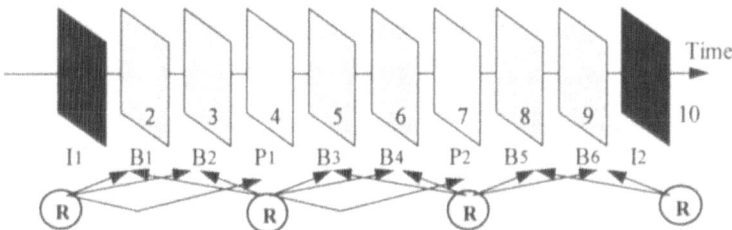

Figure 9.16. Example of a group of pictures (GOP) used in MPEG. I-frames are used as reference (R) frames for P-frames, and both I- and P-frames are used as reference frames for B-frames.

Since the B-frames are predicted both from the previous and future reference frames, the frames are reordered before motion prediction and encoding. The frames are encoded in the following order: $I_1 P_1 B_1 B_2 P_2 B_3 B_4 I_2 B_5 B_6$. Sending a frame out of sequence requires additional memory both at the encoder and the decoder, and also causes delay. Although a larger number B-frames provide superior compression performance, the number of B-frames between two reference frames is kept small to reduce cost and minimize delay. It has been found that I-frame only coding requires more than twice the bit-rate of an IBBP coding. If this delay corresponding to IBBP is unacceptable, an IB sequence may be a useful compromise.

The decoding of MPEG-1 video is the inverse of the encoding process as shown in Fig. 9.17. The coded bitstream is first demultiplexed to obtain the motion vectors, quantizer information, and entropy coded quantized DCT coefficients. An I-frame is obtained in three steps. The compressed coefficients are first entropy decoded. The coefficients are then dequantized, and inverse DCT of the dequantized coefficients is calculated. This completes the I-frame reconstruction.

The P-frames are obtained in two steps. In the first step, the corresponding error frame is calculated by decoding the compressed error frame using a procedure similar to the I-frame reconstruction (*i.e.*, entropy decoding, dequantization, and inverse DCT). In the second step, the motion predicted frame is calculated by applying motion compensation on the previous reference (I- or P-) frames. The motion predicted frames and the corresponding error frames are then added to obtain the P-frames.

A B-frame is calculated when both previous and future reference frames are calculated. Each block of the B-frame is typically predicted by motion compensation from both previous and future reference frames. The final block is then obtained by interpolating the two predicted blocks.

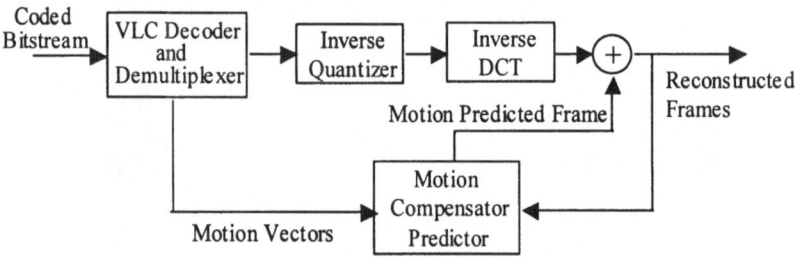

Figure 9.17. Simplified schematic of MPEG-1 Video Decoder.

9.5.3 The MPEG-2 Video Compression Standard

MPEG-2 is a generic coding standard [8] for moving pictures and associated audio in the range from approximately 1.5 to 80 Mbits/sec. The MPEG-2 standard is similar to MPEG-1, but includes extensions to cover a wider range of applications. The most significant enhancement over MPEG-1 is the addition of syntax for efficient coding of interlaced video. Although most MPEG-2 applications use 4:2:0 chroma sampling, MPEG-2 also allows 4:2:2 (which has been developed for improved compatibility for digital production equipment) and 4:4:4 chroma sampling. Several other more subtle enhancements (*e.g.*, 10-bit DCT DC precision, non-linear quantization, VLC tables, and improved mismatch control) are included, and which have a noticeable improvement on coding efficiency even for progressive video. Other key features of MPEG-2 are the scalable extensions that permit the division of a continuous video signal into two or more coded bit streams representing the video at different resolutions (spatial scalability), picture quality (SNR scalability), and picture rates (temporal scalability). Hence, the same set of signals can work for both HDTV and standard TV.

MPEG-2 standard provides the capability for compressing, coding, and transmitting high quality, multi-channel, multimedia signals over terrestrial broadcast, satellite distribution, and broadband networks. Since MPEG-2 has been designed as a transmission standard, it supports a variety of packet formats, and provides error correction capability that is suitable for cable TV and satellite links. A schematic of MPEG-2 video and audio coder and the associated packetizing scheme for transmission through a network is shown in Fig. 9.18.

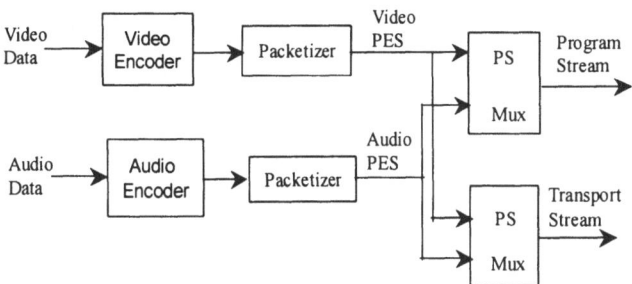

Figure 9.18. Block diagram of MPEG-2 audio-visual encoder and packetization scheme.

Profiles and Levels

MPEG-2 standard is divided into four levels, depending on complexity (see Table 9.2). Each level is further divided into six profiles, depending on the special features of the coding algorithm. Although there are 24 combinations of levels and profiles, all of these combinations have not been defined. The MPEG-2 levels and profiles for different applications are as shown in Table 9.2.

Table 9.2. MPEG-2 Levels and Profiles.

	Different Profiles					
Levels	Simple	Main	SNR scalable	Spatial scalable	High	4:2:2 Profile
Low		4:2:0 352x288 4 Mbit/s I,P,B	4:2:0 352x288 4 Mbit/s I,P,B			
Main	4:2:0 720x576 15 Mbit/s I,P	4:2:0 720x576 15 Mbit/s I,P,B	4:2:0 720x576 15 Mbit/s I,P,B		4:2:0, 4:2:2 720x576 20 Mbit/s I,P,B	4:2:2 720x608 50 Mbit/s I,P,B
High-1440		4:2:0 1440x1152 60 Mbit/s I,P,B		4:2:0 1440x1152 60 Mbit/s I,P,B	4:2:0, 4:2:2 1440x1152 80 Mbit/s I,P,B	
High		4:2:0 1920x1152 80 Mbit/s I,P,B			4:2:0, 4:2:2 1920x1152 100 Mbit/s I,P,B	

The simple profile at the main level does not employ B frames, resulting in a simpler hardware implementation. The main profile has been designed for a majority of the applications. The low level algorithm supports SIF resolution (352x288), whereas the main level supports the SDTV (standard definition TV) resolution. High-1440 supports high definition resolution (with aspect ratio of 4:3) that is double (1440x1152) to that of SDTV. The 16:9 high definition video (1920x1152) is supported in high level. Table 9.3 shows the application of this profile at different levels. The SNR and Spatial profiles support scalable video of different resolution at different bitrates. Simple, main, SNR and spatial profiles support only 4:2:0 chroma sampling. The high level supports both 4:2:0 and 4:2:2 sampling, as well as SNR and spatial scalability. A special 4:2:2 profile has been defined by MPEG-2 to improve compatibility with digital production equipment at a complexity lower than the high profile.

Table 9.3. MPEG-2 Main Profiles at different Levels

Level	Size	Pixels/sec	Bit-rate (Mbits/s)	Applications
Low	352x288x30	3 M	4	VHS quality
Main	720x576x30	12 M	15	Studio TV
High	1440x1152x60	96 M	60	Consumer HDTV
Very High	1920x1152x60	128 M	80	Film Production

9.5.4 The MPEG-4 Video Compression Standard

We have noted that MPEG-1 and MPEG-2 standards employ interframe ME to remove temporal correlation, and DCT to remove the spatial correlation of the error frames. These two standards provide good video coding performance with respect to bit-rate and subjective quality. However, one of the major drawbacks of these standards is that they do not provide content access functionality. Consequently, the ISO has recently established a new video and audio coding standard known as MPEG-4 [10]. This new standard provides techniques for the storage, transmission and manipulation of natural and synthetic textures, images and video data in multimedia environments over a wide range of bit-rates.

The design of MPEG-4 is centered on a basic unit called the audio-visual object (AVO). Here, a scene is segmented into background and foreground, which in turn is represented by video objects. Consider the scene depicted in Fig. 9.19, which can be segmented into four components (*i.e.*, the background, wall-clock, the presenter, and the computer display) as shown in Fig. 9.20. In MPEG-4, each AVO is represented separately and becomes the basis for an independent stream. Note that AVOs can also be arbitrarily composed.

2-D Animated Mesh

In addition to the object-based coding, MPEG-4 has incorporated techniques for representing synthetic images. It employs the virtual reality modeling language (VRML) to synthesize animated video. In addition, it employs animated 2-D mesh modeling to represent images such as human face [10].

Structure of the tools for representing natural video

The MPEG-4 image and video coding algorithms give an efficient representation of visual objects of arbitrary shape, and also supports content-based functionalities. They support most functionalities already provided by MPEG-1 and MPEG-2 standards, including efficient compression of standard rectangular-sized image sequences at varying levels of input formats, frame rates, pixel depth, bit-rates, as well as various levels of spatial, temporal and quality scalability.

Figure 9.19. Example of an MPEG-4 Scene.

Support for Conventional and Content-Based Functionalities

The MPEG-4 coding tools provide a trade-off between the bit-rate and functionalities. The MPEG-4 very low bit-rate video (VLBV) coding mode encodes a video using techniques similar to an MPEG-1/2 coder. A VOP is encoded using motion compensation followed by texture coding. For the content-based functionalities where the input video sequence may contain arbitrary shaped objects, this approach is extended by also coding shape and transparency information. This is known as generic MPEG-4 coding mode. Shape may be either represented by an 8-bit transparency component that allows the description of transparency if one VO is composed with other objects, or by a binary mask. The content-based functionalities may increase the bit-rate of the encoded video.

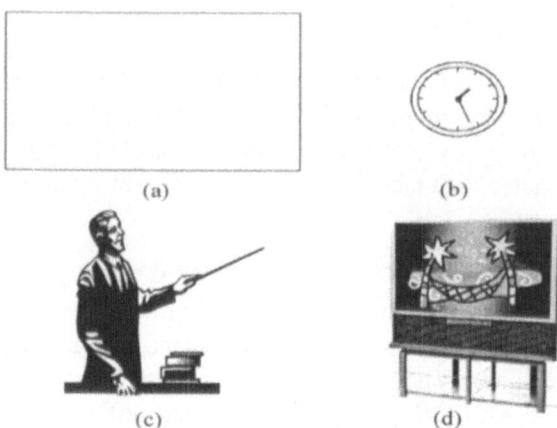

Figure 9.20. Segmenting an MPEG-4 scene into various objects. a) background,
b) wall clock, c) presenter, and d) computer display.

9.5.4.1 Video Coding Scheme

Figure 9.21 below outlines the basic approach of the MPEG-4 video
algorithms to encode rectangular as well as arbitrarily-shaped input image
sequences. The basic coding structure involves shape coding (for arbitrarily-
shaped VOs) and motion compensation, as well as DCT-based texture
coding (using standard 8x8 DCT or shape adaptive DCT). The basic coding
structure involves shape coding (for arbitrarily shaped VOs) and motion
compensation as well as DCT-based texture coding (using standard 8x8
DCT or shape adaptive DCT). An important advantage of the content-based
coding approach of MPEG-4 is that the compression efficiency can be
significantly improved for some video sequences by using appropriate and
dedicated object-based motion prediction "tools" for each object in a scene.
A number of motion prediction techniques allow efficient coding and
flexible presentation of the objects:

In certain situation, a technique known as sprite panorama in MPEG-4
may be helpful to achieve a superior performance. Consider a video
sequence where a person is walking on the street (see Fig. 9.22(a)). Since,
the person is walking slowly, several consecutive images will have similar
background, and the background may move slowly due to camera
movement or operation. A sprite panorama image can be generated using the
smaller static background (*i.e.*, the street image). The foreground object, *i.e.*,
the person's image is separated from the background. The large panorama
image (*i.e.*, Fig. 9.22(b)) and the person's image (*i.e.*, Fig. 9.22(c)) are
transmitted to the receiver separately only once in the beginning. These
images are stored in the sprite buffer in the receiver. In each consecutive
frames, the position of the person in the sprite panorama, and the camera

parameters if any, are sent to the receiver. The receiver would be able to reconstruct the individual frames from the panorama image and the picture of the person. Since, the individual frames (other than the panorama image) are represented only by the camera operation, and position of the objects, very high compression performance may be achieved with this method.

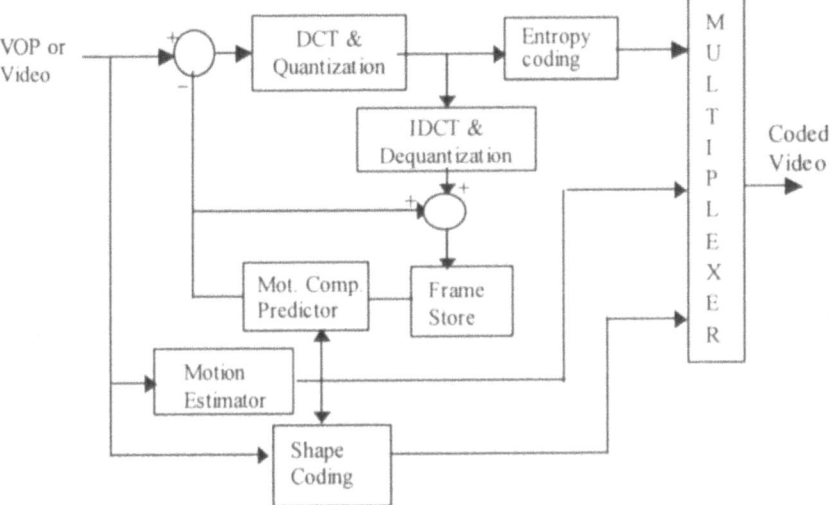

Figure 9.21. Simplified block diagram of MPEG-4 video encoder

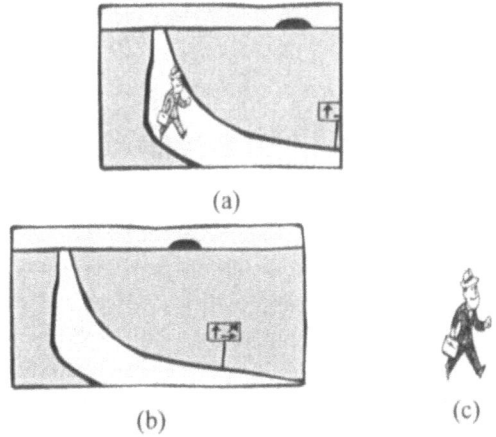

(a)

(b)   (c)

Figure 9.22. Example of sprite coding of a video sequence. a) image to be coded, b) panorama image, and c) moving object. Fig. (a) can be obtained easily from Figs. (b) and (c).

Scene description

In addition to providing support for coding individual objects, MPEG-4 also provides facilities to compose such objects into a scene. The necessary composition information forms the scene description, which is then coded and transmitted together with the media objects. Starting from the Virtual Reality Modeling Language (VRML), MPEG has developed a binary language for scene description called BIFS (BInary Format for Scenes).

In order to facilitate the development of authoring, manipulation and interaction tools, scene descriptions are coded independently from streams related to primitive media objects. Special care is devoted to the identification of the parameters belonging to the scene description. This is done by differentiating parameters that are used to improve the coding efficiency of an object (*e.g.*, motion vectors in video coding algorithms), and the ones that are used as modifiers of an object (*e.g.*, the position of the object in the scene). Since the MPEG-4 should allow the modification of this latter set of parameters without having to decode the primitive media objects itself, these parameters are placed in the scene description and not in primitive media objects.

An MPEG-4 scene can be represented by a hierarchical structure using a directed graph. Each node of the graph is a media object, as illustrated in Fig. 9.23. Nodes at the leaves of the tree are primitive nodes, whereas the nodes that are parents of one or more other nodes are called compound nodes. The tree structure may not be static; node attributes (*e.g.*, positioning parameters) can be changed, and nodes can also be added, replaced, or removed. This logical structure provides an efficient way to perform the object-based coding.

9.5.5 The H.261 Video Compression Standard

The MPEG-1 and MPEG-2 video coding standards are targeted for general purpose applications. To achieve better video coding performance, the encoders in these standards are much more complex than the decoders. This is particularly useful for applications where the number of encoders will be much fewer than the number of decoders. However, for one-to-one visual telephony, this assumption is no longer true. Both the encoder and the decoder should have low complexity in order to make the videophone products less expensive for the consumers.

The ITU-T Recommendation H.261 is a video coding standard [4] specifically developed for low bit-rate audiovisual services. Here, the video coder operates at a rate p times 64 kbits/s where $1 \le p \le 30$. Consequently, this standard is also known as $p \times 64$ standard. Typically, video conferencing using CIF format requires 384 Kbits/s, which corresponds to $p = 6$.

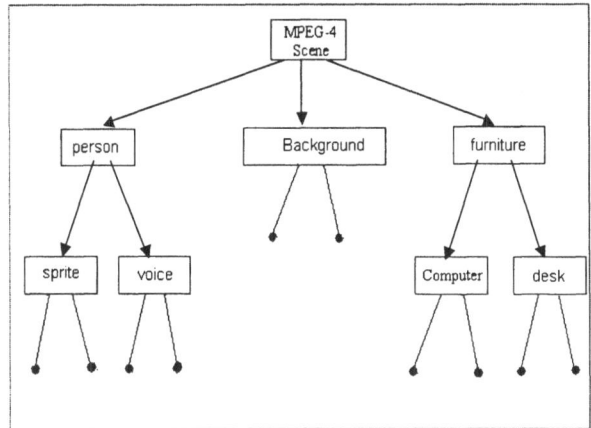

Figure 9.23. Logical structure of a scene.

The principles of H.261 coding standard is similar to that of MPEG-1. It employs motion prediction, block-DCT, quantization, and VLC coding. However, there are minor differences in actual quantization procedure, and the bitstream arrangement is also different. H.261 accepts only two frame sizes: CIF and QCIF.

9.5.6 H.263, H.263+, and H.26L Standards

The H.263 video codec [5] is an improved version of the H.261 codec, and employs the same DCT and motion compensation techniques. Several incremental improvements in video coding were added, and are mentioned below.

1. Half-pixel motion compensation to reduce the DFD energy.
2. Improved variable length coding (has option for arithmetic coding)
3. Optional modes include unrestricted motion vectors.
4. Advanced motion prediction mode including overlapped block motion compensation
5. A mode that combines a bi-directionally predicted picture with a normal forward predicted picture (PB Mode).
6. Supports wider range of picture formats (4CIF, 16CIF)

It has been found that H.263 coders can achieve the same quality as H.261 coders at about half the bit-rate.

This H.263+ standard [11] includes further improvement over Recommendation H.263 in order to broaden its range of useful applications, and to improve its compression performance. The standard accepts several new types of pictures, such as scalability pictures, and custom source formats. New coding modes have been added, including advanced intra coding, deblocking filter and reference picture selection.

Currently, the ITU-T video coding experts group is developing a new international standard for video coding known as H.26L. The primary goal of this new standard is to have a simple and straightforward video coding design capable of achieving enhanced compression performance. In addition, the standard has a provision of network-friendly packet-based video representations addressing conversational (*e.g.*, video telephony) and non-conversational (*e.g.*, storage, broadcast and streaming) applications. The H.26L has a video coding layer (VCL) that provides the core high compression video representation, and a network layer (NAL) for video delivery over a particular type of network.

The H.26L VCL has achieved a significant improvement in rate-distortion efficiency over existing standards. H.26L NAL is being developed to transport the video over a wide variety of networks such as IP networks with RTP packetization and 3G wireless systems.

9.5.7 Performance Comparison of Standard Codecs

Several standard video codecs have been examined in the previous sections. The prime objective of all codecs is to provide good coding performance. However, these codecs have different complexity and coding performance. It is difficult to compare these codecs since their target bit-rate and applications are different. Nevertheless, a brief performance comparison of a few selected codecs is presented in this section.

The performance comparison of the M-JPEG and the MPEG-2 standards has been reported by Boroczky *et al.* [12]. Figure 9.24 compares the Motion JPEG, MPEG-2 I only and MPEG-2 IP coders, at 2 Mbits/s and 8 Mbits/s, respectively, for *hallway*, *football*, and *table tennis* sequences. It is observed that the MPEG-2 I only coder provides superior performance to the M-JPEG. Moreover, use of P frames in MPEG-2 further improves the performance for all three sequences.

The performance evaluation of three public-domain video codecs is now presented. The first is the H.263 video codec developed by TeleNor [13]. This codec implements the baseline H.263 video coding scheme, as well as the optional coding modes such as unrestricted motion vector mode, syntax-based arithmetic coding mode, advanced prediction mode, and PB-frames mode. The second codec was released by MPEG Software Simulation Group (MSSG) in 1996 for MPEG-2 standard [14]. This codec can also generate MPEG-1 video bitstream, and in this test only the MPEG-1 part has been evaluated. The third codec is the MPEG-4 reference implementation provided by Microsoft [15]. Although the MPEG-4 codec software has both the frame-based coding and object-based coding options, the frame-based technique is used in this test.

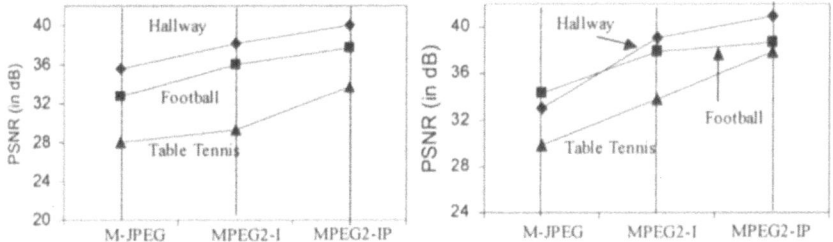

Figure 9.24. Coding performance of M-JPEG, MPEG-2 I, and MPEG-2 IP algorithms. a) for SIF size frame at 15 frames/sec and 2 Mbits/s, and b) for CCIR 601 (720x480) size frame at 30 frames/sec and 8 Mbits/s (adapted from [12]).

Coding performance is evaluated using the first 100 frames of the test sequence "bream" (as shown in Fig. 9.25) with a frame-size of 352×288. Assuming a sampling rate of 30 frames/second, the clip has duration of approximately 3.3 seconds. Note that each codec has several parameters that can be fine-tuned to achieve superior performance. In this experiment, the default parameters have been employed. Therefore, it is possible to obtain a better performance than that has been obtained in this experiment.

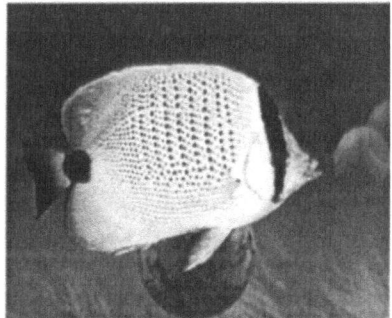

Figure 9.25. One frame of "bream" sequence

Comparison at 1.1 Mbits/second

Figures 9.26 and 9.27 shows the coding performance at 1.1 Mbits/second. In the MPEG-1 and MPEG-4 codecs, the GOP structure is IBBPBBPBBP BB, i.e., one GOP contains twelve frames. For H.263 codec, there is one I frame, and others frames are coded as P-frames. It is observed that MPEG-4 provides more than 1 dB performance improvement in this test. H.263 provides a better performance than MPEG-4 for this sequence at this bit-rate.

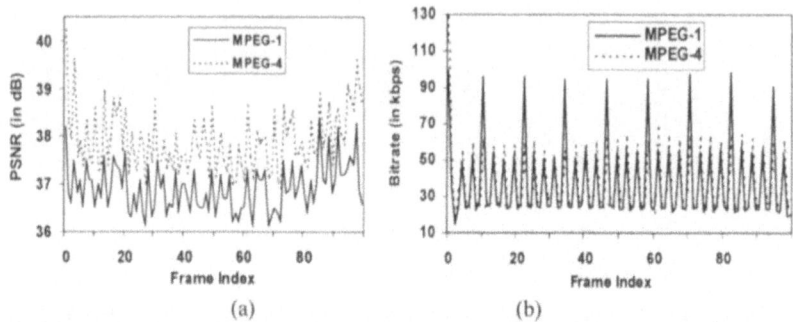

Figure 9.26. Performance of MPEG-1 and MPEG-4 codecs for bream sequence at 1.1 Mbits/s. The PSNR of the luminance component of the individual frames is shown.

Comparison at 56 Kbits/second

The performance of H.263 and MPEG-4 was evaluated at 56 kbits/s. In order to retain a good subjective quality of the encoded frames, typical video codecs drop frames at a low bitrate. When the bream sequence is encoded at 56 kbps, the H.263 and MPEG-4 codecs encode 24 and 34 frames, respectively, instead of 100 original frames.

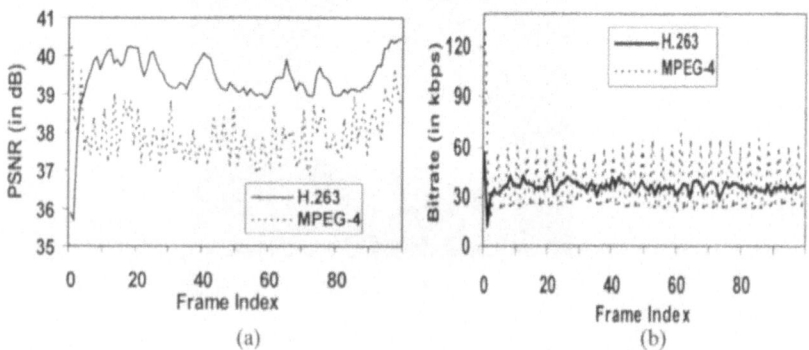

Figure 9.27. Performance of the H.263 and MPEG-4 codecs at 1.1 Mbits/s. The PSNR of the luminance component of the individual frames is shown.

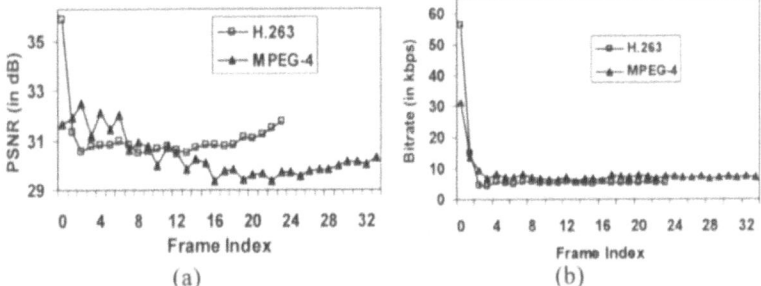

Figure 9.28. Performance comparison of H.263 and MPEG-4 codec at 56 kbps. a) the PSNR and b) the bitrate of the individual encoded (luminance) frames of the bream sequence. The average bitrate is 56 kbps.

Figure 9.28 shows the PSNR and the bitrates of the individual encoded frames. It is difficult to compare the PSNR and the size of the individual compressed frames since the encoded frames do not correspond to each other for the two codecs. However, the plots provide us some information regarding how the codec operates. The I-frame of the MPEG-4 codec uses fewer bits compared to the H.263 codec, and spends these extra bits to encode 12 extra frames (34 frames compared to 24 for H.263 codec). It has been observed that overall MPEG-4 provides superior video quality.

REFERENCES

1. A. N. Netravali and B. G. Haskell, *Digital Pictures*, Plenum Press, New York, Second Edition, 1995.

2. F. Dufaux, and F. Moscheni, "Motion estimation techniques for digital TV: a review and a new contribution," *Proc. of IEEE*, Vol. 83, No. 6, pp. 858-876, June 1995.

3. B. G. Haskell, P. G. Howard, Y. A. Lecun, A. Puri, J. Ostermann, M. R. Civanlar, L. Rabiner, L. Bottou, and P. Haffner, "Image and video coding – emerging standards and beyond," *IEEE Tran. on Circuits and Sytems for Video Technology*, Vol. 8, No. 7, pp. 814-837, Nov 1998.

4. M. Liou, "Overview of the px64 kbit/s video coding standard," *Communications of the ACM*, Vol 34, No 4, 60-63, April 1991.

5. ITU-T Recommendation H.263, *Video coding for low bitrate communications*, 1996.

6. MPEG Homepage: http://mpeg.telecomitalialab.com/standards.htm.

7. D. L. Gall, "MPEG: a video compression standard for multimedia applications," *Communications of the ACM*, Vol. 34, pp. 46-58, April 1991.

8. J. L. Mitchell, *MPEG Video: Compression Standard*, Chapman & Hall, New York, 1996.

9. R. Koenen, F. Pereira, and L. Chiariglione, "MPEG-4: context and objectives," *Signal Processing: Image Communications*, Vol. 9, pp. 295-304, May 1997.

10. ISO/IEC JTC1/SC29/WG11 N4030, *MPEG-4 Overview – (Singapore Version)*, March 2001.

11. G. Cote, B. Erol, M. Gallant, and F. Kossentini, "H.263+: video coding at low bit rates," *IEEE Tran. on Circuits and Sytems for Video Technology*, Vol. 8, No. 7, pp. 849-866, Nov 1998.

12. L. Boroczky and A. Y. Ngai, "Comparison of MPEG-2 and M-JPEG Video coding at low bit-rates," *SMPTE Journal*, March 1999.

13. K. O. Lillevold, "Telenor R&D H.263 Codec (Version 2.0)," ftp://bonde.nta.no/pub/tmn/software/, June 1996.

14. MPEG Software Simulation Group (MSSG), "MPEG-2 Video Encoder/Decoder (Version 1.2)," ftp://ftp.mpegtv.com/pub/mpeg/mssg/, July 1996.

15. Microsoft Corporation, "MPEG-4 Visual Reference Software (Version 2.3.0 FDAM1-2.3-001213)," http://numbernine.net/robotbugs/mpeg4.htm, Dec 2000.

QUESTIONS

1. A multimedia presentation contains the following types of data:
    i)   10000 characters of text (8 bit, high ASCII)
    ii) 200 color images (400x300, 24 bits)
    iii) 15 minutes of audio (44.1 KHz, 16 bits/channel, 2 channels)
    iv) 15 minutes of video (640x480, 24 bits, 30 frames/sec)

    Calculate the disk space required to store the multimedia presentation. How much space percentage does each data type occupy?

2. What is the principle behind color subsampling? Consider the video data in the above problem. Assume that the video is stored in i) 4:2:2, or ii) 4:2:0 format. How much space would be required in each case to store the video?

3. What is motion compensation in video coding? Why is it so effective?

4. Compare and contrast various motion estimation techniques.

5. While performing motion estimation for the football sequence using a full search algorithm, the displaced block difference energy of a 16x16 block was found to be as given in the following table. The energy shown is normalized for better clarity. The search range is (-7,7) in both horizontal and vertical direction. Calculate the motion vector and the motion predicted error energy if we use i) three-step, ii) 2-D logarithmic, and iii) conjugate direction search. Are these fast-search techniques able to find the global minimum?

86	91	94	95	94	89	83	76	69	65	64	68	73	76	75
75	80	86	90	93	91	84	74	63	52	46	45	50	58	65
62	65	71	78	82	83	80	71	60	48	36	28	28	35	47
54	53	56	61	68	73	75	73	64	51	36	21	13	15	28
57	53	51	54	59	65	70	70	66	56	43	27	13	9	18
65	60	54	53	57	61	64	66	65	60	53	43	30	21	21
72	71	64	60	59	60	61	63	63	62	60	58	50	39	32
75	76	73	68	64	61	61	63	64	64	64	64	60	52	43
75	79	81	79	74	68	64	63	64	64	65	66	65	61	53
75	79	82	82	78	72	66	63	62	63	65	67	67	67	63
74	77	78	77	76	70	63	60	60	63	65	68	69	68	64
74	77	78	77	76	70	63	60	60	63	65	68	68	67	64
73	74	72	72	71	67	61	60	62	65	66	68	68	67	64
73	71	69	67	67	64	60	60	63	66	67	69	71	70	68
73	70	65	61	59	59	58	58	62	66	68	71	73	72	70

6. Repeat the above problem for the following table. Do the fast search algorithms find the global minima? If we had use the search range (-15,15) instead of (-7,7), what would be the performance for the three fast search techniques?

24	30	36	42	47	51	55	57	58	59	61	62	63	62	61
25	32	39	46	51	55	59	61	62	64	66	68	68	66	65
28	34	41	48	53	59	64	66	67	69	72	73	73	71	69
30	37	43	50	53	58	66	70	70	73	77	78	78	76	74
31	38	46	53	53	55	64	71	72	76	81	84	83	81	80
31	38	47	55	56	54	57	64	71	78	83	86	86	84	84
32	38	46	55	59	57	52	51	62	78	85	86	88	88	88
34	39	45	54	59	60	53	41	47	70	83	86	88	90	90
36	41	47	53	58	61	57	45	44	59	76	85	89	91	92
37	43	49	54	57	60	59	55	54	59	70	83	91	94	95
39	43	52	56	57	57	59	64	67	68	72	82	92	98	99
43	44	50	57	56	52	54	64	73	77	77	80	89	96	99
48	47	47	55	57	50	48	60	72	80	80	79	85	94	97
51	52	49	51	54	51	48	53	64	77	81	79	82	88	90
50	53	54	50	46	49	50	49	53	66	75	76	78	77	78

7. Is DCT coding of motion compensated error frame as effective as DCT coding of natural images? Explain.

8. Explain the usefulness of the B-frames in MPEG-1 video coding standard.

9. The I-frames in the MPEG standard help to provide fast forwarding and random access while watching a digital movie. What GOP length would you choose if the movie is required to display at least one frame every 0.5 sec while fast forwarding? Assume a progressive frame rate of 30 frames/sec. What are some of the possible GOP structures (i.e., number of P- and B-frames) with this GOP length?

10. An MPEG video uses the GOP structure: IBBBPBBB. Determine the order of encoding for the first 20 frames.

11. You are designing a digital video database and looking for a good video compression algorithm. One of the main requirements is that you should be able to retrieve a video based on its content. Which video compression standard algorithm will you select? Explain the main principles of this standard algorithm.

12. Encode the first 20 frames of the Claire sequence using the image coder in Example 8.7. Plot the bit-rate versus PSNR of each frame.

13. Implement a simple inter-frame coder assuming a GOP of 20 frames. Encode the first frame using the image coder in Example 8.7, and the remaining 19 frames as P frames. Use the FSA with search window (-7,7) for the motion estimation. Calculate the DCT of the motion predicted error frame, and quantize the DCT coefficients with different step-sizes. Calculate the entropy of the motion vectors and the quantized DCT coefficients for each frame, and use the entropy as the bit-rate for the frame. Plot the bit-rate versus PSNR of each frame.

14. Compare the performance obtained in the previous two problems, and discuss the advantages of the motion compensation.

Chapter 10

Digital Audio Processing

Digital audio has become very popular in the last two decades. With the growth of multimedia systems and the WWW, audio processing techniques are becoming popular. There is a wide variety of audio processing techniques, such as filtering, equalization, noise suppression, compression, addition of sound effects, and synthesis. The audio compression techniques have been discussed in Chapter 7. In this Chapter, we present a few selected audio-processing techniques. The depth of the coverage is very narrow compared to the rich set of literature available. The readers who are interested in learning the audio processing techniques in greater details may consult the books listed in the reference section.

10.1 AUDIO FILTERING

Filtering is the process of selecting or suppressing certain frequency components of a signal by a desired amount. Digital filtering was introduced in Chapter 5. Here, we consider application of digital filters in audio processing. The following example illustrates the effect of digital filtering on an audio signal.

■ **Example 10.1**

Consider the audio signal (*bell.wav*) and its amplitude spectrum, which are shown in Fig. 10.1. The signal was sampled with a sampling frequency of 22050 Hz, and 8-bit resolution. It is observed that the signal has a sharp attack (*i.e.* rise) and a gradual delay. The power spectral density shows that the signal has frequency components across the entire 0-11025 Hz range.

Lowpass filtering

The filtering is performed using the "fir1" function, which is a generic MATLAB function for FIR filtering.

```
filt_low = fir1(64, 4000/11025) ;   % filt_low is a 64-tap filter with a normalized
% cut-off frequency 0.3628 that corresponds to 4000 Hz for a sampling frequency
%of 22.05 kHz. The same filter would have a cut-off frequency of 8 kHz if the
```

% sampling frequency is 44.1 kHz
x_lpf = filter(filt_low,1,x) ; % x_lpf is the lowpass filtered audio signal

The *fir1* function uses Hamming window by default. The cut-off frequency is chosen as 4000 Hz. Note that the fir1 function accepts normalized (with respect to the sampling frequency) cut-off frequencies. The filter is a 64-tap low pass filter with real coefficients, and has a linear phase. The filter gain is shown in Fig. 10.2(a). The power spectral density of the filtered output is shown Fig. 10.2(b). It is observed that the high frequency spectral components have been significantly reduced. The audio output is provided in the accompanying CD in *wav* format to demonstrate the lowpass effect. The complete MATLAB function is also included in the CD.

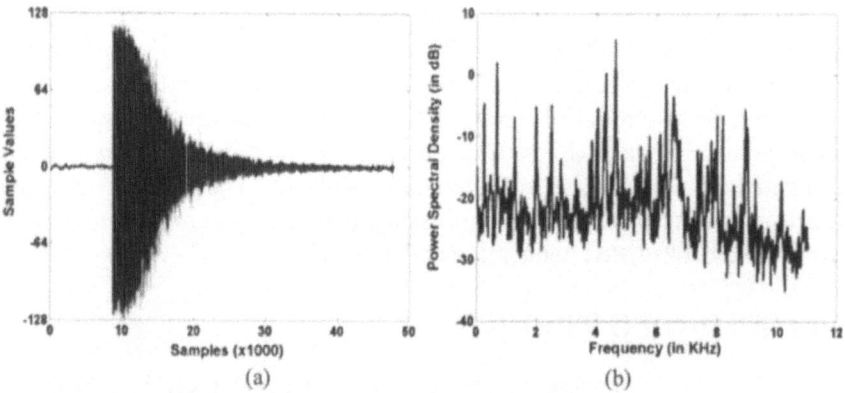

Figure 10.1. The audio signal bell.wav. a) the time domain signal, and b) it's amplitude spectrum.

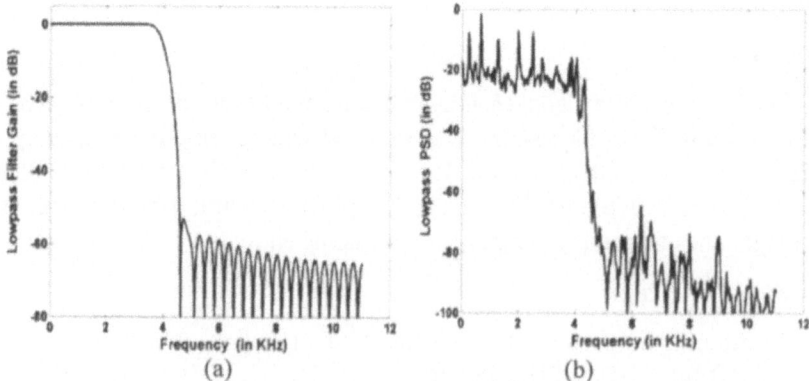

Figure 10.2. Lowpass filtering of audio signal. a) Lowpass filter, and b) power spectral density of the filter output. The filter is a 64 tap FIR filter (designed using Hamming window) with a cutoff frequency of 4000 Hz.

Bandpass filtering

The filtering is performed again using the *fir1* command.

```
filt_bp = fir1(64,[4000 6000]/11025) ;
%Bandpass filtering the audio signal
x_bpf = filter(filt_bp,1,x) ;
```

The lower and upper cut-off frequencies are chosen as 4000 and 6000 Hz, respectively. The filter is 64-tap high pass filter with real coefficients, and has a linear phase. The gain of the bandpass filter is shown in Fig. 10.3(a). The spectral density of the filtered output is shown in Fig. 10.3(a). The complete MATLAB function for bandpass filtering, as well as the audio output is included in the CD.

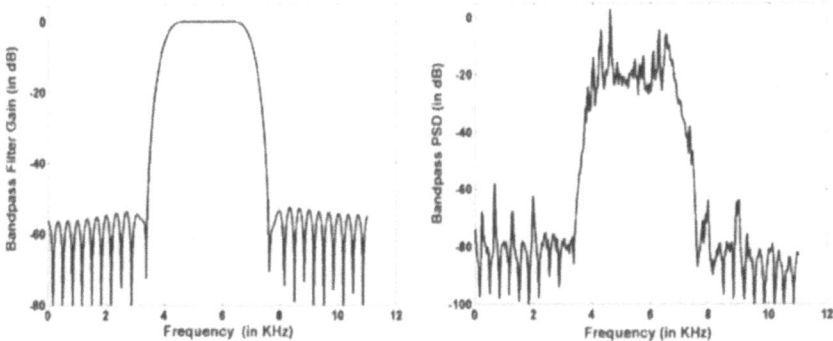

Figure 10.3. Bandpass filtering. a) Bandpass filter characteristics, and b) power spectral density of the filter output. The filter is a 64 tap FIR filter with lower and higher cutoff frequencies of 4000 and 6000 Hz, respectively.

Highpass filtering

The filtering is performed again using the *fir1* command.

```
filt_high = fir1(64,4000/11025,'high') ;
%Highpass filtering the audio signal
x_hpf = filter(filt_high,1,x) ;
```

The cut-off frequency is chosen as 4000 Hz. The filter is 64-tap high pass filter with real coefficients, and has a linear phase. The gain of the highpass filter is shown in Fig. 10.4(a). The spectral density of the filtered output is shown in Fig. 10.4(a). The audio output is again provided in the accompanying CD in *wav* format to demonstrate the highpass effect. ∎

10.2 AUDIO EQUALIZATION

Equalization is an effect that allows the user to control the frequency response of the output signal. The user can emphasize (*i.e.*, boost) or de-emphasize (*i.e.*, suppress) selected frequency bands in order to change the output sound. The amount that a selected frequency band is boosted or cut is generally indicated in decibels (dB). When an equalizers level control is set to 0 dB, no boost or cut is applied to the selected band(s).

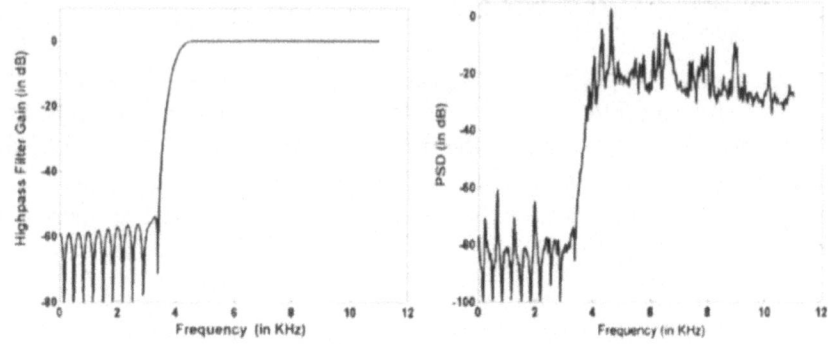

Figure 10.4. Highpass filtering. a) Highpass filter characteristics, and b) the power spectral density of the filter output. The filter is a 64 tap FIR filter with a cutoff frequency of 4000 Hz.

The equalization can be conveniently done using a filterbank. There are primarily two types of equalization systems, tone control system and graphic equalizer system, which are used in consumer audio systems. The two methods are presented below.

Tone Control

The tone control method provides a fast and simple way to adjust the sound to suit the taste of the listeners. The consumer audio systems generally have two control knobs labelled *bass* and *treble*. Each of these knobs controls a special type of filter, known as *shelving filter*. The bass controls a lowpass shelving filter while the treble controls a highpass shelving filter. Typical gain (frequency) response of these two filters is shown in Fig. 10.5. A gain larger than one boosts the audio, while a gain smaller than one suppresses the audio.

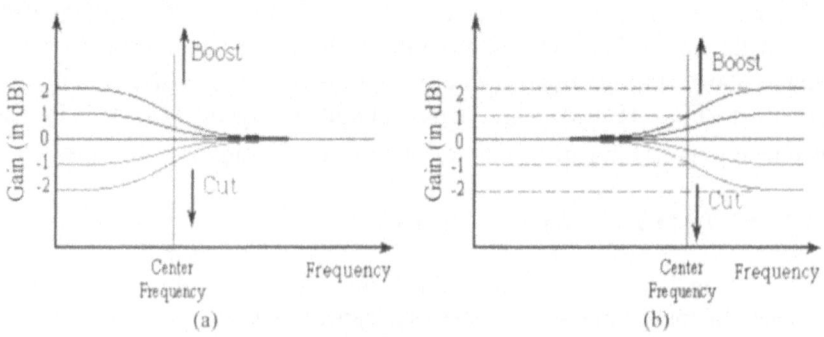

Figure 10.5. Frequency response of a) lowpass, and b) highpass shelving filter.

Note that the lowpass and highpass filters typically attempt to remove a frequency band completely. However, a shelving filter does not try to remove the frequency components completely, rather the selected bands are just boosted or suppressed while leaving the other bands unchanged.

Many audio systems have *mid* control, in addition to bass and treble. This control basically uses a bandpass filter, typically known as peaking filter, that boosts or cuts the mid frequency range. The gain characteristic of a typical peaking filter is shown in Fig. 10.6. In consumer audio systems, the passband, and stopbands of the filters are fixed during the manufacturing process, and hence the users do not have any control. The tone control systems can be implemented by placing the filters in series or parallel connection.

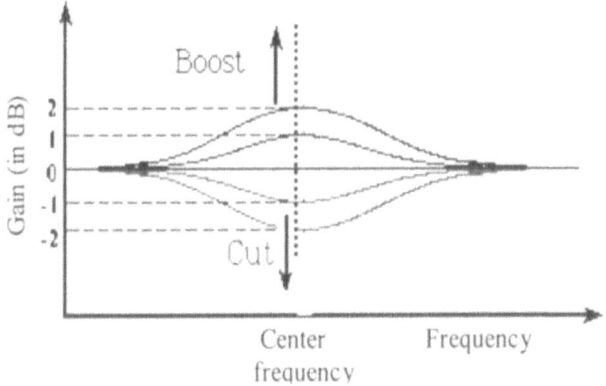

Figure 10.6. Frequency response characteristic of a peaking filter.

Graphic Equalizers

The graphic equalizers are more sophisticated than the tone control systems. The input signal is typically passed through a bank of 5-7 bandpass filters as shown in Fig. 10.7. The output of the filters are weighted by the corresponding gain factors, and added to reconstruct the signal. The filters are characterized by the normalized cut-off frequencies. Therefore, the same filters would work for audio signals with different sampling frequencies.

■ Example 10.2

Consider the audio signal *test44k* whose waveform and spectrum was shown in Chapter 4 (see Figs. 4.1 and 4.2). The sampling frequency of the audio signal is 44.1 kHz. A 5-band equalizer will be designed using a bank of filters. The cut-off frequencies of the 5 bandpass filters are chosen as shown in Table 10.1.

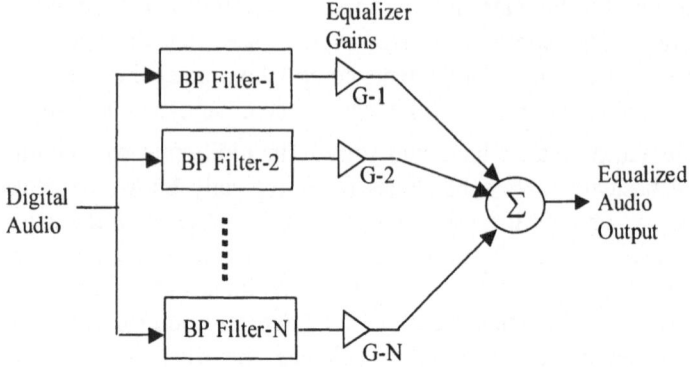

Figure 10.7. Simplified schematic of an equalizer system using a bank of bandpass (BP) filters. The output of each filter is weighted by the gain factors (G-1, G-2,...) set by the listeners.

Table 10.1. Cut-off frequencies of band pass filters

Filter#	Cut-off frequency (in Hz)		Normalized cut-off frequency	
	Lower	Upper	Lower	Upper
1	20	1200	0.0009	0.0544
2	1200	2500	0.0544	0.1134
3	2500	5000	0.1134	0.2268
4	5000	10000	0.2268	0.4535
5	10000	20000	0.4535	0.9070

Note that the frequencies are chosen in accordance with the linear octave spacing, which is appropriate due to the frequency sensitivity of the human ear (see Chapter 2). FIR filters, each of order 32 (i.e., 33-taps), are designed using the MATLAB fir1 function. For example, the third bandpass filter is designed using the following code.

```
bpf(3,:) = fir1(32, [0.1134  0.2268]);
```

Although the audio signal *test44k* contains about 30 seconds of audio samples, the first 18 seconds are considered. The audio signal is processed in blocks of 132,300 samples (*i.e.*, 3 seconds of audio). When the signal is passed through the bandpass filters, the energy of the different bands is as shown in Table 10.2.

In order to demonstrate the equalizer effect, the gains (in dB) of the different bands are selected as shown in Table 10.3. Note that the high frequency bands have lower energy, and the sensitivity of the ear is also low at these frequencies. Therefore, higher gain factors have been selected for these bands.

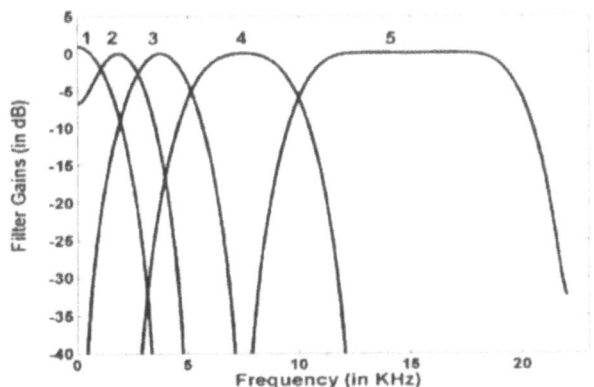

Figure 10.8. Gain response of the bandpass filters

Table 10.2. Energy variation in different bands

Time	Band Energy (in dB)				
(in sec)	Band-1	Band-2	Band-3	Band-4	Band-5
0-3	37.4610	30.8688	15.7790	10.4556	2.1273
3-6	36.1113	29.4799	15.7305	11.6237	6.9943
6-9	37.2630	30.1553	13.9430	9.2472	3.7286
9-12	34.7644	29.2201	16.6713	15.1690	5.3465
12-15	36.5304	30.2963	14.8195	9.1849	2.1991
15-18	37.3523	30.5949	14.2061	8.9226	1.8897

Table 10.3. Energy variation in different bands

Time	Band Energy (in dB)				
(in sec)	Band-1	Band-2	Band-3	Band-4	Band-5
0-3	10	0	0	0	0
3-6	0	10	0	0	0
6-9	0	0	15	0	0
9-12	0	0	0	20	0
12-15	0	0	0	0	20
15-18	10	0	0	0	0

The complete MATLAB code as well as the equalized audio output (in wav format) is provided in the CD. The audio output demonstrates that the different audio bands are boosted as per the selected gain factors. ∎

10.3 AUDIO ENHANCEMENT

In an audio storage and transmission system, the quality of the signal may degrade due to various reasons. For example, a low quality speech production system would produce a poor quality audio. The presence of background interference may cause the degradation of audio quality. The quantization noise introduced during audio compression (see Chapter 7) is another source of degradation. Audio enhancement algorithms can be used to reduce the noise contained in a signal, and improve the quality of the audio signal.

There are several enhancement techniques for noise reduction. Typical methods include:

i) Spectral Subtraction: This technique suppresses noise by subtracting an estimated noise bias found during non-speech activity. This method will be described later in detail.

ii) Wiener Filtering: It minimizes the overall mean square error in the process of inverse filtering and noise smoothing. This method focuses on estimating model parameters that characterize the speech signal. This technique requires a priori knowledge of noise and speech statistics.

iii) Adaptive Noise Canceling: This method employs an adaptive filter that acts on a reference signal to produce a noise estimate. The noise is then subtracted from the primary input. Typically, LMS algorithm is used in the adaptation process. The weights of the filter are adjusted to minimize the mean square energy of the overall output.

In this section, two techniques, namely digital filtering, and SSM, are presented.

10.3.1 Noise Suppression by Digital Filtering

If the noise component in a signal has a narrow spectrum, a straightforward digital filtering can be applied to suppress the noise components. This is illustrated with an example.

■ Example 10.3

Consider the audio signal *noisy_audio1* (the signal in *wav* format included in the CD) shown in Fig. 10.9. The power spectral density of the signal is shown in Fig. 10.10. It is observed that there is a sharp peak at 8 kHz, which is the noise (the noise can be identified easily when played).

In order to suppress the noise, a bandstop filter with a sharp transition band, also known as notch filter, is required. The filter can be designed using the MATLAB function *fir1* as follows.

```
wc =[7800  8200]/11025 ; % Normalized cutoff frequency
filt_bs = fir1(128,wc,'stop') ;   % order-128 filter, 129 tap
```

The gain characteristic of the 129-tap filter is shown Fig. 10.11(a). The power spectral density of the filtered output is shown in Fig. 10.11(b). It is observed that although the signal strength has been reduced at 8 KHZ, the noise is still there (the noise can be heard when the audio file is played).

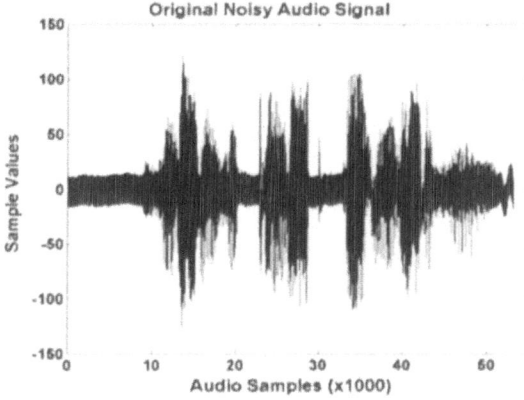

Figure 10.9. Waveform of the audio signal *noisy_audio1*.

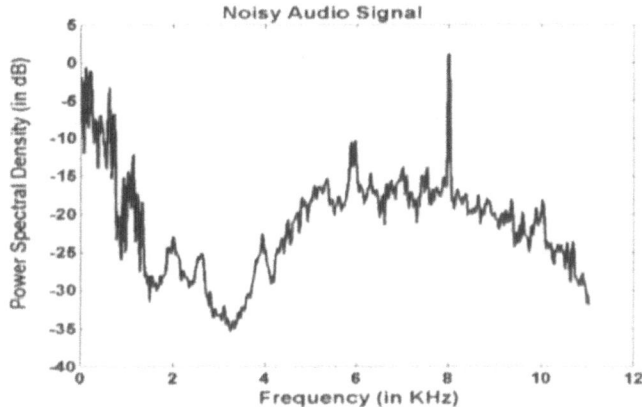

Figure 10.10. Power spectral density of the signal *noisy_audio1*.

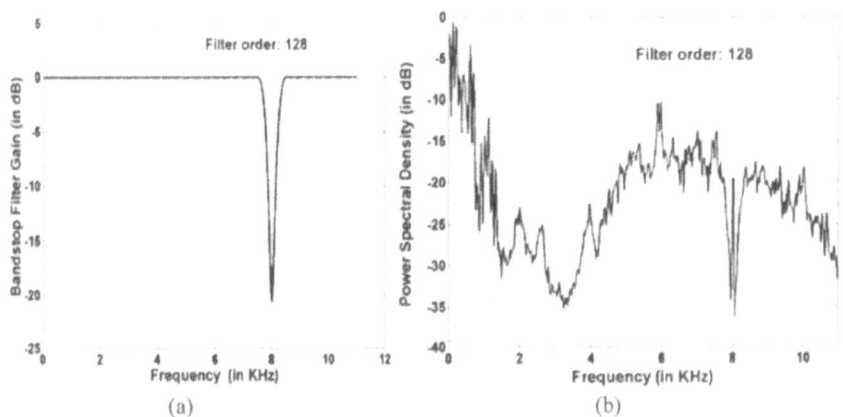

Figure 10.11. Filtering of audio signal. a) Gain response of the filter, and b) Power spectral density of the filtered signal.

In order to suppress the noise further, a 201-tap FIR filter (*i.e.*, filter order is 200) is employed whose gain characteristic is shown in Fig. 10.12(a). The filter has a sharp bandstop attenuation. The power spectral density of the corresponding output is shown in Fig. 10.12(b). It is observed that the noise component has been suppressed completely (unfortunately, the neighboring frequency components also have been suppressed). When the output file is played, the noise can no longer be detected. ■

10.3.2 Spectral Subtraction Method

It is observed in Example 10.3 that if the noise contained in a signal has a sharp spectral distribution, digital filtering can be applied to suppress the noise component. However, in many cases, the noise component has a wide spectral band. In these cases, a simple bandstop filtering may not be appropriate. In this section, another noise suppression technique, known as the spectral subtraction method (SSM), is presented.

The SSM provides a simple and effective approach to suppress stationary background noise. This method is based on a concept that the signal frequency spectrum is expressed as a sum of speech spectrum and noise spectrum. The processing is done entirely in the frequency domain.

Consider a noisy speech signal $f(k)$. The noisy signal can be assumed to consist of a noise-free signal $y(k)$ and an additive noise signal $n(k)$. In other words,

$$f(k) = y(k) + n(k) \tag{10.1}$$

In other words,

$$y(k) = f(k) - n(k) \tag{10.2}$$

Calculating the Fourier transform of both sides of Eq. (10.2), we obtain

$$Y(\Omega) = F(\Omega) - N(\Omega) \tag{10.3}$$

Eq. (10.3) states that if the noise spectrum (both amplitude and phase) is known accurately, then simply by subtracting the noise spectrum from the spectrum of the noisy speech signal, the spectrum of the noise-free signal can be obtained. However, in practice, only an estimation of the noise amplitude spectrum is available. In SSM, the spectrum of the noise-reduced signal is estimated as:

$$Y(\Omega) = F(\Omega)\frac{[|F(\Omega)| - |N(\Omega)|]}{|F(\Omega)|} = F(\Omega) \times \left(1 - \frac{|N(\Omega)|}{|F(\Omega)|}\right) = F(\Omega)g(\Omega) \tag{10.4}$$

where g(Ω) is a parameter to express the amount of noise in the signal. If there is no noise, g(Ω) =1. If the noise level is similar to that of the signal,

$g(\Omega) = 0$. However, in a highly noisy environment, the noise level may exceed the signal level, resulting in a negative signal spectrum that does not make much sense. Hence, in SSM, it is assumed that $g(\Omega) \geq \lambda$ where λ is a threshold in the range $[0.01, 0.1]$.

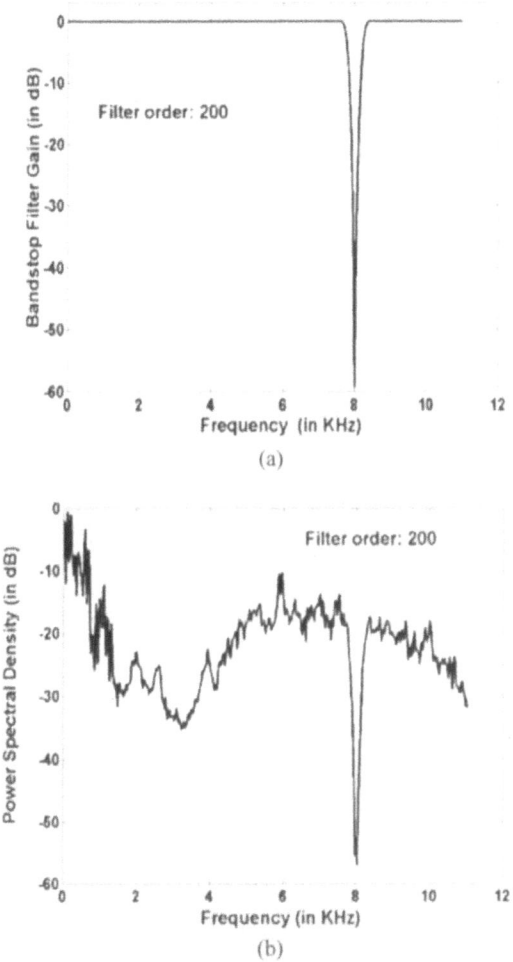

(a)

(b)

Figure 10.12. Filtering of audio signal. a) Gain response of the filter, and b) power spectral density of the filtered signal.

Eq. (10.4) assumes that the estimated amplitude of the signal at a given frequency is the real amplitude of the noisy signal modulated by $g(\Omega)$. The phase of the signal is assumed to be identical to that of the noisy signal. Fig. 10.13 shows the block schematic of the spectral subtraction method, which has three major steps [3]. First, the noise spectrum is estimated when the speaker is silent. Assume that the noise spectrum does not change rapidly.

The noise spectrum is then subtracted from the amplitude spectrum of the noisy input. Using this new amplitude spectrum, and the phase spectrum of the original noisy signal, the time-domain audio is calculated by calculating inverse Fourier transform.

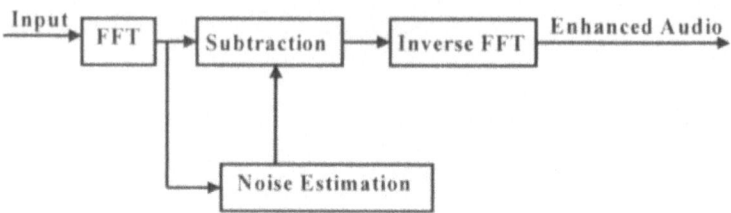

Figure 10.13. Schematic of spectral subtraction method.

■ Example 10.4

The performance of the SSM for suppressing noise is now evaluated with the audio signal *noisy_audio2*. The waveform of a noisy speech signal is shown in Fig. 10.14. The time duration of this test signal is about 2.7 seconds and the sampling frequency is 22,050 Hz. It is observed that there are gaps in the speech waveform, which represents the silence period where the noise can be heard. The power spectral density of the noise is estimated separately, and shown in Fig. 10.15. It is observed that the noise is represented by a wideband of frequencies components. Hence, a direct digital filtering is not appropriate to enhance this signal.

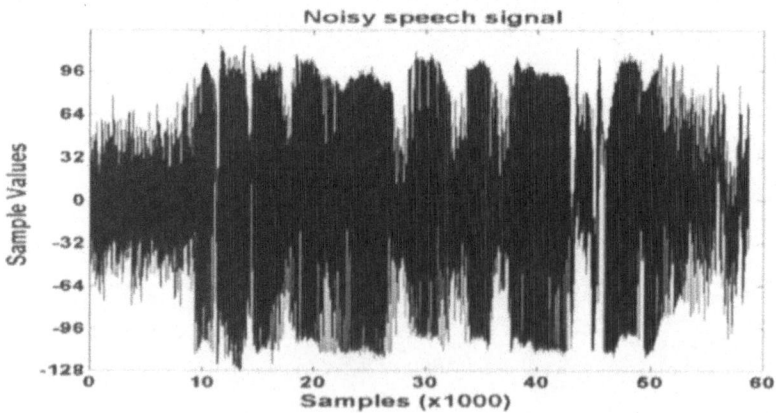

Figure 10.14. Test noisy audio signal.

The SSM is applied to enhance this signal. The gaps in the speech signal are used to estimate the background noise spectrum. A convenient way to do this is by choosing a frame length, short enough to often cover segments of the signal that do not contain speech. Assuming that we have chosen such a frame length. The spectrum of each frame is then calculated. The minimum values of the spectral components corresponding to each frequency bin over the frames are calculated. This minimum value is used as the estimate of the noise. Note that the estimate may be considered as a sub-approximation of the noise. Therefore, it is multiplied by a scale factor to obtain the noise-spectrum that is used in the SSM. For this example, good results are obtained with a scale factor of 20. Since the noise may have time-varying statistics, in some cases, the noise estimation should be restarted from time-to-time.

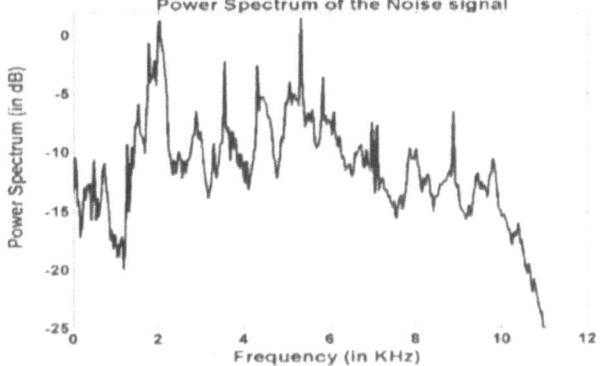

Figure 10.15. Spectrum of the noise.

The MATLAB code for the SSM processing is included in the CD. The audio frame length is chosen to be 512. There is an overlap of 50% between two consecutive audio blocks. The value of λ is set to 0.025. The window function is derived from Hamming window.

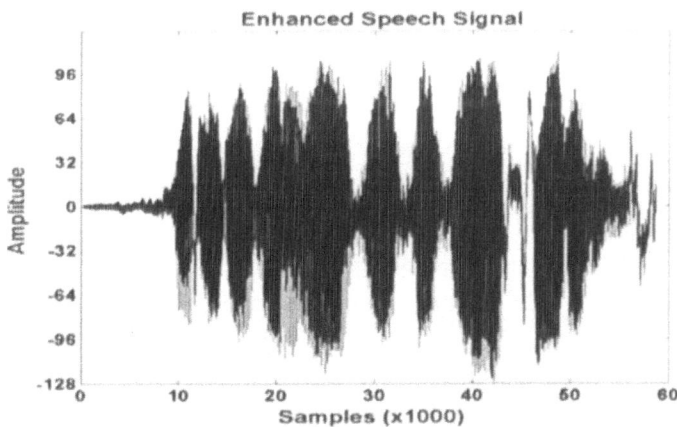

Figure 10.16. The enhanced signal in time domain

The enhanced speech signal is plotted in Fig. 10.16. It is observed that the background noise is significantly reduced. The power spectral density of the original and the enhanced signal is compared in Fig. 10.17. It is observed that the enhanced signal has reduced the high frequency components significantly in the range 2-5 kHz. The enhanced audio file has been included in the CD. It can be verified that the noise has been reduced significantly. ∎

10.4. EDITING MIDI FILES

A few selected audio processing techniques have been presented in the last few sections to enhance the quality of the digital audio signal. It was mentioned in Chapter 2 that MIDI file is becoming popular with music lovers. Unlike digital audio, MIDI is generally noise free data since it is mostly synthesized. Hence, processing techniques such as noise filtering are not applicable to MIDI files. In this section, a few small examples are provided to illustrate the creation and editing of MIDI files.

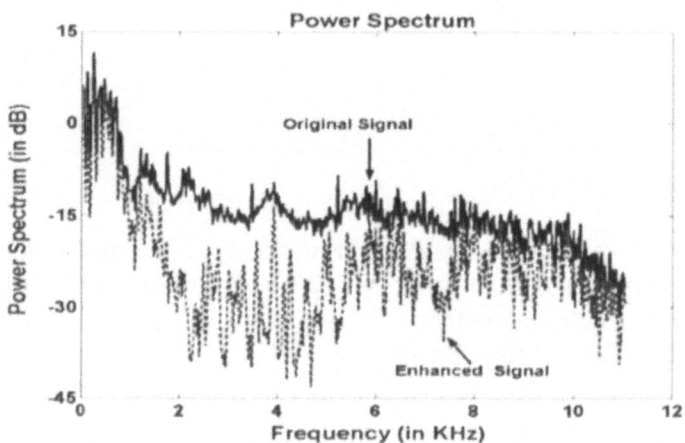

Figure 10.17. Power spectral density (in dB) of the original and enhanced signal

■ **Example 10.5**

In Example 2.2, a small MIDI file was created. The MIDI file would generate the note G3 played on an Electric Grand Piano, with velocity (which relates to volume) 100. In this example, a MIDI file will be created that generates the note E1 played on a Rhodes Piano with a velocity 68.

MIDI File in Example 2.2

```
4D 54 68 64 00 00 00 06 00 01 00 01 00 78 4D 54 72 6B 00 00 00 14
01 C3 02 01 93 43 64 78 4A 64 00 43 00 00 4A 00 00 FF 2F 00
```

As explained in Chapter 2, the third byte in the second line (0x02) corresponds to the Electric Grand Piano (see Table 2.9). In order to create a note for Rhodes Piano, the instrument number has to be changed to 0x04. The 6th byte 0x43 generates note G3. In order to create note E6, this byte should be changed to 0x64 (see Table 10.4 for the decimal equivalents of the piano note numbers). The velocity 100 is due to the 7th and 10th bytes (0x64 each) in the second line. In order to create a volume of 68, the 7th and 10th bytes should be changed to 0x44 from 0x64.

The new MIDI file can now be represented in the hex format as follows:

```
4D 54 68 64 00 00 00 06 00 01 00 01 00 78 4D 54 72 6B 00 00 00 14
01 C3 04 01 93 64 44 78 4A 44 00 43 00 00 4A 00 00 FF 2F 00
```

The MIDI file can be created using the following MATLAB code:

```
data=hex2dbytes('4D546864000000060001000100784D54726B0000001401C
30401936444784A44004300004A0000FF2F00');
fid=fopen('F:\ex10_5.mid','wb');
fwrite(fid,data);
fclose('all');
```

The *ex10_5.mid* file is included in the CD. The midi files *ex2_2.mid* and *ex10_5.mid* can be played to hear the difference between the two files. ■

Table 10.4: Note numbers for Piano. The entries are expressed in decimal format.

Octave	C	C#	D	D#	E	F	F#	G	G#	A	A#	B
-2	00	01	02	03	04	05	06	07	08	09	10	11
-1	12	13	14	15	16	17	18	19	20	21	22	23
0	24	25	26	27	28	29	30	31	32	33	34	35
1	36	37	38	39	40	41	42	43	44	45	46	47
2	48	49	50	51	52	53	54	55	56	57	58	59
3	60	61	62	63	64	65	66	67	68	69	70	71
4	72	73	74	75	76	77	78	79	80	81	82	83
5	84	85	86	87	88	89	90	91	92	93	94	95
6	96	97	98	99	100	101	102	103	104	105	106	107
7	108	109	110	111	112	113	114	115	116	117	118	119
8	120	121	122	123	124	125	126	127				

The previous example demonstrated the procedures for changing the instrument type, note type, and the velocity. The following example illustrates the addition of more tracks to a MIDI file. This would be useful for synthesizing different tracks with different time and tempo signatures, and playing different instruments simultaneously.

■ **Example 10.6**

In this example, a second track is added to the midi file created in Example 10.5 The file test3.mid, with a second track, is displayed in hex format below. Note that the text in parenthesis is only for illustration purposes.

> *(Header chunk)* 4D 54 68 64 00 00 00 06 00 **02** 00 **02** 00 78
> *(Track 1)* 4D 54 72 6B 00 00 00 14
> 01 C3 **02** 01 93 25 64 78 32 64 00 25 00 00 32 00 00 FF 2F 00
> *(Track 2)* 4D 54 72 6B 00 00 00 14
> 01 C3 **03** 01 93 43 64 78 18 64 00 43 00 00 18 00 00 FF 2F 00

When the second track is added to the file, the format is changed to 0x02 (10th byte in the first line) so that each track represents an independent sequence. The number of tracks is also changed to 02 (12th byte in the first line). A few delta-times and notes are changed to illustrate different sounds. The Electric Grand Piano is played in the first track (0x02 in line-3) and the Honky-Tonk Piano is played in the second track (0x03 in line-5). These two tracks can also be played simultaneously if format type is changed to 0x01.

■

10.5 DIGITAL AUDIO AND MIDI EDITING TOOLS

In this section, a brief overview of a few selected freeware/shareware audio and MIDI processing tools are presented. Table 10.5 lists a few audio processing tools. These tools can be used for simple to more complex audio processing. Some of these tools can be downloaded freely from the WWW.

Table 10.5. A few audio processing tools

Name of software	Operating System	Features
Glame	Linux	Powerful, fast, stable and easily extensible sound editor. Freely available from WWW.
Digital Audio Processor	Linux	Freely available from the WWW. Reasonably powerful
Cool Edit	Win95/98/ME NT/2000/XP	Very powerful, easy to use. Capable of mixing up to 128 high-quality stereo tracks with any sound card.
Sound Forge XP Studio 5.0	Microsoft Windows 98SE, Me, or 2000	Sound Forge XP Studio provides an intuitive, easy-to-use interface and is designed for the everyday user.
gAlan	Windows 98/XP/2000	gAlan allows you to build synthesizers, effects chains, mixers, sequencers, drum machines and more.

Sound and Music Editing Tools

Table 10.6 lists a few popular Midi editors that can be used to read, write and manipulate midi files. Major features of these editors are shown in the second column of the table.

Table 10.6. A few selected MIDI Processing Tools

Tools	Features
Cakewalk	It supports general MIDI, and provides several editing views (staff, piano roll, event list) and virtual piano. It can insert WAV files and Windows MCI commands (animation and video) into tracks.
Cubase VST Score	It consists of most features of Cakewalk. It allows printing of notation sheets.
MIDI Maestro	It is powerful, full-featured music software designed for use by amateur and professional music directors, conductors, and musical soloists in live musical theater and similar musical accompaniment situations. Dynamic, intuitive performance control and powerful sequencing and editing capabilities combine to make MIDI Maestro the only music software you need.
Music MasterWorks 3.62	It is a MIDI music composing program. Includes voice to note / wave to midi converter, staff / piano roll notation editor, sheet music printing, chord creation, keeping notes within a key, transpose, quantize, play/record with external MIDI devices, configurable keys, audio, and the ability to ignore the complexities of tracks and channels
DWS_Midi_Son g_Manager 1.0	Midi Song file is an efficient program, with which one can administer MIDI songs together with appropriate texts. One can input titles and interpreter, music direction, lauflaenge of the music pieces, song text, text path and MIDI path.
Pianito MicroStudio 3.0	It is an easy piano synthesizer and drum machine sequencer with 24 tracks (8 percussion, 8 midi instruments, 8 wave FXs). Real-time MIDI and WAVE recording will let you create your own songs in minutes.

REFERENCES

1. J. R. Deller, J. G. Proakis, and J. H. Hansen, *Discrete-Time Processing of Speech Signals*, Ch. 8, Maxwell Macmillan, Toronto, 2000.
2. S. F. Boll, "Suppression of acoustic noise using spectral subtraction", *IEEE Trans. on ASSP*, Vol. 27, pp. 113-120, April 1979.
3. B. P. Lathi, *Signal Processing and Linear System*, Berkeley Cambridge Press, 1998.
4. J. Eargle, *Handbook of Recording Engineering*, New York, Van Nostrand Reinhold, 1992.
5. S. Orfanidis, *Introduction to Signal Processing*, New Jersey, Prentice Hall, 1996.
6. S. Lehman, *Harmony Central: Effects Resources*, www.harmony-central.com/Effects/, 1996.

QUESTIONS

1. Name a few techniques typically used for audio processing.
2. Discuss how different filtering techniques can be used for processing audio signals.
3. Record an audio signal (or find one from the web) with sampling rate of 44100 samples/sec. Filter the signal with lowpass and highpass filters with cut-off frequencies 6000 Hz.
4. Repeat the above problem with bandpass and bandstop filters with cutoff frequencies 4000 Hz (lower cutoff) and 8000 Hz (upper cutoff frequency).
5. What is audio equalization? Which methods are typically used for audio equalization?
6. What is tone control method? Draw the typical frequency characteristics of shelving and peaking filters.
7. What is graphic equalization?
8. Record a music signal (or find one from the web), and repeat Example 10.2. Can you think of ways to improve the performance of the equalization system implemented for Example 10.2?
9. Discuss typical noise suppression techniques in digital audio.
10. When is digital filtering more suitable for noise suppression?
11. What is spectral subtraction method?
12. Find a noisy audio signal from the WWW. Apply both filtering and spectral subtraction methods to suppress the noise, and compare their performance.
13. Create a single-track MIDI file that generates note D6 on Chorused piano. Play it using a MIDI player (such as Windows Media player), and verify if it matches with the expected sound.
14. Change the note and velocity in the MIDI file created in the previous problem, and play the files to observe the effect.
15. Consider Example 10.6. Change the MIDI code such that two tracks are played simultaneously.

Chapter 11

Digital Image and Video Processing

With the growth of multimedia systems, sophisticated image and video processing techniques are now common. A typical multimedia user these days is likely to have several image and video editing tools at his or her disposal. The user can use these tools to perform several image processing tasks, such as image cropping, resizing, contrast enhancement, smoothing, sharpening, and edge detection, without knowing the intricate details of the underlying techniques. In a similar way, a video user can perform several tasks, such as video segmentation, joining multiple video clips with fade in or fade out, and zoom in and out. The purpose of this Chapter is to introduce to readers the working principles of a few select image and video processing techniques. Readers who would like to have a more detailed understanding of these techniques may consult standard textbooks on image and video processing [1]-[5].

11.1 BASIC IMAGE PROCESSING TOOLS

In this section, two frequently used image processing/editing tools are presented, beginning with image resizing techniques, followed by image cropping techniques.

11.1.1 Image Resizing

Image resizing is one of the most frequently used tools. Most current digital cameras produce images with a resolution more than 1600x1200. However, if one would like to display this image on the WWW or send it to a friend, the typical image size should be much smaller. Conversely, one may have a smaller size image and want to enlarge it to see more detail. Image resizing techniques can be employed to perform these tasks.

Image resizing is generally done by interpolating and resampling the pixels values. The most commonly used interpolation methods are

1. Nearest neighbor Interpolation
2. Linear Interpolation

3. Cubic Interpolation

Nearest Neighbor Interpolation

Nearest-neighbor interpolation (NNI), also known as *replication* or *zero-order* interpolation, is the simplest interpolation method. Here, a scan line is assumed to be a stair case function, and the heights of the stairs are assumed to be the amplitude of the pixel.

Figure 11.1 shows four pixels located at {x-1, x, x+1, x+2}, with amplitudes {A, B, C, D} respectively. Assume that we want to determine the amplitude of two pixels located at *m* and *n* on the scan line. The NNI method would simply assign A as the amplitude of pixel m, and C as the amplitude of pixel n. The same principle can be extended in the column direction.

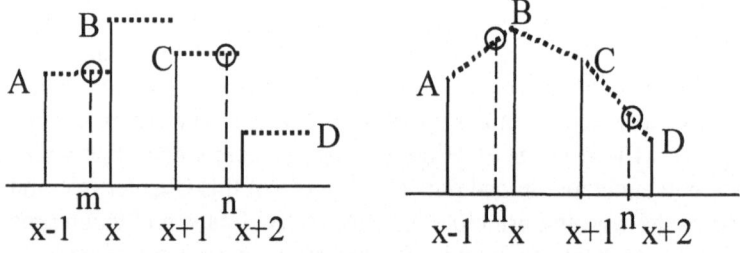

Figure 11.1. One-dimensional interpolation. (a) Nearest neighbor interpolation, and b) linear interpolation. The amplitude of the signal is known at points A, B, C, and D. The amplitude at points m and n is estimated using the interpolation method.

Although the NN method is computationally less complex, unfortunately it can produce significant image artifacts. For example, the straight edges of objects generally appear rough or jagged, which is very annoying; these effects are known as *jaggies* (see Fig. 11.2).

Bilinear Interpolation

In *Linear* Interpolation (LI), also known as first-order interpolation, the amplitude of a pixel is calculated using the weighted average of 2 surrounding pixels in a scan line. The LI method assumes that the intensity of a scan line is a piecewise linear function, and the nodes of the function are the given pixel values. Figure 11.1(b) illustrates the linear interpolation along a scan line. Assume that the amplitude of the pixel at the new location *m* is to be determined, and *m* is at a distance of α from the pixel *x*-1. It can also be said that *m* is at a distance of $1 - \alpha$ from the pixel *x*. The linear interpolation will then produce the pixel value.

$$A(1 - \alpha) + B\alpha \qquad (11.1)$$

The same interpolation principle can be extended to a column scan line. Instead of applying interpolation first row-wise, and then column-wise, 2-D linear interpolation can also performed. Here, a (new) pixel value can be calculated from four immediate neighbouring pixel values. This is known as bilinear interpolation.

■ Example 11.1

Consider the *airplane256* image of size 256x256 provided in the CD. It is difficult to read the F16 number. Hence, the NN and LI interpolation and resampling methods are used to increase the size of the image to 512x512. The interpolated images are shown in Fig. 11.2. Because of the small print area, it is difficult to judge the quality of the interpolation in Figs. 11.(a) and 11.(b). However, the tail region is blown up in Figs. 11.2(c) and (d). The jaggedness can easily be observed in Fig. (c), while the LI method produces a smoother image (Fig. (d)). ■

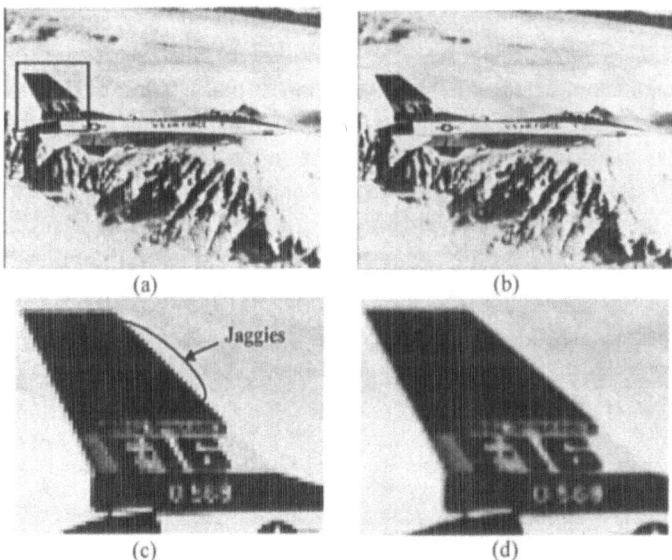

Figure 11.2. Interpolation of the *airplane image*; the image-size is increased from 256x256 to 512x512. (a) Interpolated image using NN method, b) Interpolated image using LI method, c) blow-up of Fig. (a), and (d) blow-up of Fig. (b).

Although the linear interpolation improves the quality of the interpolated image, further quality enhancement is possible with higher order interpolation. The cubic interpolation (second order interpolation) is a very popular interpolation method used in image processing. It uses 16 neighbouring pixels (4 pixels in each direction) for interpolation, and preserves the fine detail in the source image.

11.1.2 Image Cropping

Image cropping is another frequently used tool. In many instances, we are more interested on focusing on a particular region of an image. For example, consider the image in Fig. 11.3. If we would look at the people rather than the lake, image cropping may achieve this.

Figure 11.3. The image *LakeLouise*.

Image cropping is a simple process. The first step is to identify a region to be cropped. A mask can then be generated to select the area with the appropriate shape. There is a variety of masks that can be used. Figure 11.4 shows the rectangular and elliptical masks that are easy to generate. A rectangular mask is defined by four corner points. Since the pixels are arranged row-wise or column-wise, rectangular masking is equivalent to choosing the desired rows and columns of pixels. An elliptical mask on the other hand, is defined by the center and the two radiuses (major and minor). Assume that the center of the mask is (c_x, c_y), and the major (vertical) and minor (horizontal) radiuses are r_{max} and r_{min}, respectively. The mask will then include the pixels located at all (x,y) that satisfy the following condition:

$$\left(\frac{x-c_x}{r_{min}}\right)^2 + \left(\frac{y-c_y}{r_{max}}\right)^2 \leq 1 \qquad (11.2)$$

■ **Example 11.2**

In this example, we crop a few regions marked in Fig. 11.3. The original image size is 588×819 (rows $\times$ columns). For the rectangular image, the four corners of the mask are selected as (60,150), (360,150), (360,550), and (60,550). The image is shown in Fig. 11.5(a). To extract the faces using an elliptical mask, the chosen center pixels are (195,300) and (195,455). The major radius (vertical) and the minor radius (horizontal) are of length 115 and 100 pixels, respectively. The cropped images are shown in Fig. 11.5(b)-(c). ■

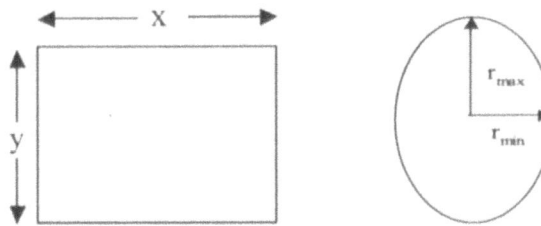

Figure 11.4. Masks for image cropping. (a) rectangular mask, b) elliptical mask.

Figure 11.5. Cropped images with a) rectangular, and b) elliptical, and c) elliptical masks.

11.2 IMAGE ENHANCEMENT TECHNIQUES

Image enhancement refers to accentuation, or the sharpening of image features such as brightness, contrast, edges, and boundaries to make the image more useful for display and analysis.

There are three broad categories of enhancement techniques: point operations, spatial operations, and transform based techniques. *Point operations* techniques perform enhancement on each pixel independently. The *spatial operations* techniques perform enhancement on the local neighborhood, whereas the transform operations techniques perform enhancement in the transform domain.

11.2.1 Brightness and Contrast Improvement

Brightness and contrast, as discussed in Chapter 3, are two important parameters of an image. The brightness reflects the average intensity of the image, and the contrast is the ability to discriminate different regions of an image. A darker image is often produced by poor lighting conditions, while low contrast images occur due to poor lighting conditions as well as the small dynamic range of the imaging sensor.

Two examples images (with 256 gray levels) are shown in Fig. 11.6(a) and (b). The first image looks darker, whereas the second image has a low contrast. The histogram of the images are shown in Figs. (c) and (d). It is observed that the first image has a large number of dark pixels (30% pixels have 0 values) mostly due to the subject's hair. The remaining pixels are distributed mostly in the gray level range [60,190]. On the other hand, the second image looks very bright, as most pixels have values between 200-255.

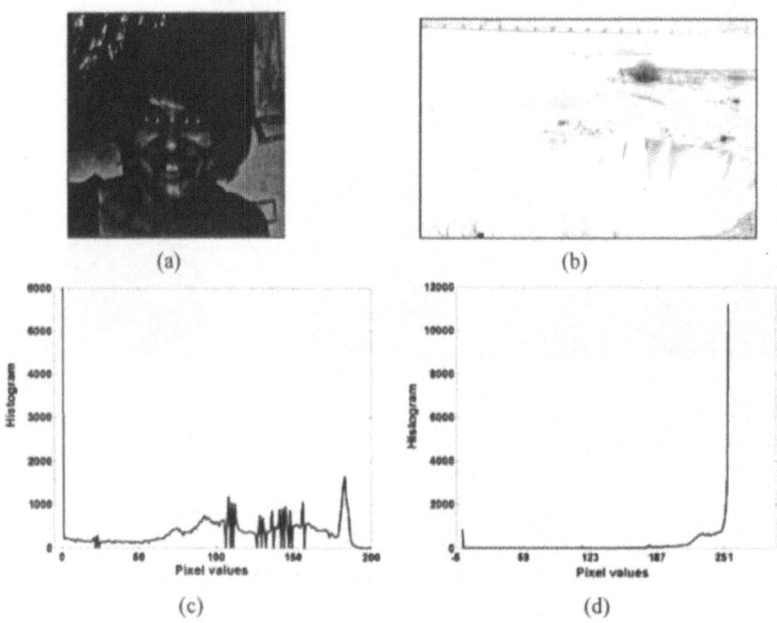

Figure 11.6. Examples of low contrast images. a) *Geeta* image, b) *Niagra* (falls) image, c) histogram of Fig. (a), and d) histogram of Fig. (b). The histogram in Fig. (c) has a sharp peak at the 0th bin, and the full dynamic range of the histogram is not shown.

In this section, two techniques are presented to change the brightness and contrast of an image. Both methods change the histogram of the image. However, this is done in two different ways.

11.2.1.1 Contrast Stretching

Contrast stretching improves an image by stretching its range of pixel intensity to make full use of the possible values. Assume that the pixels have the dynamic range $[0, N-1]$. An input pixel value $f \in [0, N-1]$ is typically mapped to another value $y \in [0, N-1]$. A frequently used contrast stretching transformation is defined as follows [1]:

$$y = \begin{cases} \alpha f & 0 \le f < p \\ \beta(f-p) + y_p & p \le f < q \\ \gamma(f-q) + y_q & q \le f < N-1 \end{cases} \tag{11.3}$$

The transformation is pictorially shown in Fig. 11.7. The horizontal axis represents the gray levels of the input pixels while the vertical axis represents the gray levels of the mapped pixel values. The input pixels values are divided into three groups: $[0,p]$, $[p+1,q]$, and $[q+1,N-1]$. Linear stretching with slopes α, β, γ are applied to the dark, middle, and bright regions, respectively. The slope of the transformation is greater than unity in the region of stretch, and less than unity in the other regions. Note that a dotted line with $45°$ slope is shown in Fig. 11.7. This mapping corresponds to no change in the histogram as the output gray level would always be equal to the input gray levels.

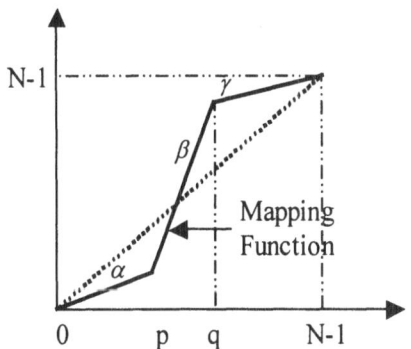

Figure 11.7. Mapping function for contrast stretching

■ Example 11.3

We want to stretch the contrast of the image shown in Fig. 11.6(a). Ignoring the peak at the first bin of the histogram, most of the pixels have gray values in the range [60,190]. If we want to stretch the contrast of the pixels corresponding to these gray levels, we may use the following mapping function.

$$y = \begin{cases} 0.17f & 0 \le f < 60 \\ 1.846(f-60) + 10.2 & 60 \le f < 190 \\ 0.1(f-190) + 250 & 190 \le f < 255 \end{cases} \tag{11.4}$$

The mapping function is shown in Fig. 11.8(a), and the mapped image is shown in Fig. 11.9(a). The output image does have an improved contrast, but the image still looks predominantly dark because the image already had a large number of dark pixels. The mapping function as defined in Eq. (11.4) increased the number of dark pixels further in the first segment, which did not help improve the overall brightness. Let the gray levels be mapped by another mapping function defined as follows:

$$y = \begin{cases} 0.17f & 0 \le f < 60 \\ 1.846(f - 60) + 10.2 & 60 \le f < 190 \\ 0.1(f - 190) + 250 & 190 \le f < 255 \end{cases} \qquad (11.5)$$

The mapping function (Mapping-2) is shown in Fig. 11.8(a). The resulting image is shown in Fig. 11.9(b). The overall brightness has clearly been improved. This is also reflected in the resultant histogram shown in Fig. 11.8(b). It is observed that the mapped image has a large number of high amplitude gray levels. ■

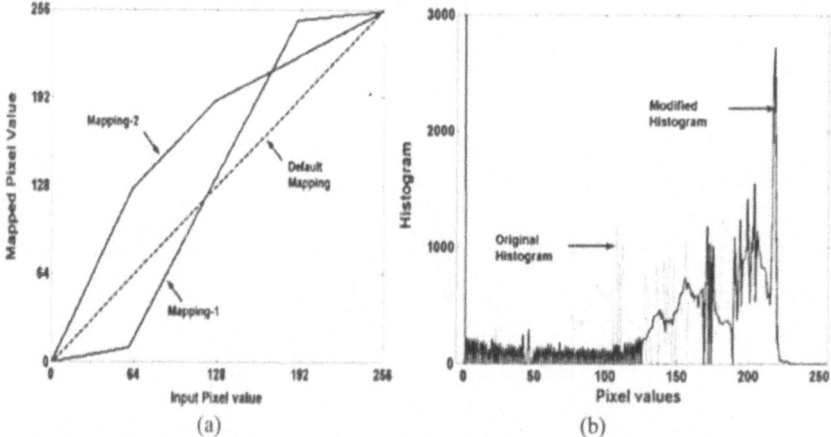

(a)   (b)

Figure 11.8. Contrast stretching of image shown in Fig. 11.6(a). a) Input-output mapping functions, and b) modified histogram with mapping-2.

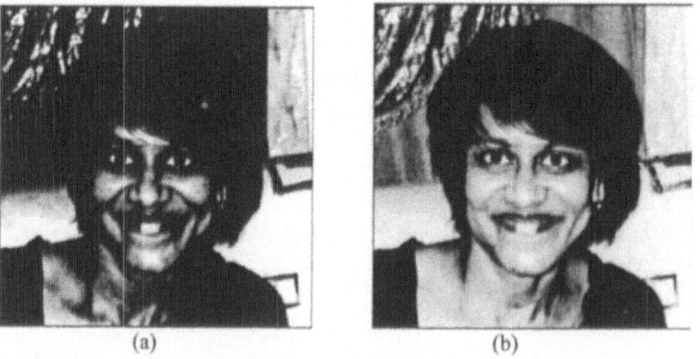

(a)   (b)

Figure 11.9. Contrast stretched images. a) Mapping-1, and b) Mapping-2.

11.2.1.2 Histogram Equalization

The contrast stretching method enhances the image by changing the histogram of the pixel values. However, the major difficulty is to select an appropriate mapping function. *Histogram equalization* is a popular method that selects its own mapping function, and provides good performance in most cases. It employs a monotonic, non-linear mapping function that reassigns the intensity values of pixels in the input image such that the output image contains a uniform distribution of intensities (*i.e.* a flat histogram).

The mapping function used in histogram equalization is an extension of a similar mapping function in continuous-domain mapping. Assume that $p(f)$ is the continuous-time probability density function of input random variable f with a dynamic range [0,1]. Then assume that another random variable y is created (as shown in Fig. 11.10) by applying the transformation $y = T(f)$. It can be shown that the output variable y will have a uniform probability density function if the following mapping is applied [2].

$$y = T(f) = \int_0^f p(w)dw \qquad (11.6)$$

If the above observation is extended to discrete domain, the transformation function becomes

$$y[k] = T(f[k]) = \sum_{j=0}^k p(f[j]) \qquad (11.7)$$

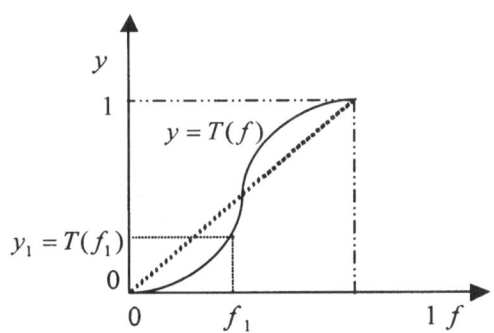

Figure 11.10. Mapping of a continuous probability density function.

It is observed in Eq. (11.5) that the mapping function is identical to the cumulative probability density function, which is also the probability distribution function of the gray levels.

■ **Example 11.4**

Consider again the images in Figs. 11.6(a) and 11.6(b). We want to generate an image whose pixel gray levels will have a uniform density function. The *histeq* function in MATLAB is used to perform the histogram equalization.

image_eq = histeq(inp_image);

The output image and its histogram corresponding to Fig. 11.6(a) are shown in Figs. 11.11(a) and 11.11(c), respectively. Similarly, the output image and its histogram corresponding to Fig. 11.6(b) are shown in Figs. 11.11(b) and 11.11(d), respectively. It is observed that the images have a higher contrast than the original images. The histogram of the equalized images are very close to the uniform histogram. ■

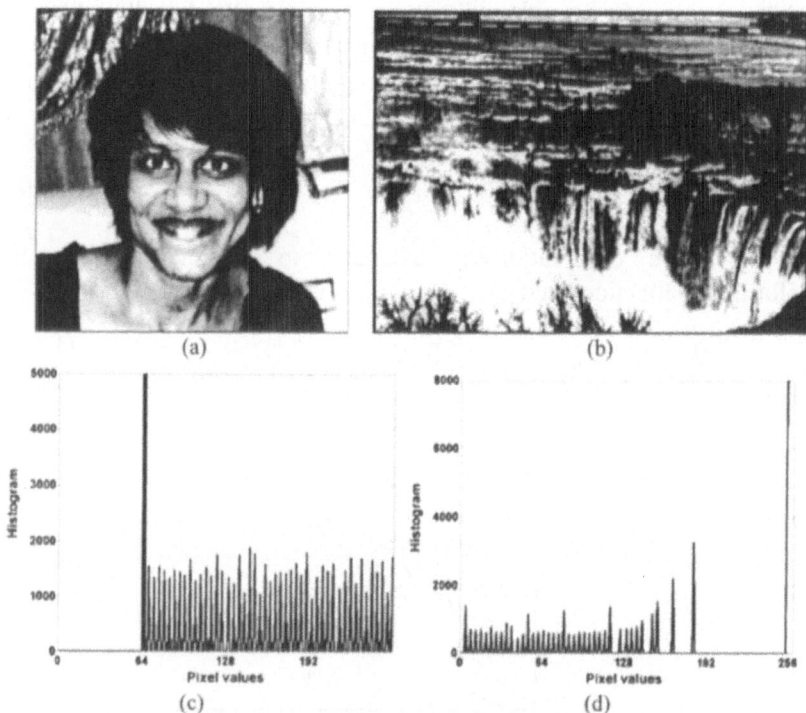

(a)   (b)

(c)   (d)

Figure 11.11. Histogram equalization. a) Histogram equalized image of Fig. 11.6(a), b) Histogram equalized image of Fig. 11.6(b), c) histogram of the image shown in Fig. (a), and d) histogram of the image shown in Fig. (b).

11.2.2 Image Sharpening

Image sharpening is the process of sharpening an image's edges. It is primarily performed by emphasizing the high frequency components.

Sharpening or highpass filtering can be performed in the frequency domain, as well as in the transform domain. Note that image blurring is the inverse process of sharpening, where the image is blurred by low pass filtering.

The impulse response of a few sharpening filters is as follows:

$$F_1 = \begin{bmatrix} 0 & -1 & 0 \\ -1 & 5 & -1 \\ 0 & -1 & 0 \end{bmatrix}, \quad F_2 = 2*\begin{bmatrix} 0 & -1 & 0 \\ -1 & 4.5 & -1 \\ 0 & -1 & 0 \end{bmatrix}, \quad F_3 = 10*\begin{bmatrix} 0 & -1 & 0 \\ -1 & 4.10 & -1 \\ 0 & -1 & 0 \end{bmatrix}$$

Note that the above filters are not strictly highpass filters. These filters can be decomposed into two components: an allpass filter and a highpass filter. For example, F_1 can be decomposed as

$$F_1 = \begin{bmatrix} 0 & -1 & 0 \\ -1 & 5 & -1 \\ 0 & -1 & 0 \end{bmatrix} = \begin{bmatrix} 0 & 0 & 0 \\ 0 & 1 & 0 \\ 0 & 0 & 0 \end{bmatrix} + \begin{bmatrix} 0 & -1 & 0 \\ -1 & 4 & -1 \\ 0 & -1 & 0 \end{bmatrix} \tag{11.8}$$

The first component outputs the original image, whereas the second component of the filter outputs a highpass image. Overall, a high frequency emphasized output is produced.

■ **Example 11.5**

A blurred Lena image is shown in Fig. 11.12(a). The image is passed through the three sharpening filters F_1, F_2, and F_3 as mentioned above. The sharpened images are shown in Figs. 11.12(b)-(d). It is observed that Fig. 11.12(d) has the most sharpened edges among the four images since its corresponding high frequency component is strongest. However, it also has maximum noise (the highpass filter also produces noise). Hence, there is a trade-off between the edge sharpening and the noise content. ■

11.3 DIGITAL VIDEO

Digital video is one of the most exciting media to work with, and there are numerous types of digital video processing techniques. The video compression techniques, which provide a high storage and transmission efficiency for various applications, were introduced in Chapter 9. In this section, we focus on the content manipulation in digital video.

A digital video is a sequence of frames that are still pictures at specific time instances. Consider a digital movie of 1-hour duration with a rate of 30 frames/sec. There would be 108000 frames. In order to perform a content-based search, the individual frames may be grouped into *scenes* and *shots*, as shown in Fig. 11.13. A *shot* is defined as a sequence of frames generated during a continuous operation and representing a continuous action in time and space. A *scene* consists of one or more shots not necessarily contiguous in time. For example, it is possible to have two shots of the same scene, one in the morning and the other in the afternoon. Consider the video sequence shown in Fig. 9.2 (in Chapter 9). The video consists of three shots (and also three scenes), with the shots having 34, 38 (approximately), and 40 (approximately) frames, respectively.

Figure 11.12. Image sharpening performance of different filters. a) image to be sharpened, b) output image produced by the filter F_1, c) output image produced by the filter F_2, and d) output image produced by filter F_3.

Each shot in a video sequence consists of frames with different scenes. There are two ways by which two shots can be joined together – i) abrupt transition (AT), and ii) gradual transition (GT). In AT, two shots are simply concatenated, while in the GT additional frames may be introduced using editing operations such as fade in, fade out or dissolve.

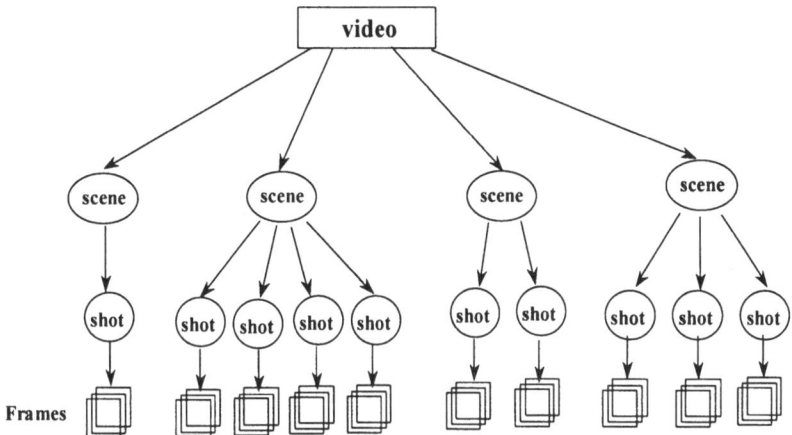

Figure 11.13. Video sequence can be grouped in terms of scenes and shots. In order to increase the granularity further, the frames can be divided into regions or objects as suggested by the MPEG-4 standard.

11.3.1 Special Effects and Gradual Transition

There is a wide variety of techniques for introducing special effects while combining two shots at the boundaries [5, 6] with gradual transition. These techniques can be primarily divided into three categories: wipe, dissolve, and fade.

In *wipe*, the images from one shot are replaced by the images from another shot, separated by a border that moves across or around the frame. There are different types of wipes depending on how the separated border is moved across the frame. In *dissolve*, the images of one shot are gradually replaced by the images of another using a weighted average of two boundary frames. In *fade*, a video shot is gradually converted to a black (or white) image, or a video frame is obtained gradually from a black (or white) frame. These three techniques are presented in more detail in the following sections.

11.3.1.1 Wipe

As mentioned above, the transition from the last frame (I_1) of the first shot to the first frame (I_2) of the second shot is obtained by a moving border as if somebody is wiping one frame and replacing it with another frame. Several wiping directions are possible such as left-to-right, right-to-left, and top-to-bottom. In left-to-right mode, the wiping starts at the left boundary of I_1, whereas in the right-to-left mode the wiping starts at the right boundary of I_1.

Assume that N_R and N_C are the number of rows and columns of the video frames, and we want to insert K wipe frames between the two boundary frames. The *r-th* transition frame in the left-to-right wipe mode can be calculated from the following equation:

$$W_r^{LR} = I_2\left(1:N_R,1:\frac{r*N_C}{K+1}\right) \oplus_H I_2\left(1:N_R,\frac{r*N_C}{K+1}+1:N_C\right) \quad (11.9)$$

where $\oplus_H$ denotes the addition of two image blocks coming from two different frames arranged in the horizontal direction. Note that Eq. (11.9) follows the MATLAB convention for specifying the rows and columns. The schematic of the above operation is shown in Fig. 11.14. Similarly, the wipe operation in the top-to-bottom mode can be expressed as

$$W_r^{TB} = I_2\left(1:\frac{r*N_R}{K+1},1:N_C\right) \oplus_V I_1\left(\frac{r*N_R}{K+1}+1:N_R,1:N_C\right) \quad (11.10)$$

where $\oplus_V$ denotes the addition of two image blocks coming from two different frames arranged in the vertical direction.

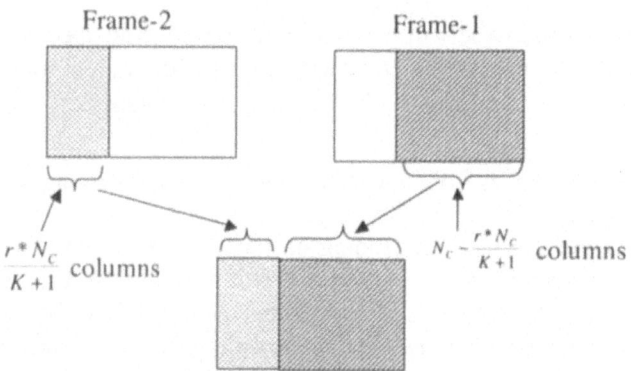

Figure 11.14.Calculation of *r*-th transition frame in the wipe operation (left-to-right). N_R and N_C are the number of rows and columns of the video frames.

■ Example 11.6

Figure 11.15 shows two frames from two different shots of the video sequence shown in Fig. 9.2. Assume that the Fig. 11.15(a) is the last frame of one shot, and Fig. 11.15(b) is the first frame of another shot. We would like to make a transition from shot-1 to shot-2 using the wipe operation.

The wipe operation is carried with K=7, *i.e.*, with seven transition frame. The wipe frames corresponding to the *left-to-right* mode are shown in Fig.

11.16, and the wipe frames corresponding to the *top-to-bottom* mode are shown in Fig. 11.17.

The transition frames have been included in the CD. In order to observe the transition frames in motion, MATLAB *movie* function can be used. ■

Figure 11.15. Two example frames. a) The last frame of the first shot, b) the first frame of the second shot.

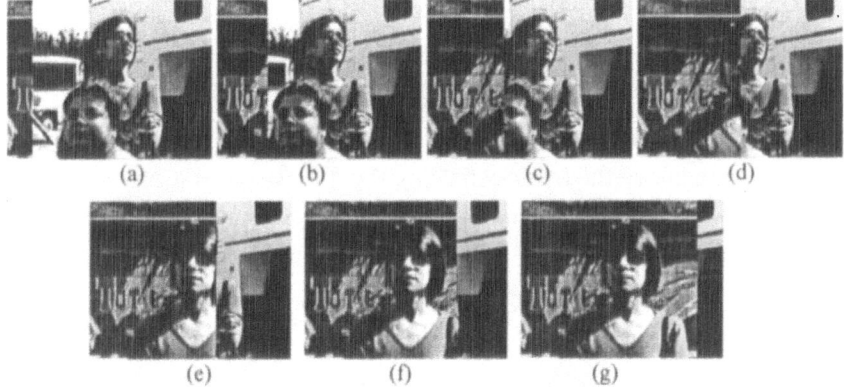

Figure 11.16. Wipe operation from left-to-right. Seven transition frames are shown in Figs. (a)-(g).

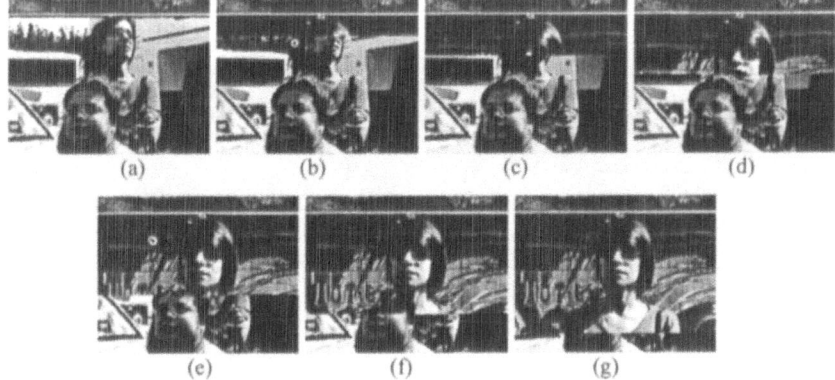

Figure 11.17. Wipe operation from top-to-bottom. Seven transition frames are shown in Figs. (a)-(g).

11.3.1.2 Dissolve

In dissolve operation, the transition frames are calculated using the weighted average of the frames I_1 and I_2. The relative weights of the two frames are varied to obtain different transition frames. Assume again that K dissolve frames would be inserted for gradual transition. The *r-th* transition frame can be calculated using the following equation:

$$D_r(1:N_R,1:N_C) = \frac{K+1-r}{K+1} * I_1(1:N_R,1:N_C) + \frac{r}{K+1} * I_2(1:N_R,1:N_C) \qquad (11.11)$$

It is observed in the above equation that for smaller r, the weight of the I_1 is higher than that of the I_2. Hence, the transition frames look closer to the start frame for smaller r. However, as r increases, the weight of the second frame increases, and the transition frames gradually move closer to I_2.

■ **Example 11.7**

Consider again the boundary frames in Example 11.6. In this example, we would like to achieve the transition using the dissolve operation.

The transition frames, shown in Fig. 11.18, are calculated using Eq. (11.11) with K=8. Note that there is no directivity involved as in wipe operation. All regions are being transformed simultaneously. ■

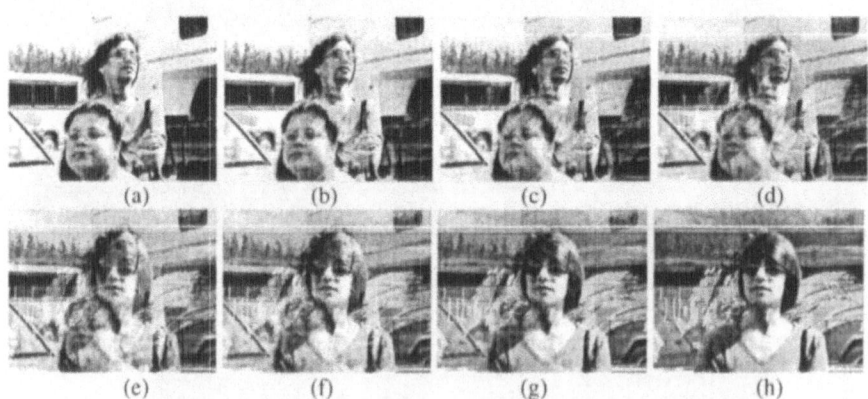

Figure 11.18. Obtaining smooth transition using the dissolve operation. Eight transition frames are shown in Figs. (a)-(h).

11.3.1.3 Fade In/Out

To achieve a gradual transition from one shot to another shot, fade out and fade in operations can be employed in sequence. The fade out and fade in operations can be performed by dissolving the frames as shown in Fig. 11.19.

Let frame-1, frame-2, and the middle frame be denoted by I_1, I_2 and I_M, respectively. Assume that K_1 dissolve frames would be inserted in the fade out, and K_2 dissolve frames would be inserted in the fade in operations.

The frames corresponding to the r-th transition frame for fade out can be calculated using the following equation:

$$F_r(1:N_R,1:N_C) = \frac{K_1+1-r}{K_1+1} * I_1(1:N_R,1:N_C) + \frac{r}{K_1+1} * I_M(1:N_R,1:N_C) \quad (11.12)$$

Similarly, the q-th fade-in frames can be calculated using the following equations:

$$G_q(1:N_R,1:N_C) = \frac{K_2+1-q}{K_2+1} * I_M(1:N_R,1:N_C) + \frac{q}{K_2+1} * I_2(1:N_R,1:N_C) \quad (11.13)$$

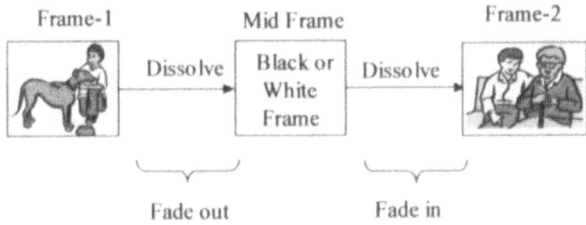

Figure 11.19. Fade out and fade in operations.

■ Example 11.8

Consider the boundary frames in Example 11.6. Here, we would like to achieve the transition using fade in and out operations.

The fade out transition frames, shown in Fig. 11.20(a)-(e), are calculated using Eq. (11.12) with K=5. The fade in transition frames, shown in Fig. 11.20(g)-(k), are calculated using Eq. (11.13) with K=5. Note that the middle frame (shown in Fig. 11.20(f)) is a gray frame, which could also be black or white. Unlike the wipe operation, no directivity is involved in the dissolve operation. All regions are being transited simultaneously. ■

11.3.2 Video Segmentation

In the previous section, we considered the video special effects that are important for creating impressive digital video. In this section, we present video segmentation, which is another important technique in video analysis. The objective of video segmentation is to segment a video into shots and scenes. As mentioned earlier, there may be an AT or GT between two shots. Consider the Beverly video sequence shown in Fig. 11.21. The video with both types of transition. In this section, a brief overview of techniques for AT detection is presented [7, 8].

Table 11.1. Shots with abrupt transition in Beverly Hills sequence

Sequence	Abrupt transition (AT) at frame #
Beverly Hills	13,17,29,46,54,80,88,125,141,155,170,187,200,221,252,269, 282,318,344,358,382,419,435,457

The video segmentation is performed by matching some characteristics of the neighbouring frames [9]. Here, we consider three types of techniques employed for AT detection: pixel intensity matching, histogram comparison, and motion vector comparison.

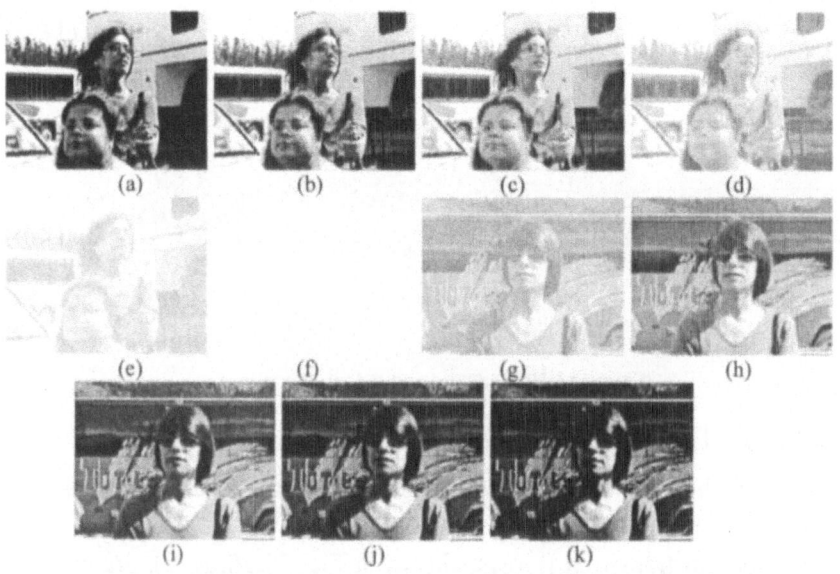

Figure 11.20. Fade out and in operations. Figs. (a)-(e) shows the fade out operation, and Figs. (g)-(k) shows the fade in operations. Fig. (f) shows the middle frame.

Pixel Intensity Matching Method

In this technique, pixel intensities of the two neighboring frames are compared. The absolute distance between the pixel values of frames i and j is calculated using the following equation:

$$PIM(i,j) = \sum_{m=0}^{M-1}\sum_{n=0}^{N-1} |f_i(m,n) - f_j(m,n)| \qquad (11.14)$$

where $f_p(m,n)$ represents the gray level of the p-th frame, with the frame size $M \times N$. If the distance is greater than a pre-specified threshold, then an AT is declared between i and j-th frames.

Although pixel intensity matching is simpler to implement, this method is sensitive to motion, and camera operations which might result in false detection.

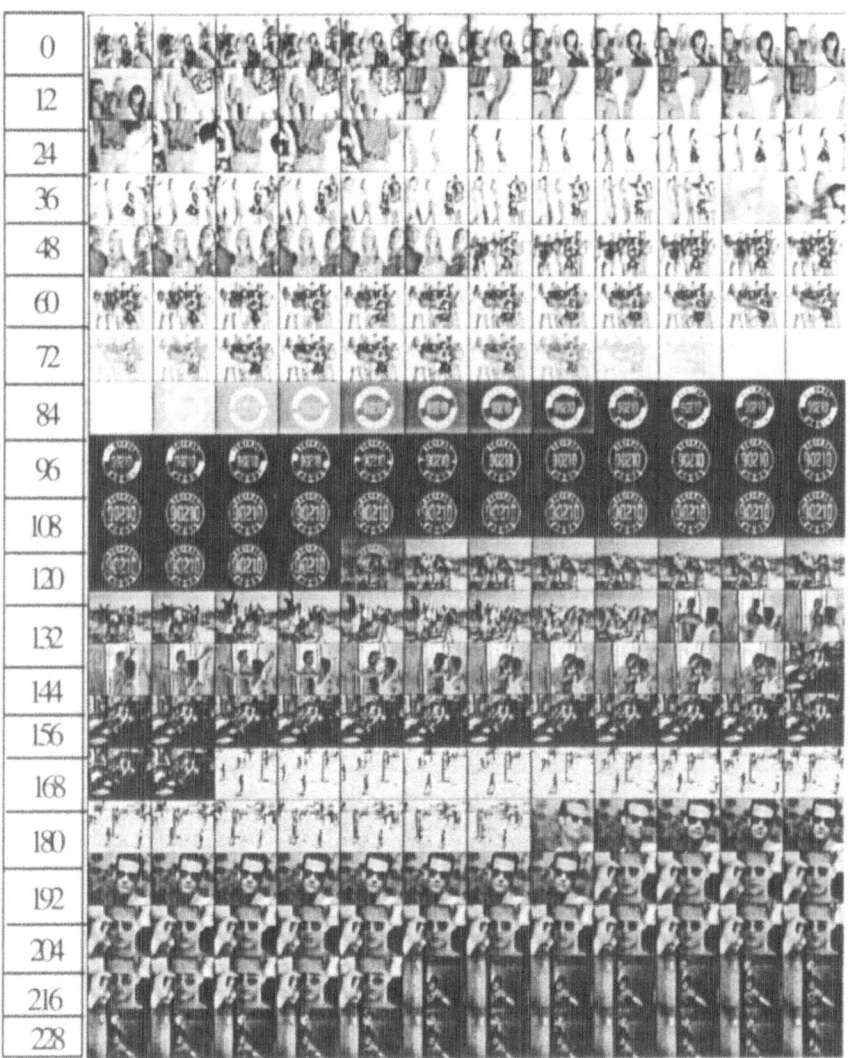

Figure 11.21. Beverly Hills sequence, frames 0-239.

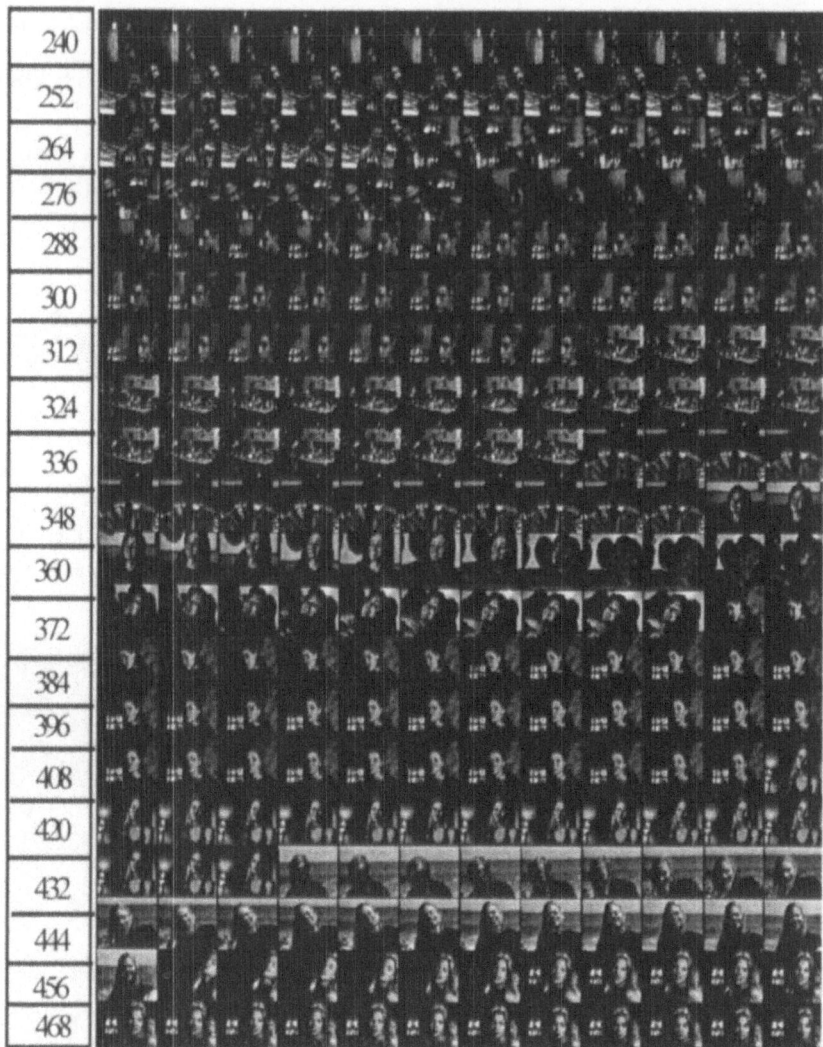

Figure 11.21 (contd). Beverly Hills sequence, frames 240-479.

Histogram-based Method

In this technique, the histograms of consecutive video frames are compared to detect shot changes. The absolute distance between histograms of frames i and j is calculated using the following equation:

$$DoIH(i, j) = \sum_{k=0}^{G} \left| H_i(k) - H_j(k) \right| \qquad (11.15)$$

where H_p is the histogram of the p-th frame, and G is the total number of gray levels in the histogram. If the distance is greater a pre-specified threshold, then an AT is declared.

Although the histogram-based technique provides good performance, it fails when the histograms across different shots are similar. In addition, the histograms within a shot may be different due to changes in lighting conditions, such as flashes and flickering objects, resulting in false detection.

Motion Vector Method

The object and camera motions are important features of video content. Within a single camera shot, the motion vectors generally show relatively small continuous changes. However, the continuity of motion vectors is disrupted between frames across different shots. Hence, the continuity of motion vectors can be used to detect ATs. This is especially useful since most of the coding techniques perform motion compensation and hence, the motion vectors are already available.

Here, we outline a technique suitable in a GOP framework (as used in the MPEG standard) employing I, P, or B frames. The GOP duration in a coded digital video is generally fixed irrespective of the temporal change. Hence, the ATs can occur at frames that can be I-, P-, and B- frames. Note that there are no motion vectors associated with I-frames. On the other hand, P-frames have only forward motion vectors, and B-frames have both forward and backward motion vectors. Hence, the criterion of deciding an AT at I-, P-, or B- frames is different. We define the following parameters that will be useful for making a decision.

For P frames:

$$R_{MP} = \frac{Number\_of\_Motion\_Compensated\_Blocks}{Total\_Number\_of\_Blocks} \tag{11.16}$$

For B- frames:

$$R_{FB} = \frac{Number\_of\_Forward\_Motion\_Compensated\_Blocks}{Number\_of\_Backward\_Motion\_Compensated\_Blocks} \tag{11.17}$$

We assume that AT does not occur more than once between two consecutive reference frames. A frame[n] may be considered a candidate for abrupt transition if the following holds true.

Case-1: frame[n] is an I-frame

    1. The $R_{FB}[n-1]$ (i.e., R_{FB} corresponding to frame (n-1), which is a B-frame) is large.

Case-2: frame[n] is a P-frame

    1. The $R_{MP}[n]$ is very small, and

2. The R_{FB} corresponding to the previous B-frame is large.

Case-3: frame[n] is a B-frame

   1. The $R_{FB}[n]$ corresponding to the current frame will be small, and

   2. If frame (n-1) is B-frame, $R_{FB}[n-1]$ is large, and

   3. If frame (n+1) is P-frame, $R_{MP}[n+1]$ is small.

Note that appropriate thresholds should be selected in the above cases to achieve a good detection performance.

■ **Example 11.9**

Here, the performance of histogram and motion vector methods for AT detection is presented. Note that a technique may correctly detect an AT, may fail to detect an AT (*i.e.*, missed AT), or may detect an AT that does not exist (*i.e.*, false AT). An ideal video segmentation technique should not provide any false AT (FAT) or missed AT (MAT). However, the detection techniques are generally not perfect and hence the FATs or MATs are almost inevitable. Threshold techniques generally reduce the number of MATs by increasing the number of FATs, and vice versa. Hence, thresholds for each technique should be selected appropriately to balance the number of FATs and MATs.

The difference of histograms between the consecutive Beverly Hills frames is shown in Fig. 11.22. The peaks in the plot correspond to abrupt changes in the histogram that likely occurs because of the scene change. Hence, the peak locations are considered as the AT position. It can be shown that [7] the histogram method produces 20 correct ATs, 5 false ATs, and 4 missed ATs. On the other hand, the motion vector method (with a GOP structure IBBBBPBBBBBPBBBBI) produces 23 correct ATs, 3 false ATs, and 1 missed AT.

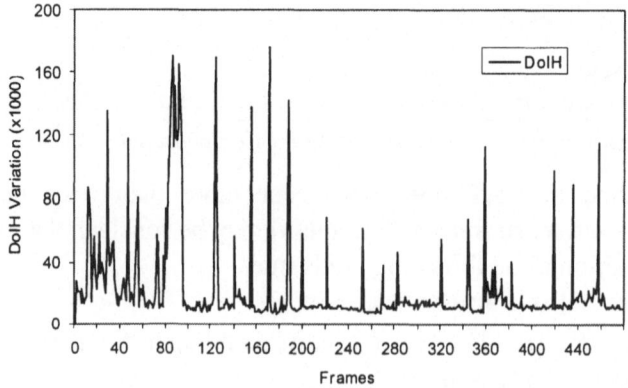

Figure 11.22. Variation of histogram difference between consecutive frames in the Beverly Hills sequence [7].

It is observed that both methods find most of the abrupt transitions. The motion vector method is particularly suitable for joint coding and indexing [7]. The motion vectors are required for video coding any way; hence extra calculation is not required. ∎

In addition to its effectiveness in determining video shots, camera operations are also effective in creating professional quality digital video. Hence, a brief overview of different camera operations is presented in the following section.

11.3.2.1 Camera Operations

Camera operation is an important technique to create professional quality digital video. There are primarily seven basic camera operations: fixed, panning, tracking, tilting, booming, zooming, and dollying [10]. The operations are shown in Fig. 11.21.

In *panning*, the camera is rotated either left or right along its vertical axis. In *tracking*, the camera is moved towards either left or right. In *tilting*, the camera is rotated either up or down along its vertical axis. In *booming*, the camera is moved up or down. In *zooming*, the size of the objects that are pointed at by the camera are magnified (zoom in) or reduced (zoom out) by controlling the camera's focal length. In *dollying*, the size of the objects in a scene are magnified or reduced by moving the camera towards or away from the object. Although both zoom and dolly change the size of the objects, there is a subtle difference between the two effects. In a zoom, the relative positions and sizes of all objects in the frame remain the same, whereas in a dolly shot these will change as the camera moves. The camera operations are generally detected using two spatial domain approaches. The first approach employs optical flow technique, whereas the second approach employs spatiotemporal patterns.

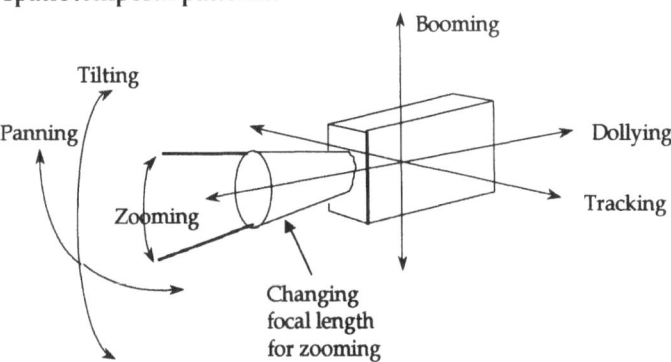

Figure 11.21. Basic camera operations.

Optical flow gives the distribution of velocity with respect to an observer over the points in an image [4]. The motion fields generated by tilt, pan and zoom are shown in Fig. 11.22. It is observed that the motion vectors due to pan and tilt point in one direction. Conversely, the motion vectors diverge or converge from a focal point in the case of zoom. Figure 11.22 shows an ideal situation with no disparity. However, in reality, some disparity might be observed due to irregular object motion and other kinds of noise. Note that the effect of object motion and camera operations on motion vectors are very similar and hence difficult to distinguish.

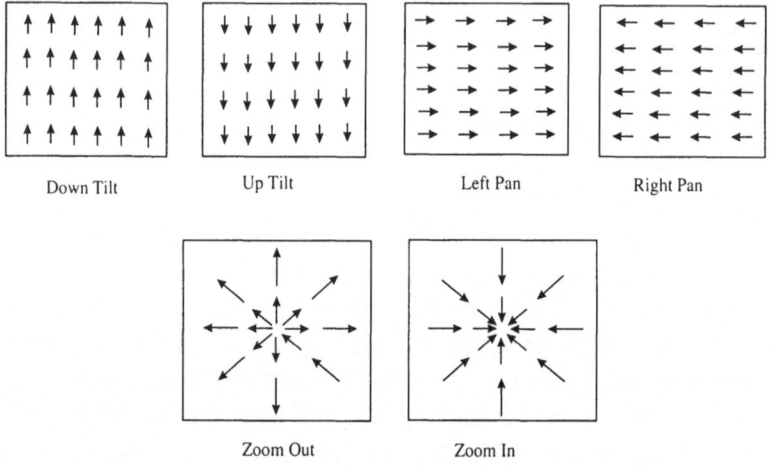

Figure 11.22. Motion vectors produced by tilt, pan and zoom.

11.4 IMAGE AND VIDEO EDITING SOFTWARES

In this section, a brief overview of a few select visual editing/processing tools (as shown in Table 11.1) is presented. The image processing tools are listed in Table 11.1, whereas the video editing tools are listed in Table 11.2. These tools can be used for simple or more complex image and video processing. Some of these tools can be downloaded freely from the WWW.

11.5 SUMMARY

A wide variety of image and video processing algorithms are available in the literature. In the first part of this Chapter, a few selected image processing algorithms – namely resizing, cropping, brightness and contrast improvement, and sharpening – were presented. This was followed by a brief overview of selected video processing algorithms. Here, the focus was again on popular algorithms. Several algorithms for shot/scene transition were discussed. This was followed by an overview of video segmentation techniques, including the effect of camera motions on the motion vector field, and how this effect can be exploited to obtain video segmentation.

Finally, a few image and video editing/processing package tools and their key features were tabulated.

Table 11.1. A few selected freeware/commercial image processing tools.

Software Tools	Operating System	Comments
Microsoft Photodraw	Windows	Provides good graphics, image processing and manipulation tools. Allows layers of images, graphics and text.
Adobe Photoshop	Windows	Provides good graphics, image processing and manipulation tools. Allows layers of images, graphics and text. Includes many graphics drawing and painting tools.
Macromedi a Freehand	Windows/ Unix	Provides text and web-graphics editing tools. Supports many bitmap, and vector formats, and FHC (Shockwave Freehand).
ImageMagi ck	Windows 95/98/2000 ,Mac,VMS, OS/2	Good collection of tools to read, write, and manipulate an image in many image formats. Available free of charge.
XITE	Unix/Wind ows NT/95	Consists of display programs with image widget and graphical user interface. Available free of charge
Gimp	Unix/Wind ows	A versatile photo editing software that is available free of charge.
XV	Unix	Provides basic photo editing tools. Available free.

Table 11.2. A few selected freeware/commercial video editing tools.

Software Tools	Operating System	Features
Adobe Premiere	Windows	An excellent tool for professional digital video editing. It supports Photoshop layers, native Photoshop files, and Illustrator artwork.
VideoWave 5.0	Windows 2000/XP	Excellent video-editing tools. Includes high-quality multimedia bundles; clear, intuitive layout.
Final Cut Pro 3	Mac OS 9/ OS X	Another excellent video editing software. Real-time effects playback on DV without PCI hardware.
VideoStudio 6.0	Windows 98/2000/ ME/XP	Video-editing software that can trim video, add soundtracks, create compelling titles and drop in stunning effects. Easy-to-learn interface

REFERENCES

1. A. K. Jain, *Fundamentals of Digital Image Processing*, Prentice Hall, 1989.
2. R. C. Gonzalez and R. E. Woods, *Digital Image Processing*, Addison Wesley, 1993.
3. W. K. Pratt, *Digital Image Processing*, John Wiley and Sons, Third Edition, 2001.

4.   A. M. Tekalp, *Digital Video Processing*, Prentice Hall, New Jersey, 1995.

5.   B. de Leeuw, *Digital Cinematography*, AP Professional, 1997.

6.   T. Wittenburg, *Visual Special Effects Toolkit in C++* , 1997.

7.   M. K. Mandal, *Wavelet Based Coding and Indexing of Images and Video*, Ph.D. Thesis, University of Ottawa, Fall 1998.

8.   B. Furht, S. W. Smoliar, and H. Zhang, *Video and Image Processing in Multimedia Systems*, Kluwer Academic Publishers, 1995.

9.   F. Idris and S. Panchanathan, "Review of image and video indexing techniques," *Journal of Visual Communication and Image Representation*, Vol. 8, pp. 146-166, June 1997.

10. A. Akutsu *et al.*, "Video indexing using motion vectors," *Proc. of SPIE: Visual Communications and Image Processing*, Vol. 1818, pp. 1522-1530, 1992.

QUESTIONS

1.   Create a gray level image or download one from the WWW. Create two images with size 60% and 140% of the original image in both horizontal and vertical directions.

2.   Select a rectangular and an elliptical region in the image from the previous problem. Crop the selected regions from the image.

3.   Create a dark gray level image or download one from the WWW. Plot the gray level histogram of the image. Using the contrast stretching method, increase the average brightness level of the image. Compare the histograms of the original and contrast-stretched images.

4.   Using the MATLAB "histeq" function, perform the histogram equalization of the image. Compare the histograms of the original and contrast-stretched images.

5.   Install a few image processing tools (several of these tools are downloadable from the WWW) listed in Table 11.1. Compare their features.

6.   Create two images (or download from the web. Calculate seven transition frames corresponding to the right-to-left and bottom-to-top wipe modes. Display the transition frames using the MATLAB *movie* function.

7.   Consider the previous problem. Calculate the transition frames corresponding to the dissolve operations.

8.   Consider the previous problem again. Calculate the transition frames corresponding to the fade out and fade in operations.

9.   What is video segmentation? Describe a few techniques that are typically used to segment a video.

10. Create a digital video sequence or download it from the WWW. Calculate and plot the difference of histograms of the consecutive frames and choose the peaks as the location of the abrupt transitions. Verify manually the correctness of the AT locations determined by the histogram method.

11. Explain the different types of camera motions that are used in a video. Draw the schematic of motion fields for different types of camera motions.

12. Consider a still image, and divide it into small blocks of size 4x4. Transform the image such that the motion vectors between two consecutive frames matches the motion vectors for the zoom operation. Display the image frames using MATLAB movie function.

Chapter 12

Analog and Digital Television

Television has been the most important communication medium in the twentieth century, and will remain important in near future. As mentioned in Chapter 1, current television systems are not truly multimedia systems, since there is no interactivity and independence between different media. However, with the advent of digital television and WebTV, the scenario is likely to change in the near future. In this Chapter, our focus is the principles of digital and high definition television. However, in order to appreciate the digital television fully, one should know the principles of analog television standards. Hence, we first present an overview of the analog television system, followed by a brief introduction to the digital and high definition television systems.

12.1 ANALOG TELEVISION STANDARDS

The existing television standards, which are analog in nature, have been around for the last six decades. There are three analog television (TV) standards in the world [1]. The NTSC standard is used primarily in North America and Japan, the SECAM system is used in France and Russia, and the Phase Alternation Line (PAL) is used in most of Western Europe and Asia. The working principles of these three systems are very similar, differing mostly in sampling frequency rate, the image resolution, and color representation.

The block schematic of a television system is shown in Figs. 12.1 and 12.2. It typically has three subsystems:

i) TV Camera: For capturing the video.

ii) Transmission System: For transmitting the video from the camera (or TV station) to the TV receiver.

iii) TV Receiver: For receiving and displaying the video.

In a television system, the video signal is captured by the TV camera. The image of a scene is focused onto a two-dimensional (2-D) photoconductive layer inside the camera [2]. The photoconductive layer stores the electric charges proportional to the illumination level of the scene. This charged image is then converted to current using an electric beam before the image is scanned from left-to-right and top-to-bottom until the scanning is complete. The electron beam is then returned to the top, and the entire imaging and scanning process is repeated. Note that the scanning produces a 1-D signal from a 2-D image. Hence, horizontal synchronization signal is inserted (by the sync adder) between the signals corresponding two horizontal lines so that the decoder can distinguish between the two lines. Similarly, a vertical sync signal is inserted between two pictures so that the decoder knows the start of a new picture. The composite video signal is then modulated with a video carrier frequency and the audio signal is modulated with the audio carrier frequency. The two modulated signals are then added to obtain the RF signal that can be transmitted through a terrestrial or cable transmission system.

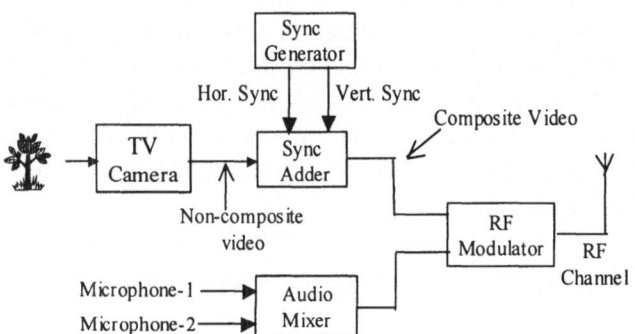

Figure 12.1. Simplified block schematic of a television camera and transmission system.

For a color television system, the color video frames are passed through red, green and blue filters (as shown in Fig. 12.2) to create three 2-D image signals. Three separate electron beams are then employed to scan each color channel. However, in many systems, a single electron beam scans all three channels with time multiplexing.

In the receiver side (see Fig. 12.3), the TV signal is captured by the antenna, and received by an RF tuner. The signal is passed through a video intermediate frequency (I.F.) system that demodulates the signal and separates the video and audio signal. This is followed by the video amplifier stage that separates the luminance and the chrominance signals (for color TV), and provides the horizontal and vertical sync pulses to the sync separator. The sync separator passes these on to the vertical and horizontal deflection systems. The video amplifier produces three color channels (R, G, and B). These signals are then fed to the corresponding electron guns in the TV receiver, which emit beams of electrons whose vertical and horizontal positions are controlled by the horizontal and vertical deflection systems. The electron beams corresponding to the R, G, and B channels hit the corresponding red, green and blue dots on the phosphor screen (more details about the display device is provided in Chapter 15). The phosphor screen then emits light and we see the moving pictures.

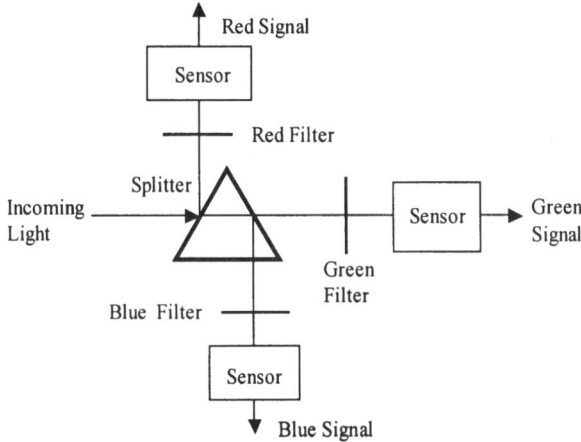

Figure 12.2. Simple color television camera. The camera image is split by three filters. Red, green and blue video signals are sent to three primary color displays where the images are combined.

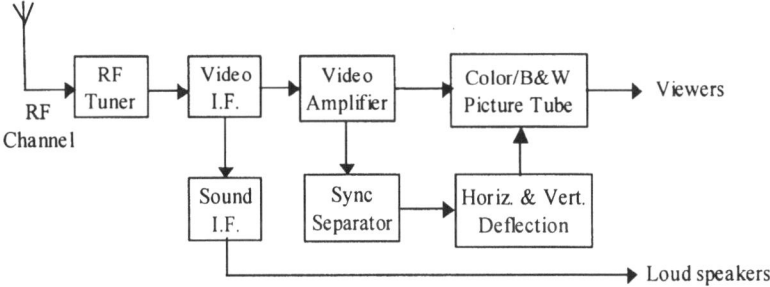

Figure 12.3. Simplified block schematic of a television receiver.

12.2 RASTER SCANNING

We have noted that the 2-D images are scanned and converted to a 1-D signal for transmission. There are primarily two scanning methods: *progressive* and *interlace*. Figure 12.4 explains these two scanning methods. Progressive scanning is generally used in computer monitors with a frame rate of at least 75 frames/second. In interlace scanning, each video frame is scanned in two steps known as *fields*. Figure 12.4(a) explains the interlace scanning for eight scan lines. In the first field, only odd lines, denoted by line# [1,2,3,4], are scanned. In the second field, only even lines, denoted by line numbers [5,6,7,8], are scanned.

The bandwidth allocated for each NTSC TV channel is 6 MHz. This limits the frame rate to 30 frames/sec. If we use progressive scanning as done in a computer monitor, we will observe unacceptable flicker with 30 frames/sec. Hence, interlaced scanning is employed in the TV system. The scanning process for NTSC television camera and receivers is shown in Fig. 12.4 (b), and the detailed scanning parameters [3] are shown in Table 12.1. It is observed that there are 512 horizontal scan lines in the NTSC television system. The scanning of each frame (*i.e.*, each image) is performed in two steps (known as *fields*). In the first field, only odd lines, denoted by line numbers [1,2,3,...262], are scanned. In the second field, only even lines, denoted by line# [263,264,...525] are scanned. This provides an effective rate of 30 frames/sec or 60 fields/sec. The picture quality improves because our eyes are fooled by the interlaced scanning, since the fields are refreshed 60 times a second. Typical sampling characteristics [3] of the television are shown in Table 12.2. It is observed that the sampling characteristics generally satisfy the Nyquist sampling criterion in the spatial coordinates.

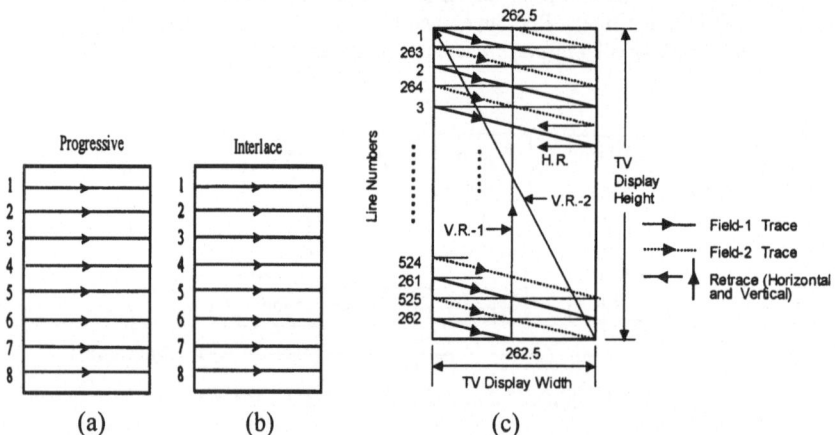

Figure 12.4. Progressive and Interlace scanning, a) Progressive scanning for 8-line display, b) interlace scanning for 8-line display, and c) Interlace scanning for NTSC television.

12.3 COLOR SPACE FOR TV TRANSMISSION

Note that the TV cameras generate the video data in $\{R, G, B\}$ color space. The data from each color channel could be scanned and transformed into 1-D sequence and sent independently. However, in practice, the video is transformed from $\{R,G,B\}$ to $\{Y,C_1,C_2\}$ color space where Y is known as the luminance channel and C_1 and C_2 are the two channels representing the chrominance signals. This conversion is primarily for two reasons – i) to make the color TV signal compatible with the monochrome (black and white) television, and ii) to save the bandwidth for the television transmission channel.

Table 12.1. Parameters for NTSC and PAL broadcast television.

Parameters	NTSC	PAL
Lines/frame	525	625
Active lines/frame	485	575
Aspect ratio (width/height)	4/3	4/3
Frames/second	30	25
Line interlace	2:1	2:1
Line frequency	15750 Hz	15625 Hz
Nominal video bandwidth	4.2 MHz	5.5 MHz
Line period	63.5 μs	64 μs
Active line period	52.1-53.3 μs	51.7-52.2 μs
Line blanking interval	10.2-11.4 μs	11.8-12.3 μs
Synchronizing pulse width	4.19-5.7 μs	4.5-4.9 μs
Field period	16.67 ms	20 ms
Active field period	15.3-15.4 ms	18.39 ms

Table 12.2. Sampling characteristics of broadcast television.

Parameters	NTSC	PAL
Angle subtended at eye (width x height)	$12.7^0 \times 9.5^0$	$12.7^0 \times 9.5^0$
Picture size (samples wide x samples high)	443 x 485	572 x 575
Hor. Sampling frequency (samples/deg)	35	45
Ver. Sampling frequency (samples/deg)	51	61
Temporal sampling frequency (samples/sec)	60	50

There are two types of TV receivers – monochrome and color receivers. When color video is transmitted, it is expected that both color and monochrome receivers would be able to display the signal, although the monochrome receiver will display only the intensity image. Even though, $\{R,G,B\}$ colors can be employed for TV transmission, they cannot be used directly by the monochrome receiver. In addition, the bandwidth required for $\{R,G,B\}$ transmission is approximately three times of one color channel. It is possible to save bandwidth by employing a more efficient color space. Hence for NTSC transmission, $\{Y,I,Q\}$ color space is used for transmission while $\{Y,U_I,V_I\}$ color space is used for the PAL system.

Let us assume that a video camera produces normalized component signals $\{R_N, G_N, B_N\}$. Here, normalization means that $R_N = G_N = B_N$ will produce the reference white C (Fig. 3.18 in Chapter 3) for the NTSC system, and reference white D_{6500} for PAL and SECAM systems. The luminance signal Y is generated using the following equation in all TV systems.

$$Y = 0.299R_N + 0.587G_N + 0.114B_N \tag{12.1}$$

However, the two color channels are generated differently in NTSC, PAL and SECAM systems. The generation of the chrominance signals in different TV systems is presented below.

12.3.1 NTSC System

In the NTSC system, the two color channels are generated as follows:

$$U_I = \frac{B_N - Y}{2.03} \tag{12.2a}$$

$$V_I = \frac{R_N - Y}{1.14} \tag{12.2b}$$

Note that for the reference white, $U_I = V_I = 0$ since $R_N = B_N = Y$. Because most real world scenes contain pastel and saturated colors, the chrominance signals $\{U_I, V_I\}$ will be much smaller than the luminance signals. In order to reduce the bandwidth further, a $33°$ rotation is carried out to produce $\{I, Q\}$ signals.

$$I = V_I \cos 33° - U_I \sin 33° \tag{12.3a}$$

$$Q = V_I \sin 33° + U_I \cos 33° \tag{12.3b}$$

The angle $33°$ has been chosen to minimize the bandwidth requirement for the Q signal. In summary, the $\{Y, I, Q\}$ components for the NTSC TV transmission system can be calculated from the normalized camera output using the following matrix multiplication:

$$\begin{bmatrix} Y \\ I \\ Q \end{bmatrix} = \begin{bmatrix} 0.299 & 0.587 & 0.114 \\ 0.596 & -0.274 & -0.322 \\ 0.211 & -0.523 & 0.312 \end{bmatrix} \begin{bmatrix} R_N \\ G_N \\ B_N \end{bmatrix} \tag{12.4}$$

The I and Q signals are quadrature amplitude modulated according to the following relationship:

$$C(t) = I(t)\cos(2\pi f_c t + 33°) + Q(t)\sin(2\pi f_c t + 33°)$$

where f_c is the color sub-carrier frequency equal to 3.58 MHz.

The envelope of the chroma signal $C(t)$ relative to the luminance signal $Y(t)$, i.e. $\sqrt{I^2+Q^2}/Y$ approximately provides the saturation of the color with respect to the reference white. On the other hand, the phase of the $C(t)$, i.e.,

$$\tan^{-1}\frac{Q(t)}{I(t)}$$

approximately provides the hue of the color. In addition to the decreased bandwidth, there is another advantage of using the $\{I,Q\}$ signal. The chroma signal is generally distorted during the transmission process. However, the HVS is less sensitive to saturation distortion than the hue distortion.

The composite video signal for the NTSC transmission is formed by adding the luminance and the chroma signal as follows:

$$V_{NTSC}(t) = Y(t) + C(t)$$

The $\{I,Q\}$ coordinates are shown in Fig 12.5 [4]. The choice of the $\{I,Q\}$ coordinates relates to the HVS characteristics as a function of the field of view and spatial dimensions of the objects being observed. The color acuity decreases for small objects. Small objects, represented by frequencies above 1.5 MHz, produce no color sensation at the HVS. Conversely, intermediate sized objects, represented by frequencies in the range 0.5-1.5 MHz, are observed satisfactorily if reproduced along the orange-cyan axis. Large objects, represented by frequencies 0-0.5 MHz, require full three colors for good color reproduction. The $\{I,Q\}$ coordinates have been chosen accordingly to achieve superior performance.

12.3.2 PAL System

As in the NTSC system, Eqs (12.1)-and (12.2) are used to generate signals for the PAL TV transmission. The transformed signal can be represented as

$$\begin{bmatrix} Y \\ U_t \\ V_t \end{bmatrix} = \begin{bmatrix} 0.299 & 0.587 & 0.114 \\ -0.147 & -0.289 & 0.436 \\ 0.615 & -0.515 & -0.100 \end{bmatrix} \begin{bmatrix} R_N \\ G_N \\ B_N \end{bmatrix}$$

Note that the chrominance signals $\{U_t,V_t\}$ are not rotated in the case of the PAL system (see Fig. 12.6). As a result, the $\{U_t,V_t\}$ requires a larger bandwidth. In practice, each chrominance channel requires approximately 1.5 MHz bandwidth.

$$C(t) = U_t\cos(2\pi f_c t) + V_t\sin(2\pi f_c t)$$

$$V_{PAL}(t) = Y(t) + C(t)$$

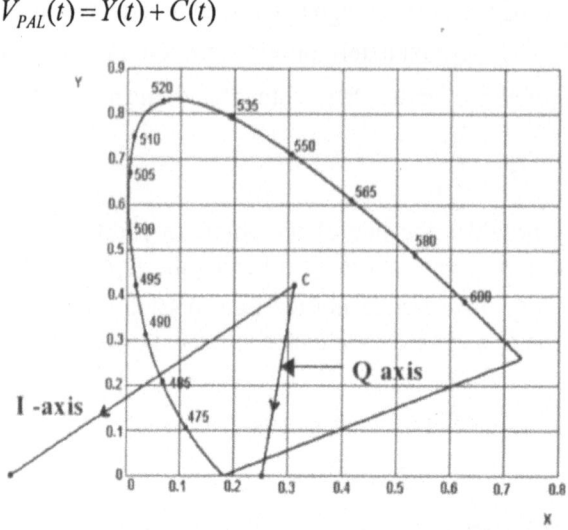

Figure 12.5. Color coordinates used for NTSC TV transmission system. The color difference coordinates $\{U_I, V_I\}$ are rotated to generate the $\{I, Q\}$ signals.

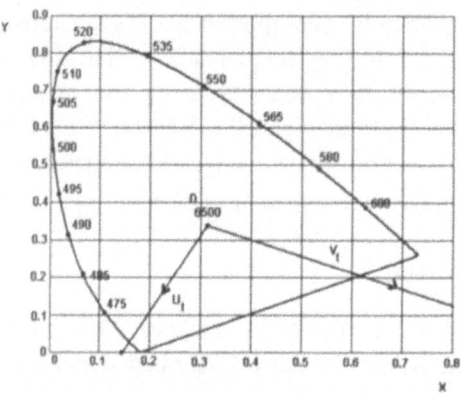

Figure 12.6. Color coordinates used for PAL and SECAM TV transmission systems.

■ Example 12.1

In this example, the advantage of $\{Y, I, Q\}$ and $\{Y, U, V\}$ channels is demonstrated. The original color Lena image is represented in $\{R, G, B\}$ space, and the color image is decomposed to $\{Y, I, Q\}$ and $\{Y, U, V\}$ components. The average energy of each component is shown in Table 12.3. It is observed that the energy is distributed among the R, G, and B components. However, in $\{Y, I, Q\}$ and $\{Y, U, V\}$ color space, most of the energy is compacted in the Y component. Figure 12.7 shows the Y, I, and Q images. It is observed that I and Q components have a smaller magnitude. The histogram of the different components is shown in Fig. 12.8. The

MATLAB code used to obtain these results is included in the CD. The readers can expect similar results with most images.

Table 12.3. Energy compaction ability of different color system for Lena image. The entries are the percentage of the total energy in each component. Signals are made zero mean before calculating the energy.

	First component (R/Y)	Second component {G/I/U}	Third component {B,Q,V}
RGB	37.82	43.95	18.24
YIQ	85.40	11.04	3.56
YUV	87.36	5.40	7.24

The entropy of the different color components is shown in Table 12.4. It is observed that {R, G, B} components have a higher entropy than the components of {Y, I, Q} and {Y, U, V} systems. ∎

(a)   (b)   (c)

Figure 12.7. The $\{Y,I,Q\}$ components of the color Lena image. a) Y component, b) I component, c) Q component.

12.4 NTSC TELEVISION SYSTEM

So far we have discussed various parameters and the color spaces corresponding to the NTSC and PAL color TV systems. In this section, we present the channel assignment, encoder, and decoder for the NTSC system. The principles are very similar for PAL and SECAM, despite some differences.

12.4.1 Channel Assignment

We have noted that the Y channel requires the largest channel bandwidth in a NTSC transmission followed by $\{I,Q\}$ channels. Experimental studies have shown that among I and Q, I channel requires a larger bandwidth than the Q channel. Hence, in NTSC standard, the *I* signal is lowpass filtered to 1.3 MHz, and *Q* signal is lowpass filtered to 0.5 MHz.

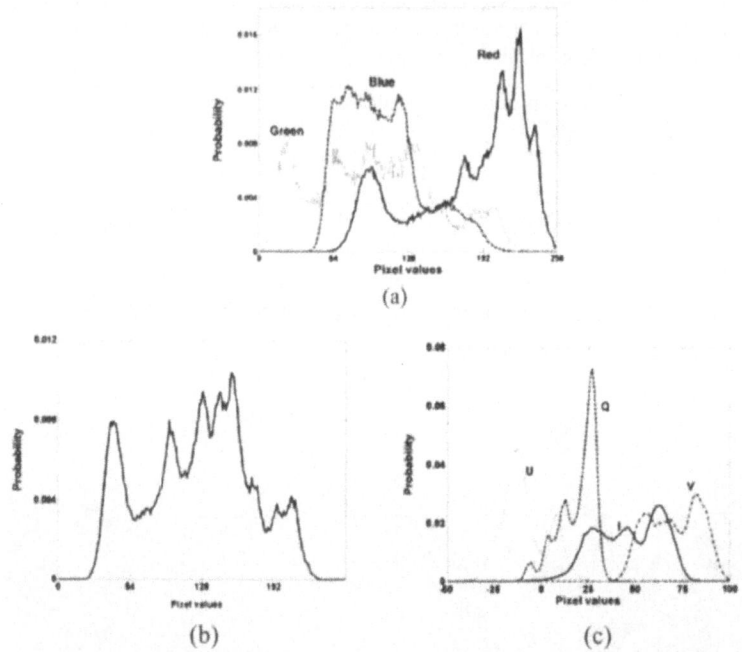

(a)

(b)   (c)

Figure 12.8. Probability density functions of different color components corresponding to the Lena image. (a) {R,G,B}, (b) Y, (c) {I,Q,U,V}. The dynamic range of the {R,G,B} components is greater than that of the {I,Q,U,V} components.

Table 12.4. Entropy of different color components for Lena image. The entropy is expressed in bits/pixel/component

	First compo. (R/Y)	Second compo. {G/I/U}	Third compo. (B, Q, V)
RGB	7.25	7.60	6.97
YIQ	7.45	6.00	5.05
YUV	7.45	5.46	5.69

In composite video, the chroma is superimposed on the luminance signal as shown in Fig. 12.9. The NTSC channels have a typical bandwidth of 6 MHz. The detailed bandwidth assignment is shown in Fig. 12.10. The luminance component is transmitted as amplitude modulation of the video carrier, while the chroma components are transmitted as the amplitude modulation sidebands of a pair of suppressed subcarriers in quadrature. The frequency of the chroma subcarrier is 3.58 MHz above the video carrier at 1.25 MHz. Sound transmission is achieved by frequency modulation. The frequency of the sound carrier is 4.5 MHz above the frequency of the video carrier.

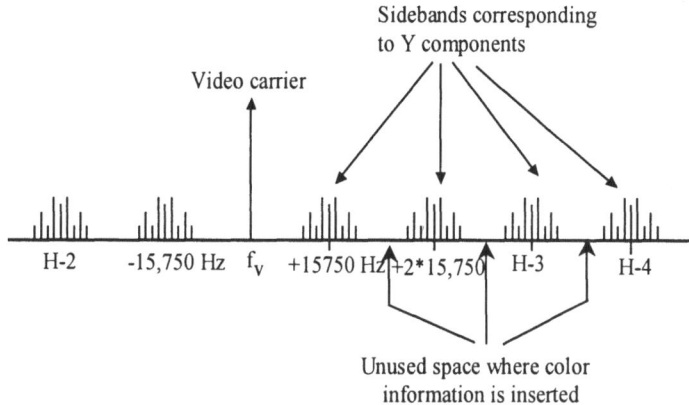

Figure 12.9. Interleaving of luminance and chrominance signals in the frequency domain. The harmonics (H) are 15750 Hz apart in the frequency domain. There is unused space between two harmonics, which can be used to represent chrominance components.

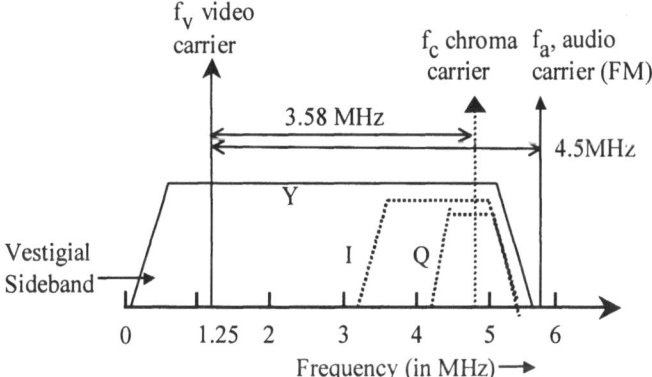

Figure 12.10. NTSC signal spectrum with 6 MHz channel assignment. The Y (luminance) signal is modulated by AM vestigial sideband using a video carrier of 1.25 MHz frequency. The I and Q signals have a double-sided bandwidth of approximately 2.6MHz and 1 MHz, respectively.

12.4.2 NTSC Encoder and Decoder

A simplified block schematic of an NTSC encoder is shown in Fig. 12.11. The $\{R,G,B\}$ signal is converted to a $\{Y,I,Q\}$ signal by a matrix multiplication. The I and Q signals are then filtered using lowpass filters of 1.3 MHz and 0.5 MHz bandwidth, respectively [1]. The filtered I and Q

signals are then combined to produce the chroma signal. In composite video, the chroma signal is superimposed on the luminance signal as shown in Fig. 12.9. As mentioned in the previous section, the NTSC signal has an overall bandwidth of 6 MHz.

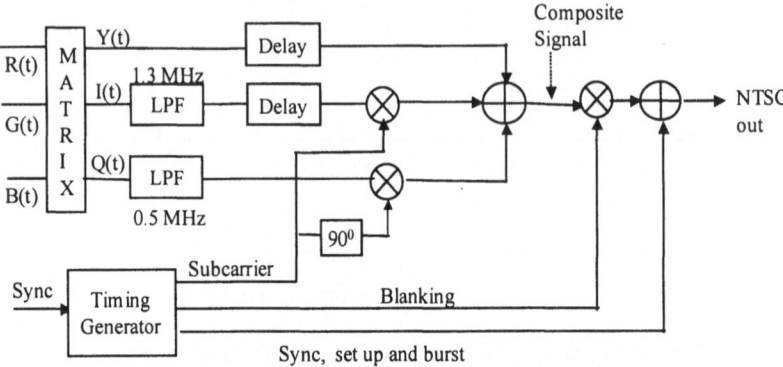

Figure 12.11. A typical NTSC encoder.

Although a chroma signal can be generated using the procedure shown in Fig. 12.11, a simpler method is generally used in practice. The chroma is generated in two stages as shown in Fig. 12.12. In the first stage, $\{Y, R-Y, B-Y\}$ signals are generated from the $\{R, G, B\}$ signals. The $\{R-Y, B-Y\}$ components are then used to produce the chroma signal.

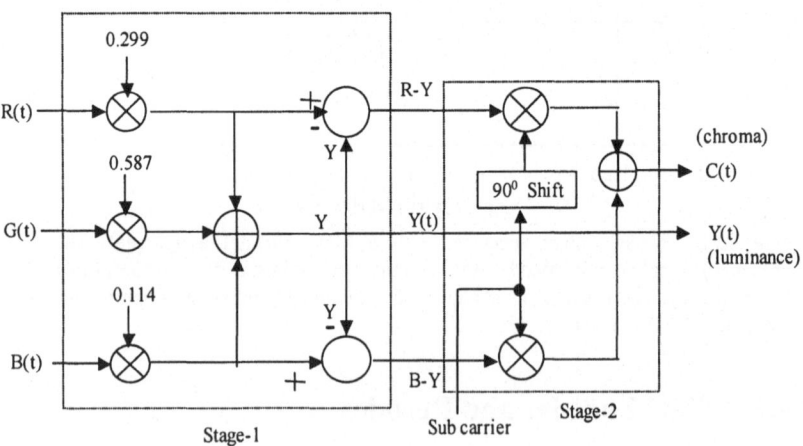

Figure 12.12. Color components are converted to color difference signals.

A simplified schematic of an NTSC demodulator is shown in Fig. 12.13. The composite signal is passed through a comb filter that separates the chrominance and luminance components. This is followed by synchronous demodulation of $C(t)$ to recover $I(t)$ and $Q(t)$. The R, G, and B signals are then obtained from the {Y,I,Q} signals by a matrix multiplication.

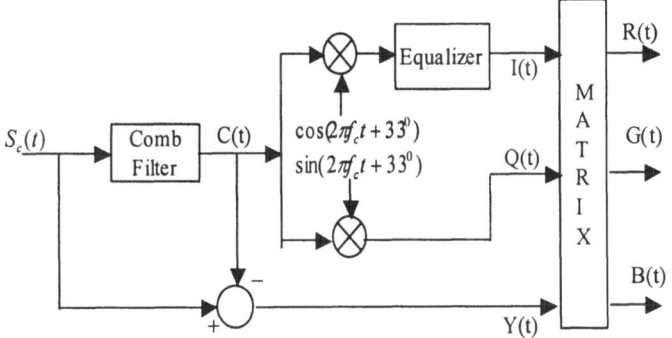

Figure 12.13. Simplified schematic of NTSC signal demodulation

12.5 COMPONENT AND S-VIDEO

It has been noted that in television video, the chroma signal is put on top of the luminance signal and transmitted as a single signal. This is known as composite video. Although the composite video requires less bandwidth, its demodulator is difficult to implement, especially the design of an efficient comb filter. For reasons of economy, simple bandpass filters are often used for luma-chroma separation, and equalization is often disregarded. Suboptimal demodulation results in visible, but not often objectionable, degradation in the reproduced picture. For example, chrominance-to-luminance cross-talk will occur if the chrominance components below 3 MHz are present, as these components will be mistakenly considered for the luminance signal. Similarly, luminance-to-chrominance cross-talk will occur due to the luminance components near the color subcarrier frequency.

There are two alternative methods to improve the picture quality. In the first method, the three color channels {R, G, B} {Y, I, Q} are transmitted separately. This method is known as component analog video. Since there is no mixing of color and luminance components, there is no need for comb filters. The video produced in this method is of very high quality. However, perfect synchronization of the three components is required for reproduction. This method needs about three times more bandwidth; therefore, this system is not used in practice. However, in the case of digital video, digital compression techniques can reduce the bandwidth significantly. As a result, many digital systems may afford to provide component video.

In the second method, known as the Y/C method or S-VHS, the chrominance and luminance signals are kept separate (as shown in Fig. 12.14). Since the chroma and luminance spectrums are not overlapped, the two components can be separated easily. This results in a better reproduction of color, although not as superior as component video. But, the Y/C method does require a slightly larger bandwidth compared to composite video.

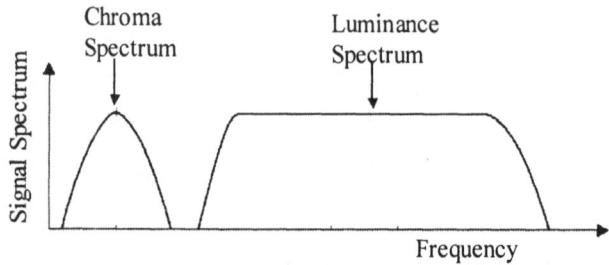

Figure 12.14. In the Y/C model, the resultant chroma is kept separate from the luminance component. It is much easier to separate the two components, and better picture quality is often achieved (compared to composite video).

12.6 DIGITAL TELEVISION

It was observed in section 12.1 that current analog monochrome television standards – namely, NTSC, PAL and SECAM – were adopted in the 1940s. The color standards were adopted in 1950s. Television standards have served their purpose well, and the quality of the TV video has improved significantly over the years mainly due to the advancement of camera and display technology. However, the primary obstacle to improved picture quality is existing TV standards. In the last 15-20 years, significant work has been done to develop and introduce digital and high-definition television standards [5, 6].

In digital TV (DTV), the video frames are represented and transmitted as digital data. Once a video is digitized, it can be treated like any other multimedia data. As shown in Chapter 9, a digital video signal requires a large space for storage, or a high bandwidth for transmission. Hence, digital TV signals are compressed using an MPEG-2 compression algorithm. The compressed signals are then modulated and broadcasted through terrestrial, cable, or satellite networks.

Why Digital TV?

There are several advantages of DTV over analog TV standards. The first and foremost advantage is flexibility. Inside a DTV, there is a powerful computer (for decoding the signal). Digital processing brings to a typical digital system significant advantage over analog system. The picture quality is much better compared to analog TV. In addition, multilingual closed captioning is possible with DTV. Finally, DTV paves the way for eventual computer-TV integration, and it is highly likely that in a few years of time, there will not be much difference between a computer and a television.

Digital TV has two main formats: standard definition and high definition. The standard definition TV (SDTV) provides a resolution almost two-times compared to analog TV quality. Even more impressive, the high-definition TV (HDTV) provides a resolution almost four-times compared to analog TV. Note that the working principles of the SDTV and HDTV standards are same; the only difference is the higher resolution for the HDTV standard.

Some of the features that have been incorporated in the various HDTV standards are as follows:

- Higher spatial resolution
- Higher temporal resolution (up to 60 Hz progressive)
- Higher aspect ratio (16:9 compared to existing 4:3)
- Multichannel CD-quality surround sound with at least 4-6 channels
- Reduced artifacts as compared to analog TV by removing composite format and interlace artifacts

Digital TV Transmission

Figure 12.15 shows a generic block schematic of a terrestrial DTV/HDTV system, which is adopted by the ITU-R. The DTV system has three main subsystems:

1. Source coding and compression
2. Service multiplex and transport
3. RF/Transmission

The *source coding and compression* subsystem employs bit rate reduction methods to compress the video, audio, and ancillary digital data. The term "ancillary data" generally refers to the control data, conditional access control data, and data associated with the program audio and video services, such as closed captioning.

The service multiplex and transport subsystem divides the digital data stream into packets of information. Each packet is labeled properly so that it can be identified uniquely. The audio, video, and ancillary data packets are multiplexed to generate a single data stream.

The RF/Transmission subsystem performs the channel coding and modulation. The channel coder takes the input data stream (which has been compressed), and adds additional information so that the receiver is able to reconstruct the data even in the presence of bit errors due to channel noise. In the receiver side, the compressed bit-stream is received and decoded. The decoded signal is then displayed at the DTV/HDTV monitor.

Different DTV Standards

Like analog television standards, there are also three major DTV/HDTV standards in the world [7]. In Japan, the HDTV standard employs the MUSE (multiple sub-Nyquist sampling encoding) technique. In Europe, the standard is known as DVB (digital video broadcasting). In North America, the system is known as ATSC (advanced television systems committee) DTV standard.

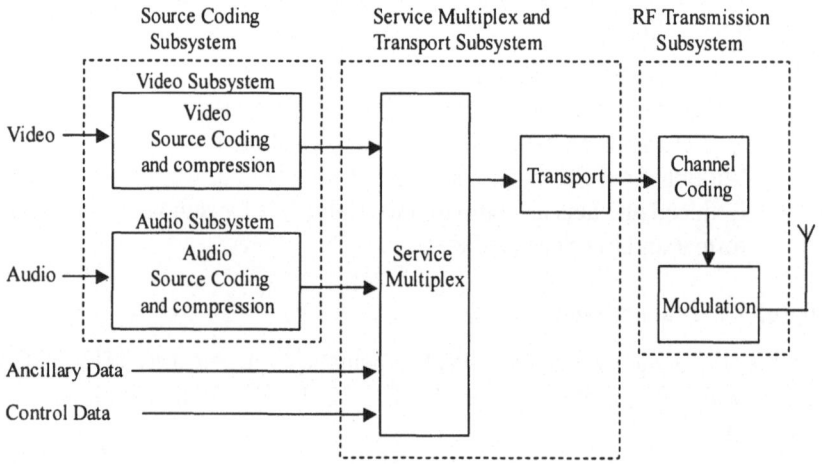

Figure 12.15. A generic digital terrestrial television broadcasting model.

The MUSE system in Japan employs a wide band frequency modulated 1125-line, 60 field/s system. The overall bandwidth of the TV signal is 8.1 MHz/channel. Note that MUSE employs a mixture of analog and digital technologies (hence it is not completely digital like the DVB and ATSC standards).

The DVB standard in Europe is based on a standard-resolution TV with 625 lines and 50 fields/s pictures. It employs MPEG-2 standard algorithms for video and audio compression. The HD mode in DVB has a resolution of 1250 lines, resulting in superior quality digital video. There are three variants of DVB: DVB-T (for terrestrial), DVB-S(for satellite), and DVB-C (for cable). The compression technique is similar in all three variants. The main difference is that they employ different modulation techniques appropriate for each type of channel. DVB-T employs the COFDM (coded orthogonal frequency division multiplex) technique, DVB-S employs the

QPSK (quadrature phase shift keying) technique, and the DVB-C employs the QAM (quadrature amplitude modulation) technique.

The development of HDTV in North America was started by the Federal Communications Commission (FCC) in 1987. It chartered an advisory committee (ACATS) to develop an HDTV system for North America. Testing of proposed HDTV systems was performed 1990-1993. The Grand Alliance [8] was formed in 1993 to develop an overall superior HDTV standard combining these proposed systems. An overview of the ATSC HDTV standard is presented below.

12.6.1 Grand Alliance HDTV Standard

The Grand Alliance (GA) HDTV standard follows the schematic shown in Fig. 12.15. The HDTV system employs an MPEG-2 algorithm to compress the video data and the AC-3 standard algorithm to compress the audio data. For data transport, the HDTV system employs the MPEG-2 transport stream syntax for the multiplex and transport subsystem. For RF/Transmission, the HDTV system employs two modes in the modulation scheme: a terrestrial broadcast mode (8 VSB), and a high data rate mode (16 VSB).

Input Video Formats

There are four major video resolutions in the DTV system as shown in Table 12.5. Most video formats have square pixels for interoperability with computer display systems. A degree of compatibility with NTSC is retained with the support of 59.94, 29.97, and 23.98 frames/sec temporal resolution. Note that the display format is independent of the transmission formats since the display is not expected to switch between formats instantaneously. Hence, the decoder should have the capability to change the temporal and spatial resolution of incoming video.

Table 12.5. Video resolution formats supported by the ATSC DTV standard.

Format	Frame Size	Aspect Ratio	Frame Rate Interlaced/Progressive
HDTV	1920H×1080V	16:9	30 f/s Interlaced 30 and 24 f/s Progressive
HDTV	1280H×720V	16:9	60, 30 and 24 f/s Progressive
SDTV	704H×480V	4:3 or 16:9	30 FPS Interlaced 60, 30 and 24 f/s Progressive
SDTV	640H×480V	4:3	30 FPS interlaced 60, 30 and 24 f/s Progressive

The interlaced formats are supported in the standard for smooth transition from interlaced NTSC to progressive HDTV standard. The progressive scan in HDTV allows a trade-off between spatial and temporal resolutions, and also between resolution and coding artifacts. For 24 or 30 Hz film, the spatial resolution can be 1280×720 or 1920×1080 depending on scene complexity. If scene complexity is high, 1280×720 can be chosen to reduce coding artifacts.

With 1920×1080 frame size, and 30 frames/sec (progressive scanning), there will be 1920×1080×30 = 62.2 millions pixels per second. If each pixel requires 24 bits for representation, the bit-rate is 1.49Gbits/s. However, in order to transmit the digital video over a 6 MHz Channel, the maximum data rate can be at most 20 Mbits/s. Therefore, the raw digital video data needs to be compressed with a high compression ratio (about 60:1) before transmission.

Encoder and Decoder

The schematic of the Grand Alliance encoding system, which is based on the MPEG-2 standard, is shown in Fig. 12.16 [8]. The input video can come from a wide variety of sources. The adaptive preprocessor may change the format of the input video to improve the picture quality. For example, the interlaced video may be changed to progressive video. Also, the motion estimation and compensation is a major part of the encoding system. A superior motion prediction is generally achieved using a combination of both coarse and fine motion vectors (e.g., fractional-pixel estimation). The forward analyzer typically calculates the perceptual significance of the DCT coefficients, and controls the quantization step sizes in order to achieve the targeted bit-rate. The video encoder loop calculates the motion predicted error frames (for P and B only). The DCT of the error frames are then calculated. The DCT coefficients are quantized and entropy and run-length coded.

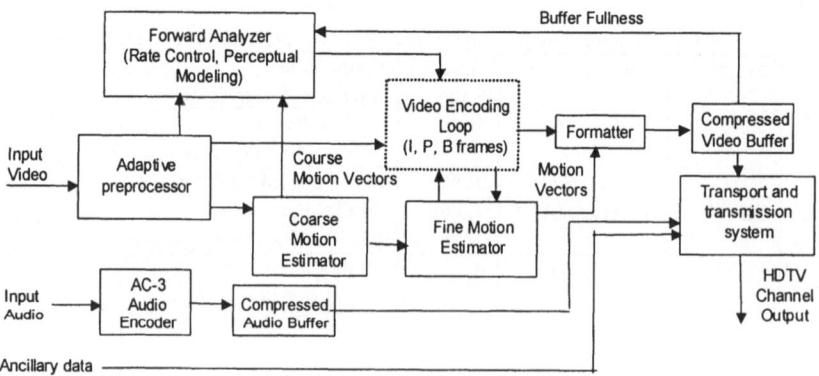

Figure 12.16. Simplified schematic of the ATSC (or GA) HDTV system.

The audio data in the GA HDTV system is encoded using the AC-3 coding algorithm. The audio, video, and ancillary data are then multiplexed and modulated for transmission through terrestrial, cable or satellite systems.

The decoding of HDTV video is the inverse of the encoding process. First, the HDTV signal is demodulated. The demodulated bitstream is then separated into audio, video, and ancillary bitstream. The audio decoder (see Chapter 7) and video decoder (see Chapter 9) are then applied to reconstruct the audio and video signals, which are then fed to the loudspeakers and the display system (e.g., cathode ray tube) for watching a channel on the TV screen.

HDTV Transmission System

The compressed videos must be organized into a format which can be reliably transmitted through an imperfect channel and allow the decoder to recover quickly from the loss of data in the channel. Robust transmission is achieved by dividing the compressed video bit-stream into fixed length blocks and applying error correction processing to each block.

In packet communication, the size of the packets has a significant impact on the efficiency of the data transmission. If the packet size is shorter, more packets will be required to transmit the same information, and the overhead will increase. But if the packets are large, the recovery time from a lost packet will increase. As a compromise, it has been decided that the GA transport packets will be of length 188 bytes. The organization of a packet is shown in Fig. 12.17.

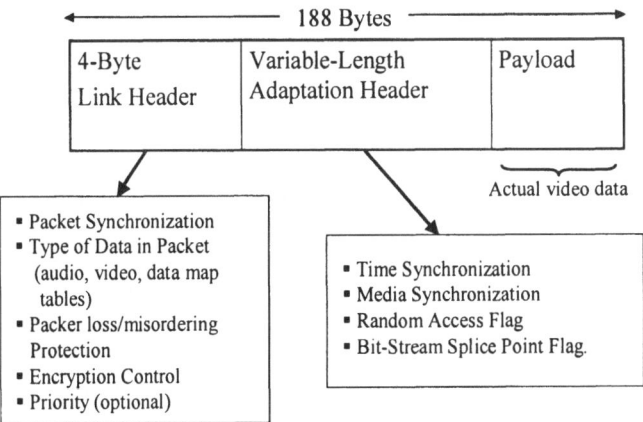

Figure 12.17. HDTV transport packet.

The program clock reference in the transport stream contains a common time base to synchronize the audio and video. For lip synchronization between audio and video, the streams carry presentation time stamps that instruct the decoder when the information occurs relative to the program clock.

The HDTV signal is required to fit inside the same 6-MHz channel spacing as is currently used for today's NTSC transmissions. Hence, the challenging task in the current HDTV standard is to reliably deliver approximately 20 Mb/s data through the terrestrial TV broadcasting environment using one 6 MHz channel. The following two techniques are employed to achieve the targeted transmission rate:

  i) Vestigial sideband modulation with Trellis coding
  ii) Reed-Solomon coding for error detection/correction

Note that trellis coding and its associated bit-rate overhead are not needed for cable TV (since the cables have much less noise in the channel). Thus, the HDTV system can deliver 43 Mb/s over one 6 MHz channel through cables. The terrestrial transmission system is an 8-level vestigial sideband (VSB) technique. The 8-level signal is derived from a 4-level AM VSB and then trellis coding is used to turn the 4-level signals into 8-level signals. Additionally, the input data is modified by a pseudo-random scrambling sequence that flattens the overall spectrum. Cable transmission is by a 16-level VSB technique without trellis coding. Finally, a small pilot carrier is added (rather than the totally suppressed carrier as is usual in VSB). This pilot carrier is placed so as to minimize interference with existing NTSC service.

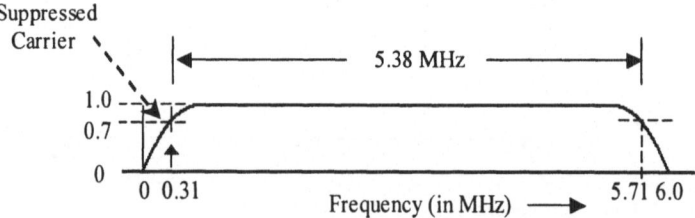

Figure 12.18. Spectrum of HDTV signal.

The spectrum of the HDTV signal is shown in Fig. 12.18. Note that the spectrum is flat except at the two edges where a nominal square root raised cosine response results in 620 kHz transition regions. The suppressed carrier of the VSB is a small pilot tone, located 310 kHz from the lower band edge where it can be hidden from today's NTSC television receivers. The VSB signal can be protected from interference from the strong energy of NTSC carriers by comb filters at the receiver.

In order to preserve the integrity of received data, error corrective coding is employed. The basic concept is to add redundancy to a message in order to eliminate the need for retransmission of the data. In the case of HDTV, Reed-Solomon (RS) coding is applied to perform the forward error correction (FEC). The RS coder can correct multiple errors, and has been used in many practical applications such as magnetic and optical storage systems, and space and mobile communications.

An RS code is described as an (n,k) code, where the code words consist of n symbols of which k are message symbols. In this case, the amount of redundancy is $(n-k)$. It can be shown that the maximum number of symbol errors that can be corrected is $(n-k)/2$.

Trellis Coding

After channel coding is performed, the signal has to be modulated (*e.g.*, frequency shift keying, phase shift keying) in order to efficiently transmit it through a channel. The GA HDTV employs trellis coding to perform the modulation. Here, modulation and coding are treated as a combined entity. The number of signal points in the constellation is larger than that required for the modulation format. Convolution coding is used to introduce a certain dependency between successive signal points. Soft decision decoding is performed in the receiver where the permissible sequence of signals is modeled as a trellis structure.

Current Status

The FCC mandated that all TV stations in the USA have to begin broadcasting digital TV by 2002. Several TV networks such as CBS, HBO, SHO, DISH, PBS, ABC, NBC broadcast HDTV signal on a regular basis. As per the FCC mandate, the analog NTSC transmission should cease at the end of 2006. Hence, the TV industry is expected to undergo immense transition in the next few years. Although HDTV offers a much better quality video, the cost of HDTV is becoming an obstacle to the adoption of the new system. As an intermediate solution, people can buy set-top boxes that will convert the HDTV/DTV signal to analogue NTSC format, which in turn can be viewed on NTSC receiver. PC plug-in cards are also available to watch HDTV on computer screens.

REFERENCES

1. J. Watkinson, *Television Fundamentals*, Focal Press, Boston, 1996.
2. M. Robin and M. Poulin, *Digital Television Fundamentals: Design and Installation of Video and Audio Systems*, McGraw-Hill, New York, 1998.

3. D. E. Pearson, *Transmission and Display of Pictorial Information*, Pentech Press, 1975.

4. A. N. Netravali and B. G. Haskell, *Digital Pictures*, Second Edition, Plenum Press, 1994.

5. Special Issue on Digital Television Part 1: Enabling Technologies, *Proc. of IEEE*, Vol. 83, June 1995.

6. Special Issue on HDTV, *IEEE Spectrum*, April 1995.

7. L. Zong and N. G. Bourbakis, "Digital video and digital TV: a comparison and the future directions," *Real-Time Imaging*, Vol. 7, pp. 545-556, 2001.

8. E. Petajan, "The HDTV Grand Alliance System," *IEEE Communications Magazine*, June 1996.

QUESTIONS

1. How many television systems have been developed worldwide? Compare and contrast these standards.

2. Explain horizontal and vertical scanning.

3. What is the advantage of interlaced video over progressive video? What is the disadvantage?

4. How long are the horizontal and vertical retrace periods in NTSC and PAL systems? What happens during the horizontal and vertical retrace period?

5. What are the advantages of using {Y, I, Q} and {Y,U,V} color spaces over {R, G, B} color space for TV transmission?

6. Choose an image of your choice. Repeat Example 12.1 and determine the energy compaction performance of the {Y, I, Q} and {Y,U,V} color spaces. You may use the MATLAB code provided in the CD.

7. Sketch the NTSC signal spectrum. Why do I and Q channels require less bandwidth compared to Y channel?

8. How is composite video generated in the NTSC system? How is the chroma signal generated?

9. Why is a good comb filter necessary for a good quality TV receiver?

10. Compare and contrast component, composite, and S-video.

11. What are the advantages of digital television over analog television?

12. Draw the schematic of a generic digital TV system. Explain the operation of different subsystems.

13. How many HDTV systems have been developed worldwide? Compare them.

14. Draw the schematic of the GA HDTV encoding system.

15. Draw the schematic of the GA HDTV decoding system.

16. What is the packet length of the HDTV bitstream? What are the advantages and disadvantages of shorter and longer packets?

17. Draw a typical HDTV signal spectrum. Explain the different parts of the spectrum. Where is the pilot carrier located?

18. Explain the channel coding and modulation techniques of GA HDTV system.

19. Explain the principles of the Reed Solomon code. What is the maximum number of errors that can be corrected by (p,q) RS code?

20. Explain the principle of the Trellis code.

Chapter 13

Content Creation and Management

With the rapid growth of multimedia systems, the creation, management and presentation of multimedia content have become popular. As well, the computing power has increased significantly in recent years, making complex multimedia presentations more feasible. Multimedia information can be represented in a variety of ways such as traditional documents, hypertext, and hypermedia. In this Chapter, the main focus is to illustrate the content creation process, and its efficient representation.

The Chapter begins with an introduction to multimedia authoring for creating multimedia content. A few select authoring tools and their working principles are then explained. The limitation of traditional documents is discussed followed by the introduction to hypertext and hypermedia systems. Finally, a brief overview of the current and upcoming multimedia standards is presented.

13.1 MULTIMEDIA AUTHORING

Authoring is the process of creating multimedia applications [1]. A typical authoring system consists of a set of software tools for creating different applications. It is actually a speeded-up form of programming where one does not need to know the intricacies of a programming language. It is only essential to understand how to use the tools to develop a presentation. Authoring systems vary widely in orientation, capabilities, and learning curve.

Several studies have found that development time (and hence the development cost) of a large project can be significantly reduced with interactive authoring environment, as opposed to a conventional programming environment. In addition, the re-use of code is likely to be increased if another project (with some overlap with the old project) employs the same environment. However, the content creation (graphics, text, video, audio, animation) is not generally affected by the choice of an authoring system; any production time gains result from accelerated prototyping, not from the choice of an authoring system over a compiled language.

Tools are individual computer programs that perform one or more tasks needed to create an application. Authoring usually requires special hardware for audio and video, and it also places special demands on the operating system software in your computer. Figure 13.1 is a simplified schematic of a typical authoring environment.

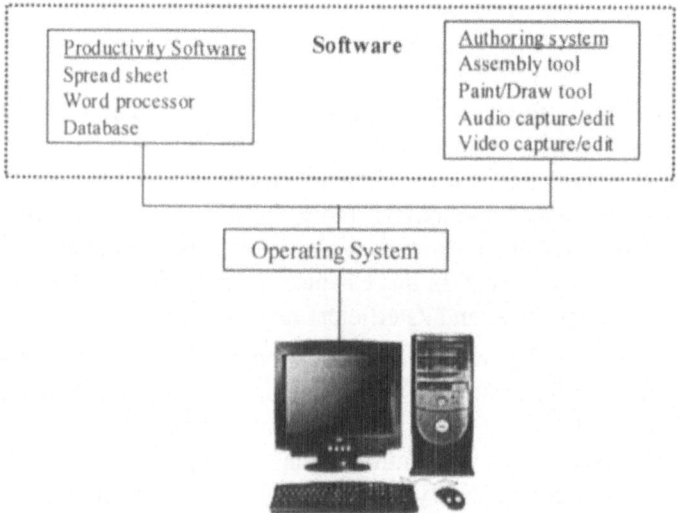

Figure 13.1. A Generic Environment for Multimedia Authoring.

13.1.1 Authoring Steps

There are several steps for multimedia authoring. Figure 13.2 shows the major steps, with the function of each block explained below.

Concept: The objectives for the project are defined at the concept stage. This includes identifying the application audience, the type of application (*e.g.*, presentation, interactive, *etc*), the purpose of the application (*e.g.*, to inform, to entertain, to teach, *etc*), and the general subject matter. Ground rules such as style, application size, target platforms for the design stage are generally set at this time.

Design: The content design is a core process in multimedia authoring. The purpose of the design stage is to detail the project architecture, the styles, and the necessary content material. Authoring software becomes valuable at the design stage. There are primarily five ways [2] a message can be delivered, as shown in Table 13.1. A message can be expressed through scripts, audio, graphics, animation and interaction in a presentation. The different content types should complement each other's role in order to be effective.

Table 13.1. Content design principles.

Content type	Comments
Scripting or writing	The message should be kept as simple as possible so that the intended audience is able to grasp it. A typical way of scripting is to first write the full message, and then shorten it.
Audio (hearing)	Audio is an effective method to convey a message. It can be a speech, music, or a sound effect. A speech can be used to convey a direct message from a person (*e.g.*, President of a country, director of a company) to the intended audience. Appropriate music can be used to set the mood of a presentation.
Graphics (illustration)	Pictures are more effective than words in conveying messages, which justifies the old adage: "A picture is worth a thousand words." Hence, graphics, especially color graphics, should be used as much as possible. However, one should be careful while selecting colors,; they should be consistent with the content.
Animation or video	Animation or video is a very effective way to illustrate a complex phenomenon (*e.g.*, how does the human heart work?). In addition, it keeps the attention of the audience, and prevents the audience from falling asleep during a long presentation. There are several forms of animation such as moving text, cartoons, and digital video. With consumer digital video cameras, digital video is now easy to obtain. However, it requires a large storage space, and may require specialized hardware or software for the compression and decompression.
Interaction	Interactivity is a very effective method to convey a message. People tend to forget most text messages during a presentation. Audio and graphics messages are remembered in greater detail. However, if the audience is allowed some interaction, the message is remembered for a long time.

Collecting Content Material: From the content lists created in the design stage, the author must collect all the content material for the project. Content material is obtained either by selecting it from available internal sources such as libraries or creating it in-house for the project.

Several Multimedia creation tools are available commercially for developing content. Tools can be categorized according to their operations performed on the basic multimedia objects. There are about five different categories of multimedia creation tools:

i)   Painting and Drawing Tools
ii) Audio Editing Programs (see Chapter 10)
iii) 3-D Modeling and Animation Tools
iv) Image Editing Tools (see Chapter 11)
v)   Animation, Video and Digital Movies (see Chapter 11)

A few select audio, image and video editing tools have been presented in Chapters 10 and 11. More tools can be found in the references [3].

Assembly: The entire application or project is put together in the assembly stage. Presentation packages (for example, PowerPoint) do their assembly while you are entering the content for various screens. Once all the screens are assembled and placed in order, the presentation is ready.

Testing: Once the application is built and all content materials are assembled, the project is tested to be sure that everything works and looks accordingly. An important concern in testing is to make sure the application works in the actual environment of the end user, not just in the authoring machine.

Distribution: If the application is intended to run by someone else on a different machine, the authoring system needs to have an option to make this possible.

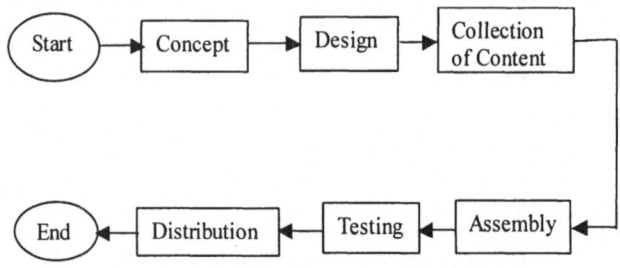

Figure 13.2. Flow diagram of the authoring process.

13.2 MULTIMEDIA AUTHORING TOOLS

Multimedia authoring tools provide the framework for organizing and editing the elements of multimedia project, including graphics, sounds, animations, and video clips. Authoring tools are used for designing interactivity and the user interface for presenting your project on screen, and for assembling multimedia elements into a single cohesive project. Authoring tools consists of two types of interfaces: text based and graphical based. In a text-based interface, the author simply enters information into the system by typing it from the keyboard while the display shows a text screen. This type interface may also use a mouse for things like menu selection, but a mouse is not a primary tool. In a graphical interface, the author uses pointing devices such as a mouse, touch screen or graphical tablet to position objects on the screen.

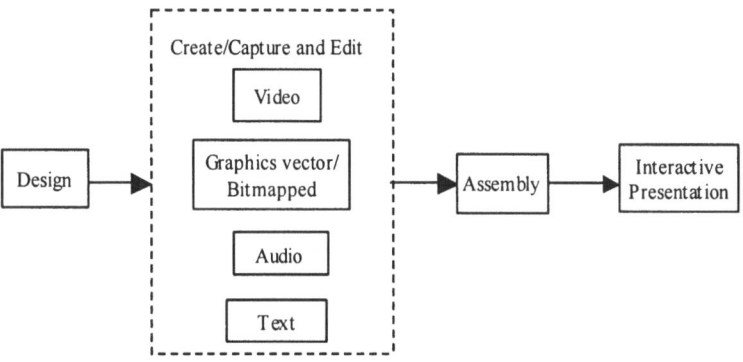

Figure 13.3. Typical flow diagram of processes in an authoring tool.

There is a wide variety of authoring tools that have been designed on different paradigms [4]. It is important to know their paradigms so that the appropriate tools can be employed for developing an application. The authoring tools can be broadly divided into four categories:

i)   Card or page based tools
ii)   Icon-based, event driven tools
iii) Time-based and presentation tools
iv)   Object oriented tools

A brief discussion on each of the above categories is presented below.

13.2.1 Card/Page Based Tools

In these authoring tools, the elements are organized as pages of a book, or a stack of cards. Thousands of books or cards may be available in the book or stack. These tools are more convenient when the bulk of the content consists of elements that can be viewed individually, like the pages of a book or cards in a card file. The authoring system lets you link these pages or cards into organized sequences. You can jump, on command, to any page you wish in the structured navigation pattern. Examples of a few card-based tools are

ToolBook (Windows)- developed by Asymetrix [5]
Metacard (Mac/Unix/Windows)
HyperCard, developed by Apple

Figure 13.4 shows the view of the ToolBook design window for a
Windows environment. This tool is conceptually easy to understand and has
a library of basic behaviors, *e.g.*, mouse-click and transitions. It has its own
programming language – Open Script – that is easy to learn. It can import a
variety of different types of media, and is convenient for applications such
as computer-based training, marking and assessment. The multimedia
projects developed by ToolBook are called Books. The Book is divided into
pages and the pages can contain text fields, buttons and other multimedia
objects. In Fig 13.4 you can see the buttons *previous* [A] and *next* [B]; by
clicking on these buttons you can navigate through the pages of the Book.
Different objects [C] in the catalog can be placed on the pages of the Book
to create an interactive presentation.

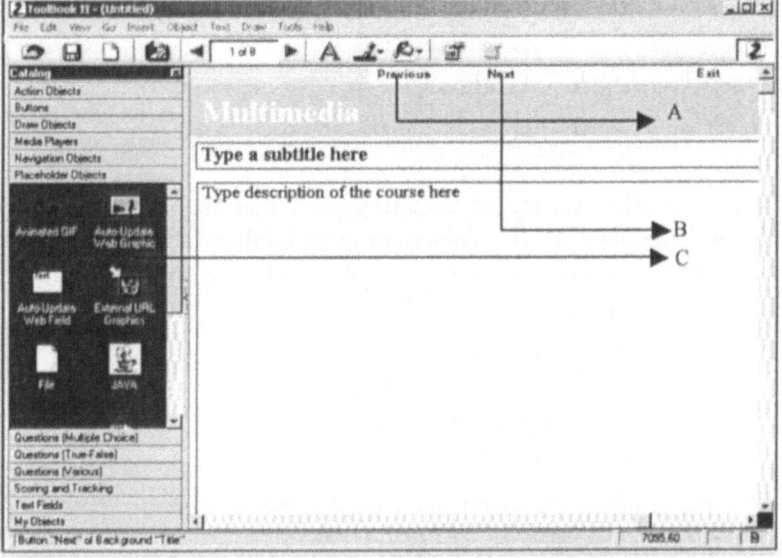

Figure 13.4. View of ToolBook for Windows. A: previous button, B: next
button, and C: objects in the catalog.

13.2.2 Icon-Based Tools

In these tools, the multimedia elements are arranged like objects in a
structural framework. The arrangement requires the creation of a flow
diagram of activities displaying the logic of the program by linking icons
that symbolize events, tasks and decisions. When the structure of the

application has been created, it is enhanced by text, graphics, sound, animation and video. This process of diagram creation is considered very useful during the design and development, especially in large navigational structures. Icon-based tools are also called *event driven* tools. Examples of a few icon-based tools are:

> Authorware 6.0(Mac/Windows), developed by Macromedia [6]
> Icon Author 7.0 (Windows)- developed by AimTech
> Quest 6.0 (Windows)- developed Allen Communication

Authorware is an authoring environment for creating cross-platform interactive multimedia systems. It provides tools for producing interactive learning and training applications that use digital movies, sound, animation, text and graphics. The creation of applications is simply a case of dragging icons onto a flowchart. Authorware is used by many corporations for imparting professional training. Figure 13.5 shows the different user options in Authoware tool. The icons [A] in the icon palette [B] specify the flow of the project and sequence different events and activities.

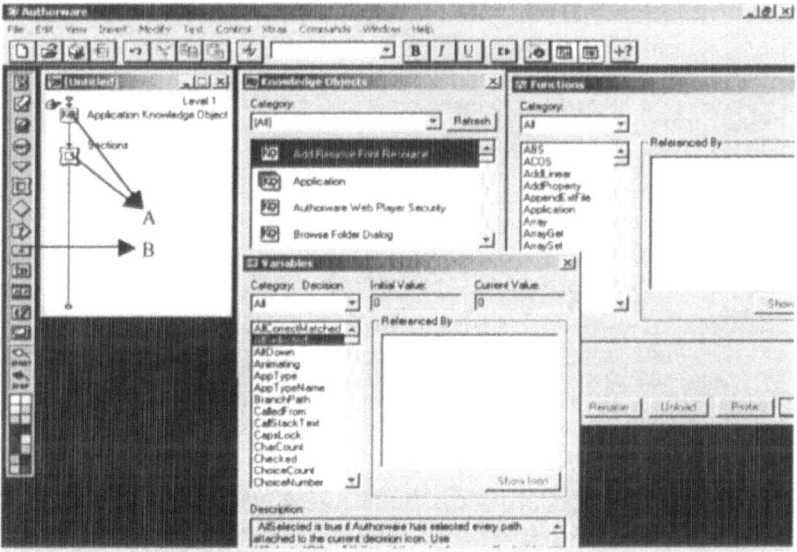

Figure 13.5.View of Authorware. A: Icons in the design window, and B: Icon palette.

13.2.3 Time-Based Tools

In time-based tools, the elements and events are organized along a timeline, with resolutions as high as 1/30 second. Time-based tools are best to use when you have a message with a beginning and end. Sequentially organized graphic frames are played back at a speed that can be set. Other elements (such as audio events) are triggered at a given time or location in the sequence of events.

Macromedia Director is a popular time-based tool, developed by Macromedia for the Macintosh/Windows environment. It has a library of basic behaviors, *e.g.*, mouse click, and transitions. The Director has its own programming language called Lingo, which is easy to learn and can import different types of media. In Director, the assembling and sequencing of the elements of the project are done by Cast [A] and Score [B]. The cast contains multimedia objects such as still images, sound files, text, palettes, and Quick Time movies. Score is a sequencer for displaying, animating and playing cast members. Figure 13.6 shows the cast window and score window of Macromedia Director, which are important for developing a multimedia presentation.

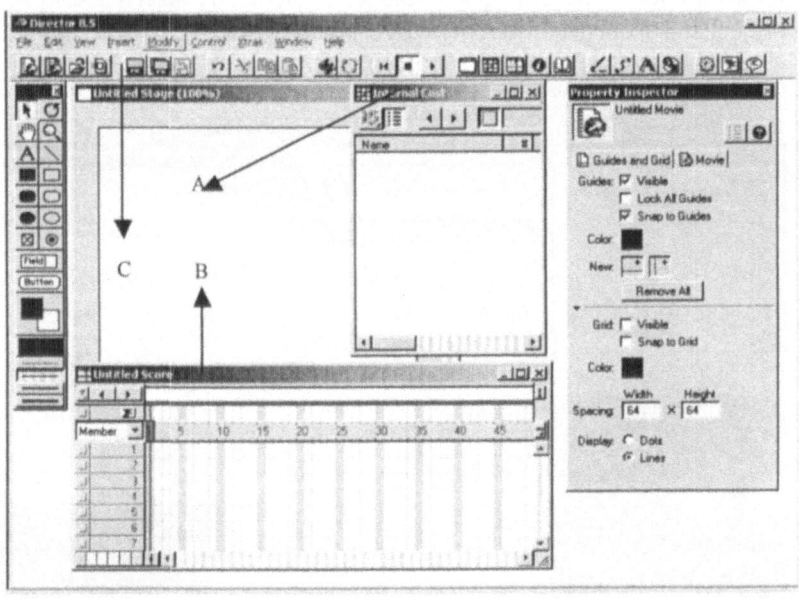

Figure 13.6. View of Macromedia Director. A: Cast window, B: Score Window, and C: Tool Bar

13.2.4 Object-Oriented Tools

In these tools, different elements and events are treated as objects that live in a hierarchical order of parent and child relationships. Messages are passed among these objects instructing them to perform activities according to their assigned properties. Objects typically take care of themselves. When a message is sent to them, the objects perform their jobs without external input. Object-oriented tools are particularly useful for games, which contain many components with many personalities. Examples of a few object oriented tools are:

> mTropolis (for Mac/Windows)- developed by Quark
> Apple Media Tool, (for Mac/Windows)- developed by Apple
> MediaForge (for Windows)-developed ClearSand Corporation

13.3. MULTIMEDIA DOCUMENTS

The authoring tools discussed in the last two sections are used to create a multimedia presentation or document. Unfortunately, a major weakness of current authoring tools is that the documents or presentations created by different tools are highly incompatible. For example, if a document is created in MS Word in a Windows environment, it is very difficult to read the document in a Unix environment. Even other word-processing documents in the same operating system may not be able to read the document.

The incompatibility among different tools can be solved if a document is represented by a set of structural information that is independent of the operating system environment or the tools that create the document [7]. During the presentation, a given tool will use the structural information and produce the final presentation. An ideal scenario is shown in Fig. 13.8.

In order to achieve the above flexibility, all documents should conform to a standard document architecture so that the content of documents can be exchanged. Document architecture describes the connections among the individual elements represented as *models*. There are three main models for document architecture [8]. The manipulation model describes the operations for creation, change, and deletion of multimedia information. The representation model defines the protocols for exchanging the information among different computers, and the formats for storing the data. The presentation model describes the overall presentation format.

Manipulation of Multimedia Data

To date, several document architectures have been developed. However, the most popular architecture used by publishers is the *Standard Generalized Markup Language* (SGML). Figure 13.9 shows the schematic of SGML document processing. There are two major steps in the processing: parsing and formatting. Note that the formatting information is expressed using SGML tags. Parsing is the analysis of a string or text into logical syntactic components in order to validate the conformability of the document to logical grammar. On the other hand, the formatter is a defined structure for the processing, storage and display of a document. Note that only the formatter knows the meaning of the tags and it transforms the document into a formatted document. The transformation from the actual information to its final representation is carried out according to the rules specific to the document architecture.

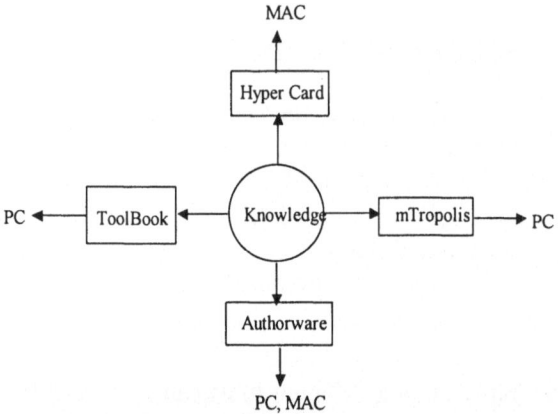

Figure 13.8. Transferring knowledge between platforms.

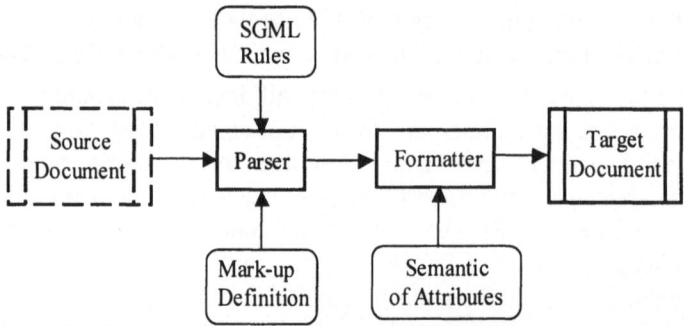

Figure 13.9. SGML document processing: from the information to the presentation.

The SGML document specifies a document in a machine-independent way. For example, it records how different units of information are structured, how these units are related to each other, and the ways in which they are interlinked. A series of software conversion tools may be used to process the marked-up multimedia resources and port them to various machine-dependent platforms for delivery to the users.

13.4. HYPERTEXT AND HYPERMEDIA

A document is a convenient method for transmitting information. The readers read a document and reconstruct the knowledge associated with it. In ordinary documents, such as books and periodicals, the information is presented sequentially. Most authors assume a sequential reading, and present the information accordingly. Generally, it is difficult to read a Chapter without reading the previous ones.

Although sequential knowledge representation is very effective, it is not without limitations. Humans think in a multidimensional space. For example, when we think about a person, several apparently unrelated thoughts come to our mind such as the person's attributes, his or her strengths/weaknesses/hobbies, or the last meeting with the person. Hence, information presented in a document is generally multidimensional in nature. The sequential presentation can be considered as a linearization process that converts multidimensional knowledge to one-dimensional knowledge. The readers, while reading the document, delinearize the information mentally, and reconstruct the multidimensional knowledge (see Fig. 13.10). In the process of reconstructing knowledge, there is often loss of knowledge, and thus, the linear representation may not be very efficient. It is intuitive that a nonlinear information chain would be able to provide a better representation of information.

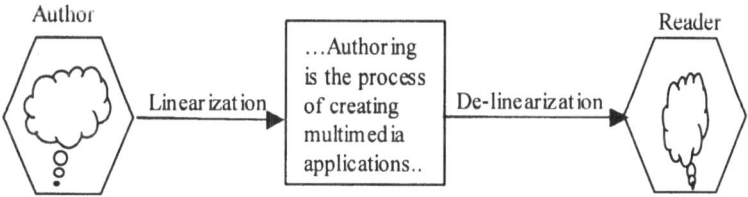

Figure 13.10. Information transmission from the author to the reader.

13.4.1 Nonlinear Information Chain

The essential idea underlying a linear organization [7] is that a body of knowledge is organized into several smaller units designed to follow each other in sequence, although some jumping back or skipping ahead is allowed (see Fig. 13.11(a)). It is observed that each of the basic units from which a linear structure is composed has just one entry point and one exit. Another important point is that the material embedded within each module that makes up a linear information structure is intended to be processed in a strictly sequential fashion.

In contrast, the essential features of a non-linear approach are illustrated in Fig. 13.11(b). In this diagram, the basic units that form the knowledge are joined in many complex ways, allowing them to be processed in a variety of non-linear pathways. Here, each module has one entry point, but can have any number of exit points. Therefore, from within the body of one unit it is possible for the user to jump to any number of other units depending upon the reader's interests and requirements.

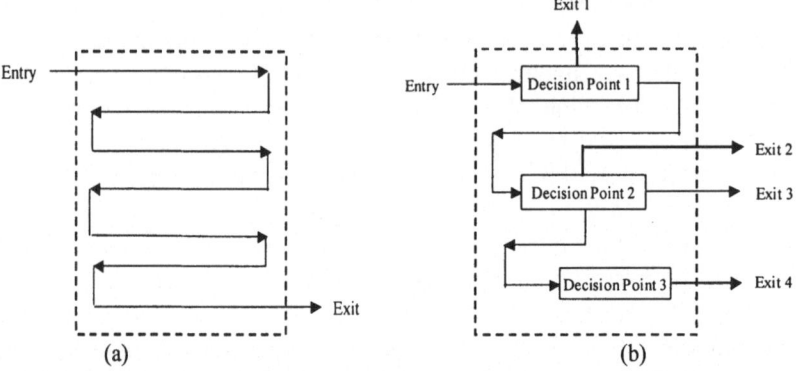

Figure 13.11. Organization of information/knowledge units. a) linear, and b) nonlinear organizations.

13.4.2 Hypertext and Hypermedia Systems

The nonlinear information chain introduced in the previous section can be implemented using a hypertext system, which was introduced in the 1960s. These systems typically use a nonlinear information representation whereby a reader can decide his or her own reading path.

A hypertext system is made of *nodes* (concepts) and *links* (relationships). A *node* generally represents a single concept or idea, and may contain text, graphics, animation, audio, video, images or programs. Node*s* are connected to other nodes by link*s*. The node from which a link originates is called the *reference,* and the node at which a link ends is called the *referent.* The links are highlighted on the screen or monitor by reactive areas (or hotspot or button) that can be selected by the user by means of a point and click operation. The forward movement in a hypertext system is called *navigation.* Figure 13.12 shows how a piece of textual information can embed several reactive areas. In this diagram, these areas are denoted by the underlined keywords.

When a document includes the nonlinear information links of hypertext systems as well as the continuous and discrete media of multimedia systems, it is called a *hypermedia* system (see Fig. 13.13). For example, if a nonlinear link consists of text and video data, then this is a hypermedia, multimedia, and hypertext system. The World Wide Web (WWW) is the best example of a hypermedia application.

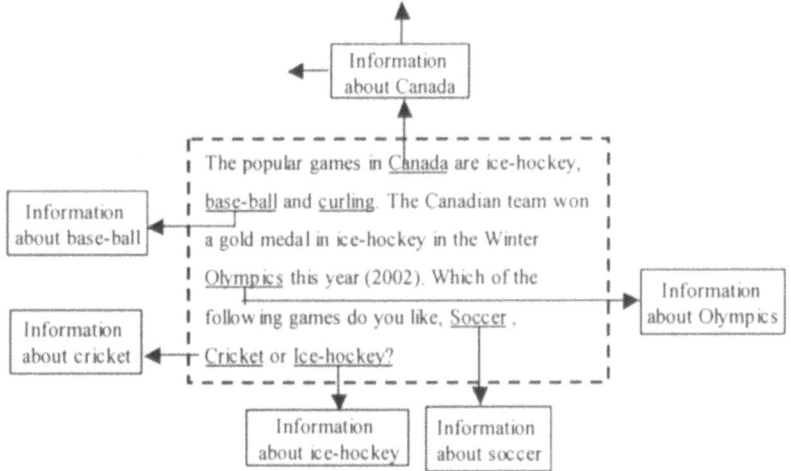

Figure 13.12. Reactive areas within a section of hypertext.

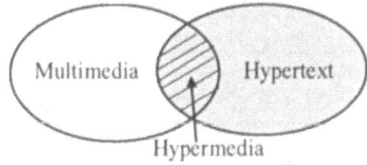

Figure 13.13. The hypertext, multimedia and hypermedia relationship.

13.4.3 Mark-up Languages

The concept of today's WWW was proposed more than half a century ago by Vannevar Bush. The concept was later refined by several researchers. A brief history of hypertext and web development is provided in Table 13.2. Only in the early 1990s, did it become feasible to implement the hypertext system, resulting in the World Wide Web. The web documents are generally written using the hypertext markup language (HTML) that is understood by all web browsers. However in the late 1990s, several other markup languages, such as XML and SIML have been proposed for different applications. A brief introduction to these markup languages is now presented.

13.4.4 HTML

The hypertext mark-up language (HTML) is the primary language [9] of all web pages on the WWW today. It can be considered as a (simplified) subset of the SGML, with emphasis on document formatting. HTML consists of a set of tags that are converted into meaningful notations and formats when displayed by a web browser. An example of an HTML document is given in Table 13.3, demonstrating that every start tag has an end tag. These two tags instruct the browser where the effect of a specific tag begins and ends.

Table 13.2. Brief history of the hypertext/WWW development.

Year	Events
1945	Vannevar Bush wrote about Memex (an imaginary device similar to today's multimedia workstation) in the landmark paper, "As We May Think."
1960	Ted Nelson started the Xanadu hypertext project. The terms "hypertext" and "hypermedia" for non-sequential writings and branching presentations were coined.
1969	IBM developed Generalized Mark-up Language (which is not specific to a particular system).
1973	Kahn and Cerf presented the ideas for the structure of Internet.
1986	SGML (ISO 8879) was established.
1987	Apple introduced Hypercard (an authoring system) for building interactive hypermedia documents.
1990	Tim Berners-Lee proposed the World Wide Web. HTML standard was proposed.
1993	Mosaic was developed in NCSA (National Center for Supercomputing Applications). Netscape was developed by the same team in 1994.
1995	JAVA was introduced for platform independent application development.
1998	The W3C consortium developed XML that that would embrace the WWW ethics of minimalist simplicity while retaining SGML's major virtues.

Example 13.1

A small HTML document is shown in Table 13.3. When the HTML document is viewed through a browser, a table will be displayed showing the names and courses taken by a group of students (see Fig. 13.14). A HTML document employs a large number of tags to generate a formatted document. A brief explanation of the meaning of different tags is as follows.

Typical HTML documents [9] start with an <HTML> tag and ends with an </HTML> tag. Between these two tags is the body of the document. The content between the <title> and </title>tags specifies the title of the document. The <head> tag provides the header information of a webpage. The content between the <h2> and </h2> tags display the headings of the document. The <table> tag generates a table with the border width specified in the border entity. The tags <TR> and </TR> create a table row. The content between <TH></TH> tag specifies the table headings. The content between <TD>and </TD> tags is used to specify the content in each cell of the row.

When this file is saved as *"Example13_1.html"*, and opened using a typical web browser such as Internet Explorer or Netscape, the table as shown in Fig. 13.14 will be displayed by the browser.

Table 13.3. XML code for the database in Example 13.1.

```
<HTML>
<title>Table in HTML</title>
<head>
<h2><font color="blue" >This table is created using HTML</font></h2>
</head>
<body>
<TABLE border="2">
<CAPTION> <h3><font color="red">Names of Students and Courses
Taken</font></h3></CAPTION>
<TR>
  <TH>Student <br>Number</TH>
  <TH>Name</TH>
  <TH>Multimedia Systems<br>by M. Mandal</TH>
  <TH>Digital Logic Design <br>by X. Sun</TH>
</TR>
<TR>  <TD>12345</TD><TD>Victor</TD><TD>x</TD><TD>x</TD>  </TR>
<TR>  <TD>32455</TD><TD>John</TD> <TD>x</TD><TD>-</TD>  </TR>
<TR><TD>32412</TD><TD>Andrew</TD> <TD></TD><TD>x</TD>  </TR>
<TR>  <TD>32851</TD><TD>Jack</TD><TD>x</TD><TD>x</TD>  </TR>
</TABLE>
</body>
</HTML>
```

Limitations of HTML

Although HTML has served its purpose, and has made the WWW very popular, it has several limitations. The foremost limitation is that it is not intelligent. It displays the data with the desired format without knowing what is being displayed. Consider the table created in Example 1. If we want to redisplay the table with the student IDs in an increasing order, the browser will not be able to do it. The browser does not know that the first column is displaying the student IDs of a group of students. It simply treats all data as dumb data without any meaning.

Second, the layout of an HTML document is static in the browser's window. The readers or the browsers do not have any control over the format in which the document is viewed. The control on the browser's side might be desirable in order to satisfy the user's requirements. For example, people with weak eyesight may prefer larger font sizes.

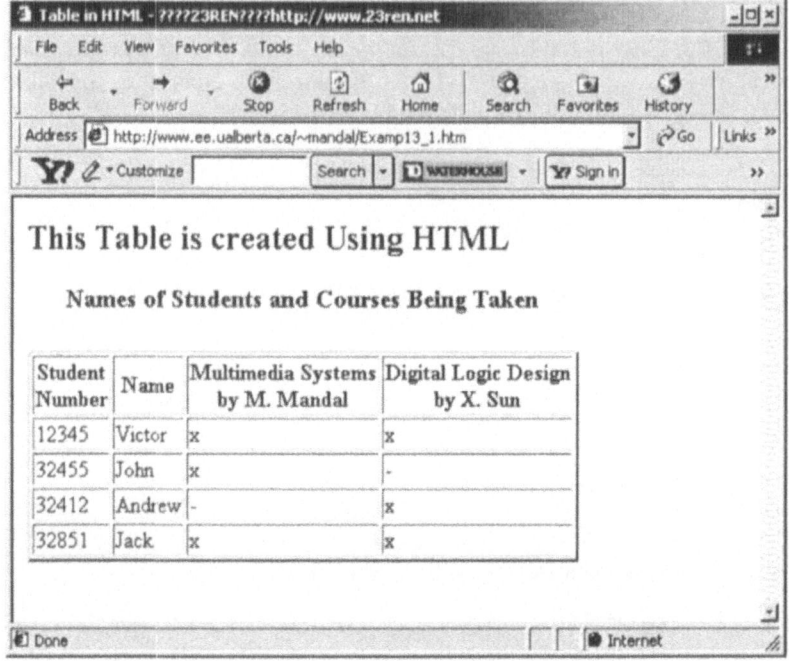

Figure 13.14: A Student Database is shown, generated by using HTML

Third, precisely marking information for all types of documents may require a large number of tags that are just not present in HTML. For example, an electrical engineer probably would want special symbols for electrical formulas, and measurement data. A car manufacturer may want to be able to talk about engines, parts and models. Satisfying the needs of all trades and people would require a large amount of elements, which may be difficult to manage by both developers and users.

Finally, the HTML has very little internal structure. One can easily write valid HTML that does not make sense semantically. One of the main reasons for this is that the contents of BODY have been defined such that one can place the allowed elements in almost any order. For example, one may consider H1 tag for the book title, H2 tag for the part title, and H3 tag as the Chapter title. Ideally, the H3 tags should be inside the H2 tags, and H2 tags should be inside the H1 tags. However, the HTML standard does not demand any such ordering.

13.4.5 XML

In order to address the limitation of the HTML, a new markup language was developed in 1996 by the WWW consortium's (also called w3 consortium) XML working group [10]. This is known as XML (eXtensible Markup Language). XML is a simplified version of the SGML, but contains the most useful features.

A major feature of XML is that a content developer can define his own markup language, and encode the information of a document much more precisely than is possible with HTML [11]. The programs processing these documents can "understand" them better, and therefore process the information in ways that are not possible with HTML. Imagine that you want to build a computer by assembling different parts from different companies according to a data type definition (DTD) that has been tailored for different computer parts. You can write a program that would choose parts from the specified companies, and give you an overall price. Conversely, you can write a program that would select the least expensive parts from all quoted prices, and give you the minimum price of a computer with a given configuration. In another case, given a price budget, the XML code will tell you the best computer configuration you can buy. The possibilities are almost endless because the information is encoded in a way that a computer can understand.

Comparison of XML and HTML

Although XML and HTML are both subsets of SGML, XML provides distinct advantages over HTML. HTML is a Markup Language for describing how the content of a document is rendered, *i.e.*, how the document would look in a browser. XML is a markup language for describing structured data, *i.e.*, content is separated from presentation. Because an XML document only contains data, applications decide how to display the data. For example a Personal Digital Assistant may render data differently than a cell phone or a desktop computer.

A document written in HTML and XML may appear the same in a browser, but the XML data is smart data. The HTML tells the browser how the data should be formatted, but XML tells the browser what it means. This is demonstrated in the following example.

Example 13.2

Consider the problem in Example 15.1. The same table can also be created using XML such that intelligent processing can be done. The code is provided in Table 13.4.

Table 13.4. XML code for the database in Example 13.2.

Student Database in XML	Table Generation

```<!doctype html public>```
```<html>```
```<body>```
```<xml ID ="xmlDoc">```
    ```<Students>```
        ```<student>```
        ```<Number>12345</Number>```
          ```<name> victor</name>```
          ```<ee635>x</ee635>```
          ```<ee601>x</ee601>```
        ```</student>```
        ```<student>```
        ```<Number>32455</Number>```
          ```<name> John</name>```
          ```<ee635>-</ee635>```
          ```<ee601>x</ee601>```
        ```</student>```
        ```<student>```
        ```<Number>32412</Number>```
          ```<name> Andrew</name>```
          ```<ee635>x</ee635>```
          ```<ee601>-</ee601>```
        ```</student>```
        ```<student>```
        ```<Number>32457</Number>```
          ```<name> jack</name>```
          ```<ee635>x</ee635>```
          ```<ee601>x</ee601>```
        ```</student>```
    ```</Students>```
```</xml>```
	```<table border ="1" datasrc="#xmlDoc">```
	```<h3><font color="green">This Table is created```
	```using XML embedded in HTML</font></h3>```
	```<caption><h3><font color ="red">Name of students```
	```and courses taken</font></h3> </caption>```
	```<thead>```
	 ```<tr>```
	   ``<th> Student
	 Number</th>```
	   ```<th> name </th>```
	   ``<th> Introduction to Multimedia
	by M.Mandal</th>```
	   ``<th> Digital Logic Design
	by X.Sun```
	```</th>```
	 ```</tr>```
	```</thead>```
	```<tr>```
	  ```<td><span datafld ="Number"></span></td>```
	    ```<td><span datafld ="name"></span></td>```
	    ```<td><span datafld ="ee601"></span></td>```
	    ```<td><span datafld ="ee635"></span></td>```
	```</tr>```
	```</table>```
	```</body>```
	```</html>```

The same table was generated in Example 13.1 using HTML and now it is generated using XML. Note that the XML does not have any predefined tags, and hence the browser does not know how to display the XML document. The XML document is generally displayed using an XSL (extensible style sheet language) document that describes how a document should be displayed. The XSL is a standard recommended by the World Wide Web Consortium. In this example, the XML is embedded in a HTML document, which describes the structure of XML document.

The flexibility of the XML in defining the tags can be observed in the code. Here, we can define user-friendly tags that make the document easier to understand. It is also easier for the search engines to look for specific tags if they are looking for some relevant information. The code in Example 13.2

is also easy to expand; if a new student arrives, we just need to add the code in column 1 of Table 13.5. The table will be expanded automatically and the new entry will be accommodated in it. In comparison, we consider Example 13.1, we have to add the code given in column 2 of Table 13.5. Such processes are more complicated for the user, and make searches more difficult.

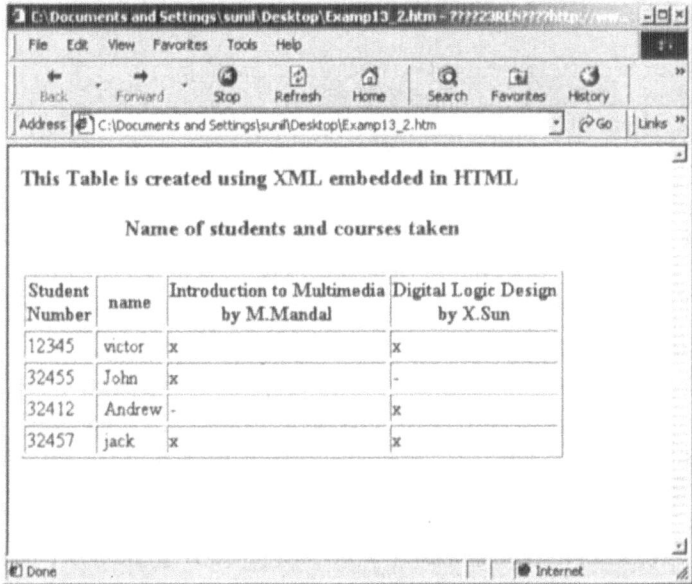

Figure 13.15: A student's database generated using XML embedded in HTML.

Table 13.5. XML code for adding a new student in the database in Example 13.2.

XML code	HTML code
<student> <Number>32412</Number> <name> </name> <ee635>x</ee635> <ee601>x</ee601> </student> <student>	<TR> <TD>32455</TD> <TD>John</TD> <TD>x</TD> <TD>-</TD> </TR>

13.5 WEB AUTHORING TOOLS

Web authoring has become very popular with the rapid growth of the Web. These tools [12] develop structured multimedia documents that can be accessed through the World Wide Web. There are primarily five types of tools available for web authoring: Markup Editing Tools, Multimedia

Creation Tools, Content Management Tools, Programming Tools and Conversion Tools. A brief description of each of these tool types is presented below.

Markup Editing Tools: These Tools assist authors to produce markup documents. These include text-based and WYSIWYG (what you see is what you get) editors for markup language such as HTML, XHTML, and SMIL, and word processors that save as markup formats. Examples of a few markup editing tools are

> Macromedia Dreamweaver, developed by Macromedia [6]
> Microsoft FrontPage, developed by Microsoft
> AdobeSiteMill, developed by Adobe [13]
> GoLiveCyberStudio – developed by GoLive Systems.Inc

Figure 13.7 shows the view of Macromedia Deamweaver, which is one of the most popular commercial markup editing tools. It provides an integrated site management tools, and file transfer protocol (FTP) is also incorporated within the application.

Multimedia Creation Tools: Such tools assist authors to create multimedia Web content without allowing access to the raw markup or code of the output format. These include multimedia production tools outputting SMIL or QuickTime, as well as image editors, video editors, and sounds editors. A popular multimedia creation tool is Corel Draw by Corel Corporation.

Content Management Tools: These tools assist authors to create and organize specific types of Web content without the author having control over the markup or programming implementation. Note that if the tool allows the author to control the markup that is actually used to implement the higher-order content, then that functionality would be considered to be a Markup Editing Tool. Rational Suite Content Studio-developed by Rational Inc is a popular content management tool.

Programming Tools: These tools create different kinds of Web Applications such as Java applets, Flash, server and client-side scripts. The programming tools can also assist authors to create markup languages (*i.e.* XML) and to create user interfaces (*i.e.* user interfcace markup language- UIML). For examples, Jbuilder developed by Borland Inc. is a programming tool which can develop java applets and standalone java applications. Macromedia Flash developed by Macromedia [6] is another popular programming tool.

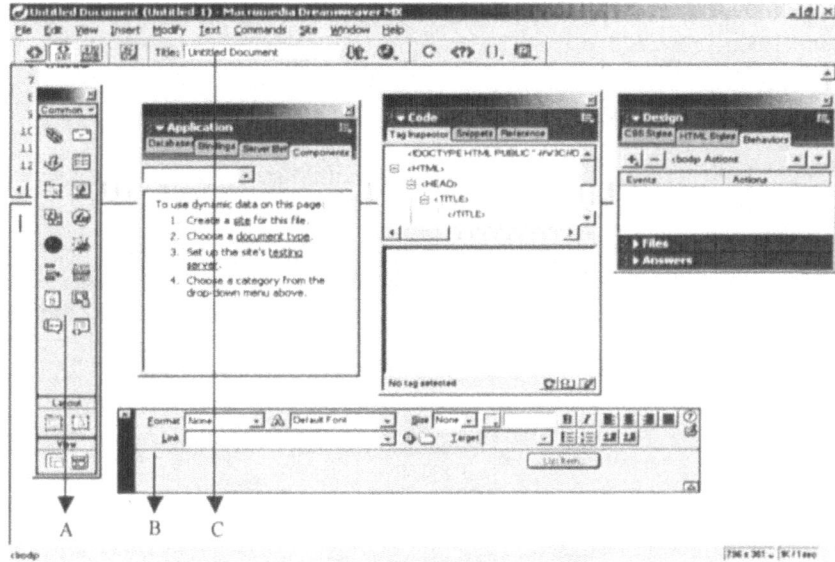

Figure 13.7. View of a Macromedia Dreamweaver. A:HTML objects, B: property window, and C: ToolBar.

Conversion Tools: These tools convert content from one format to another. This includes tools for changing the format of images, for conversion of other document formats to XHTML, and for importing document formats. For example, BeyondPress is a conversion tool used to convert QuarkXPress document to HTML/CSS. RTF2HTML is another conversion tool used to convert RTF documents to HTML documents.

13.6. MULTIMEDIA STANDARDS

Due to the rapid growth of the multimedia industry, several standards are being developed for creating and managing multimedia content. ISO/IEC JTC1/SC 29 is responsible for developing standards for representing the multimedia and hypermedia information. Figure 13.16 shows a schematic of different standard bodies, and their activities. There are three main working groups under it. The WG 1 is responsible for developing still image coding standards such as JBIG, JPEG, and JPEG-2000. The WG11 takes care of primarily the video content while the WG 12 takes care of coding the multimedia and hypermedia information. The first two working groups mainly focus on the information content whereas the last working group focuses on the information structure. In this section, a brief review of these standards is presented.

13.6.1 The MPEG-7 Standard

With the increasing popularity of multimedia, the volume of multimedia data is growing at a very rapid rate. Currently, a large amount of data is available in digital form for various applications such as digital library, World Wide Web, and personal databases. Proper identification and management of content is essential if the vast amount of digital data is to be efficiently used.

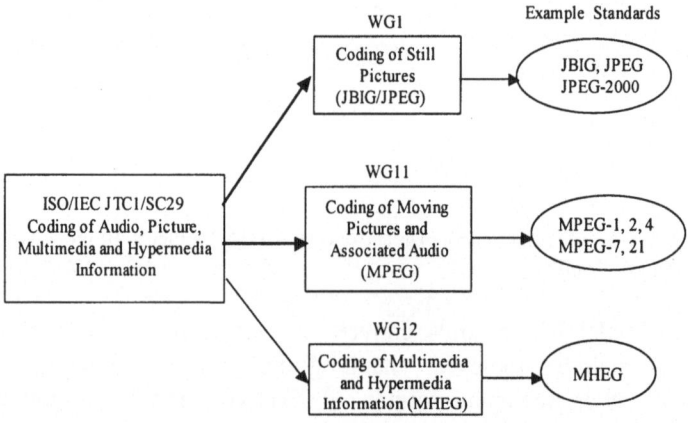

Figure 13.16. Relation between JPEG, MPEG and MHEG standards.

The database design has so far concentrated on the text data. Efficient search engines are available for searching data based on text query. However, multimedia data is significantly different from the text data. The content-based audio and visual retrieval systems seemed to be more appropriate for the multimedia data query. Block diagram of a typical multimedia data archival and retrieval system is shown in Fig. 13.17. Data matching and retrieval techniques are generally based on feature vectors. For example, images can be searched based on its color, texture, and shape. Generating feature vectors for efficient content-based retrieval is an active area of research [14]. A detailed review content-based retrieval techniques for images and video is presented in [15] (a copy of the thesis is included in the CD-ROM for interested readers).

In order to manage the multimedia (*e.g.*, audio, still image, graphics, 3D models, video) content efficiently, MPEG group of ISO is developing the MPEG-7 standard. Note that the MPEG has so far developed MPEG-1, MPEG-2, and MPEG-4 that are primarily video compression standards.

MPEG-7 is a break from this tradition. The MPEG-7 standard, formally known as "Multimedia Content Description Interface", provides a rich set of standardized tools to describe multimedia content [16]. The major emphasis here is to develop a standard set of descriptors that can be used to describe various types of multimedia information. The multimedia content is described in such a way that it can be understood by human users as well as automated systems, such as a search engine. The scope of MPEG-7 is schematically shown in Fig. 13.18. It is observed that MPEG-7 does not specify the algorithms for feature extraction, but standardizes only the description that provides a common interface between the *feature extraction* and the *search engine* subsystems.

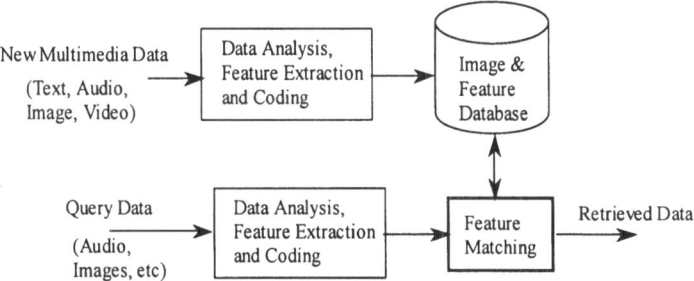

Figure 13.17. Schematic of a multimedia archival and retrieval system.

The MPEG-7 framework has three major components: i) a set of descriptors, ii) a set of description schemes and iii) a language to specify the description schemes. A *descriptor* is a representation of a feature. For example, color, texture, shape are important feature of an image whereas motion, camera operations are important features of a video sequence. Note that multiple descriptors can be assigned to a single feature. For example, the color feature can be described in several ways such as the most dominant color, and color histogram.

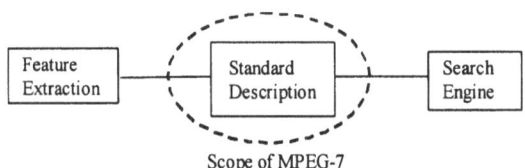

Scope of MPEG-7

Figure 13.18. Scope of the MPEG-7 standard.

A *description scheme* specifies the semantics and structure of the relationship between various components (these components may be other descriptors or other description schemes). Finally, a *data description language* (DDL) is required, which will allow the creation of new description schemes and descriptors. In addition, it would allow the extension and modification of the existing description schemes. Because of the flexibility of the XML schema language, the MPEG-7 group has decided to adopt it as the DDL for the MPEG-7 standard.

Although, MPEG-7 is not a coding standard, it builds on the other standard representations such as PCM, MPEG-1, MPEG-2 and MPEG-4. For example, a motion vector field in MPEG-1 may be used to describe the content of a scene, and the MPEG-4 shape descriptor of an object may be used in the content representation. MPEG-7 allows different granularity in its descriptions, offering the possibility to have different levels of discrimination. Even though the MPEG-7 description does not depend on the (coded) representation of the material, MPEG-7 can exploit the advantages provided by MPEG-4 coded content. If the material is encoded using MPEG-4, which provides the means to encode audio-visual material as objects having certain relations in time (synchronization) and space (on the screen for video, or in the room for audio), it will be possible to attach descriptions to elements (objects) within the scene, such as audio and visual objects.

13.6.2 The MPEG-21 Standard

Although, MPEG-7 is able to describe a multimedia object using metadata, it is not support the exchange (transaction) of objects between two users. The MPEG-21 is the latest (as of 2002) standard being developed by the MPEG, where an interoperable framework is developed in which the users would be able to exchange multimedia content in the form of digital items across various networks and devices [17].

The MPEG-21 framework and its relationship with other compression standards are shown in Fig. 13.19. It is observed that a digital item (DI), which is a package of digital work or compilations, is the unit of transaction in MPEG-21. Some key parts in MPEG-21 are: i) DI declaration, ii) DI identification, iii) Rights expression language (REL), iv) rights data dictionary (RDD), and v) DI adaptation.

The DI declaration includes identifiers, metadata and rights expression. The DI identification specifies an identifier naming and resolution system that allows digital items and their parts to be identified uniquely. The REL specifies a language for declaring rights and permission associated with a DI. The RDD defines the terms to be used in the rights expressions. The DI adaptation specifies an XML-based language for describing user environment, media resource adaptability and other concepts for adapting a DI to the users environment.

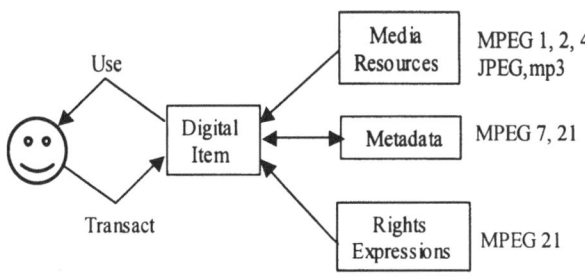

Figure 13.19. Transaction in MPEG-21 framework with digital item as the basis of transactions.

13.6.3 The MHEG Standard

MHEG [18] is an ISO/IEC International Standard developed by the Multimedia and Hypermedia Information Coding Expert Group (MHEG). It specifies a coded representation of final-form multimedia/hypermedia information objects. These objects define the structure of a multimedia/ hypermedia presentation in a system-independent way so that they can be easily exchanged as units within or across various systems from storage devices to telecommunication networks. The objectives of the MHEG standard are

- to provide a very simple but useful, and easy method to implement structure for multimedia applications requiring minimum system resources.
- to define a final digital format for presentations, making presentations easily expandable and machine independent.

There are two main levels of presentation in MHEG: *classes*, and *objects*. The MHEG classes represent the formal specification of the data structures used in an application. The semantics of the MHEG classes define the functionality and requirements for an MHEG runtime environment. MHEG defines a number of classes from which instances are created when a presentation is designed. In MHEG, these instances are known as MHEG objects.

13.7 SUMMARY

The first part of this Chapter presented an overview of the principles of multimedia authoring. Each multimedia project has its own underlying structure and purpose and will require different features and functions. Hence, the tools should be chosen according to one's needs. Note that today's word-processing software is powerful enough for most applications.

There is no reason to buy a dedicated multimedia-authoring package if your current software can perform the job; you can save money as well as the time spent on learning new authoring tools.

In the second part, the concept of hypertext/hypermedia was introduced. Similarities of the two most popular mark-up languages (HTML and XML) was discussed. Finally, a brief introduction presented the existing and upcoming multimedia standards, namely MPEG-7, MPEG-21, and MHEG.

REFERENCES

1. A. C. Luther, *Authoring Interactive Multimedia*, AP Professional, 1994.
2. D.E. Wolfgram, *Creating Multimedia Presentations*, QUE, 1994.
3. N. Chapman and J. Chapman, *Digital Multimedia*, John Wiley & Sons, 2000.
4. T. Vaughan, *Multimedia: Making it work*, McGraw-Hill, 1998.
5. Asymetrix Website, http://www.asymetrix.com
6. Macromedia Website, www.macromedia.com
7. P. Barker, *Exploring Hypermedia*, 1993.
8. R. Steinmatz and K. Nahrstedt, *Multimedia: Computing, Communications and Applications*, Prentice Hall, 1996.
9. Hypertext Markup Language, http://www.w3.org/MarkUp/
10. Extensible Markup Language, http://www.w3.org/TR/REC-xml
11. C. F. Goldfarb and P. Prescod, The XML Handbook, Prentice Hall, Second Edition, 2000.
12. Web authoring tools, http://www.w3.org/WAI/AU/tools.
13. ADOBE website, http://www.adobe.com.
14. V. Castelli and L. D. Bergman, *Image Databases: Search and Retrieval of Digital Imagery*, John Wiley & Sons, 2002
15. M. K. Mandal, *Wavelet Based Coding and Indexing of Images and Video*, Ph.D. Thesis, University of Ottawa, Fall 1998.
16. ISO/IEC JTC1/SC29/WG11 N2207, *MPEG-7 Context and Objectives*, March 1998.
17. ISO/IEC JTC1/SC29/WG11 N3500, *MPEG-21 Multimedia Framework PDTR*, July 2000.
18. J. R. Smith, "Multimedia content management in MPEG-21 framework," *Proc. of the SPIE: Internet Multimedia Management Systems III*, Vol. 4862, Boston, July 31-Aug 1, 2002.
19. Multimedia and Hypermedia Information Coding Expert Group, http://www.mheg.org/.

QUESTIONS

1. What is meant by multimedia? List the various components of multimedia.
2. Explain briefly a generic environment required for multimedia authoring.
3. Describe steps of multimedia authoring.

4. Discuss the principles of multimedia content design.

5. Discuss with examples different multimedia authoring tools.

6. Compare with examples different web authoring tools.

7. What are the three main models of multimedia document architecture?

8. Provide a schematic of SGML document processing.

9. What is the limitation of information representation in ordinary documents? How do hypermedia systems overcome this limitation? Explain the navigation process in hypertext systems.

10. What is HTML? Explain the navigation process supported by the HTML.

11. Explain the limitation of the HTML. How does XML solve some of the limitations of HTML?

12. Present an overview of the current web-authoring tools.

13. Provide an overview of the different multimedia standards that are being developed.

14. Explain the framework of MPEG-7 and MPEG-21 standards. How are these standards related to the other compression standards such as JPEG, and MPEG-{1,2,4}?

15. What are the objectives of the MHEG standard?

Chapter 14

Optical Storage Media

It was noted in the previous Chapters that multimedia data are voluminous. A high-density floppy disk (with 1.4 Mbytes storage) can store only a few images, while a short video can easily require a storage space up to tens of Mbytes. Large capacity storage space is thus required for recording and distribution of multimedia data. There is a variety of high capacity storage media available today. They are primarily based on magnetic (*e.g.*, hard disk, zip drive) and optical (*e.g.*, CD, laser disc, DVD) or a combination of both of these technologies (e.g., magneto-optical disks). The optical storage has become the most popular media for multimedia applications in the last two decades. In fact, in the early 1990s, computer retailers used to flaunt a PC with a CD drive as a *multimedia PC*. However, as CD drives are available in almost all PCs today, the term "multimedia PC" has lost its status.

There are primarily two popular types of optical storage media available on the market – the compact disk (CD) and the digital versatile disc (DVD). The CD, developed by Sony and Phillips, was introduced in 1983. It was initially used for music applications, and later extended to other applications, such as computer data and video storage. DVDs were introduced in the early 1990s. Both CD and DVDs come with different flavors: i) read-only (CD/DVD-ROM), ii) once writable (CD/DVD-R), and iii) rewritable (CD/DVD-RW).

In this Chapter, the fundamental principles of CD and DVD technology are presented. The first section covers the physical mechanism and characteristics, while the second section covers the multimedia data storage principles of the CD and DVD. There are several good books on CD and DVD technology, with a few listed in the reference [1-6]. The readers may consult theses references for more detailed information.

14.1 PHYSICAL MEDIUM

In this section, the physical characteristics of CDs and DVDs are presented. First, the physical dimensions of CDs and DVDs are detailed, followed by the material characteristics of CD/DVD media.

14.1.1 Cross-section of a CD

A CD is a flat circular disc with a thickness of approximately 1.2 mm. The physical dimensions of a typical CD are shown in Fig. 14.1. The outer diameter of a CD is 12 cm, and its inner hole diameter is 15 mm. However, the maximum and minimum recording diameters are 11.7 cm and 5.0 cm, respectively. Note that CDs are also available with an 8 cm outer diameter. However, these CDs have a lower storage capacity, and hence are not popular.

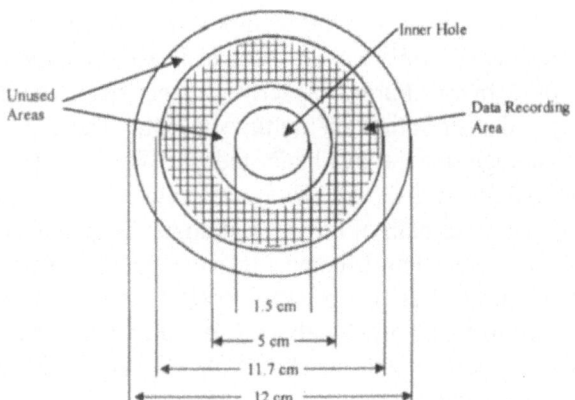

Figure 14.1. Physical dimensions of a CD (not drawn to the scale). Some CDs (although not as popular) have an outer diameter of 8 cm.

The CD data storage principle is very similar to a gramophone recording principle whereby the audio data pattern is recorded similar to grooves on a gramophone record. In a CD, the record pattern is pressed on one side of a transparent plastic substrate. The non-grooved regions are known as *lands* or *flats*, and the grooved regions are known as *pits*. The lands and pits are covered with a reflective and protective layer. This becomes the label side of the disc. Thus, there are primarily three layers in a CD - i) protective layer, ii) reflective layer, and iii) substrate layer (see Fig. 14.2(a)).

The data recorded on a CD is read by a laser beam that detects the lands and pits. The laser beam is focused through the transparent plastic substrate (see Fig. 14.2) and the reflected light is detected by a photodetector. Note that the pits appear as bumps from the substrate layer side. When the laser beam is focused on the lands, most light will fall on the surface, and will be returned in the same direction. However, when it falls on the pits, the light will be scattered, and only a small portion of the light will be returned in the same direction. In other words, the intensity of the laser light reflected from the lands is larger than that reflected from the pits, and thus the lands and pits (*i.e.*, the recorded binary data pattern) are detected. When the laser beam is focused on the reflective layer, the spot size of the beam is about 1 mm at the surface of the CD, but is about 1.7 μm at the reflective layer. Hence, if a small scratch or dust particle (smaller than 1 mm) is present on the CD surface, the laser beam can still be focused and the data can be read. However, the focus control must be very accurate for correct reading.

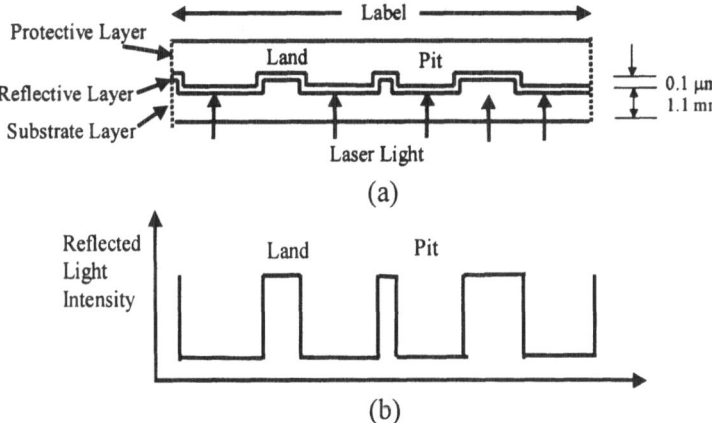

Figure 14.2. Different layers in a CD. a) The lands and pits in a CD, b) Reflected light intensity. The laser light is focused from the transparent substrate side.

14.1.2 Digital Versatile Disc (DVD)

The principles of the CD and DVD are very similar. Both CD and DVD have identical outer physical dimension (8/12 cm). However, DVD tracks are denser, and pits and land lengths are smaller (see Fig. 14.3), resulting in a much larger storage capacity. Table 14.1 shows the different parameters for CD and DVD. In the case of CD, the pits have a width of 0.6 μm, and a depth of 0.1 μm. The distance between the spiral tracks, known as *track pitch*, is 1.6 μm, while the pit spacing (*i.e.*, the distance between the center lines of two consecutive tracks) is about 2.0 μm. In the case of DVD, the dimensions are smaller as shown in Fig. 14.3(b) and Table 14.1.

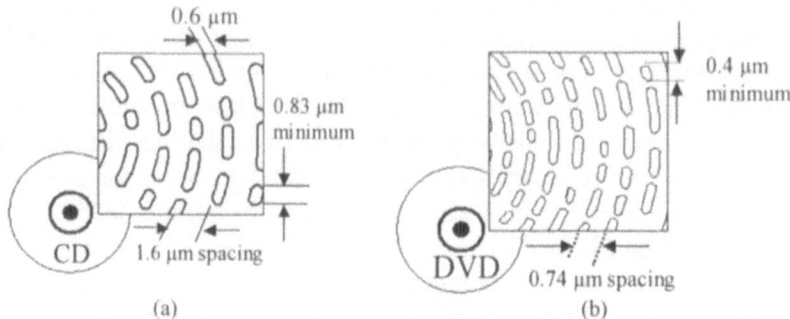

Figure 14.3. Tracks and pitch dimensions of a) CD and b) DVD.

Table 14.1. Physical Parameters of CD and DVD

	CD	Single-Layer DVD	Double-Layer DVD
Disc Diameter	12/8 cm	12/8 cm	12/8 cm
Track Pitch	1.6 μm	0.74 μm	0.74 μm
Minimum Pit Length	0.83 μm	0.4 μm	0.44 μm
Maximum Pit Length	3.05 μm	1.87 μm	2.05 μm
Laser Wavelength	780 nm	650 or 635 nm	650 or 635 nm

Layers of DVD

A typical (12 cm) CD has a storage capacity of 650 Mbytes, whereas a typical (12 cm) DVD has a storage capacity of 4.38 GBytes. However, in order to increase the storage capacity, a DVD can have multiple layers. Depending on the number of sides and layers, DVDs can be classified into four categories: i) single sided and single layer, ii) single-sided and dual layer, iii) double-sided and single layer, and iv) double-sided and dual-layer. Fig. 14.4 shows different layers of a DVD. Table 14.2 provides the storage capacities of different DVD types.

14.1.3 Physical Formats and Speeds

The physical format of CD/DVD is important to achieve the desired characteristics. In the last several decades, storage engineers have focused on designing efficient systems for erasable random access disks. The design goal of the CD/DVD format is different from that of hard disks. Here, large storage space is more desirable even at the cost of slower random access.

The entire storage space in a hard/floppy disk is divided into tracks that are concentric circles. The bits are packed closer in the inner tracks than the outer tracks. This variable data density allows the hard/floppy disks to spin with constant angular velocity as the head moves over the disc. This reduces the seek and latency times for data access.

Table 14.2. DVD Storage capacity ($1\ GB = 1.0737 \times 10^9\ bytes$)

DVD Type	Diameter	Capacity	Sides	Layers
DVD-1	8 cm	1.36 GB	1	1
DVD-2	8 cm	2.48 GB	1	2
DVD-3	8 cm	2.72 GB	2	1
DVD-4	8 cm	4.95 GB	2	2
DVD-5	12 cm	4.38 GB	1	1
DVD-9	12 cm	7.95 GB	1	2
DVD-10	12 cm	8.75 GB	2	1
DVD-18	12 cm	15.9 GB	2	2

CDs and DVDs, however, have a single spiral track similar to gramophone records. With each spin of a disc, the track advances outward from the center. The data density is uniform all over the disc. Hence, the outer edge packs more data compared to an equivalent magnetic media, resulting in a larger storage space. Since the data density is uniform, CD/DVDs spin with constant linear velocity. For example, if the CD spins at 1200 RPM at the inner edge, it will probably spin at 600 RPM at the outer edge. However, this variable spin rate increases the seek and latency times. The reading speeds of CDs and DVDs are shown in Table 14.3. The basic raw data rate (1x) of a CD is 4.32 Mbits/sec. After demodulation, decoding, error correction, and formatting, the net data rate is approximately 1.2 Mbits/sec (=150 Kbytes/sec).

Table 14.3. Reading speed of CD and DVD

	CD	Single-Layer DVD	Double-Layer DVD
Read Velocity	1.2 m/sec	3.49 m/sec	3.68 m/sec
Basic Read Rate	1.41/1.23 Mbits/s	11.08 Mbits/s	11.08 Mbits/s
Basic Read Rate	4.32 Mbits/s	26.16 Mbits/s	26.16 Mbits/s

Note that CD and DVD drives are available which can read data much faster. Table 14.4 shows the raw data rate of CD and DVD drives with different speed factors. A 6x CD (DVD) spins 12 times faster than a 1x CD (DVD), and reads data 12 times faster. Note also that the speed factors are meaningful for reading raw data. When we play audio from CDs or movies from DVDs, higher speed drives and x-factors make no difference. The audio or the video data would always be played back at 1x speed.

14.1.4 Playback of CD and DVD

The playback of a CD/DVD is the reverse process of the CD/DVD recording. Here, the objective is to read the raw data from a prerecorded CD/DVD, then demodulate the bitstream and reconstruct the signal or data.

The core of the playback is the optical pickup mechanism. Fig. 14.5 shows a popular optical readout system known as the three-beam method. In this method, a laser beam is produced by a laser diode. When the laser beam is passed through the diffraction grating, three beams are produced. The laser beams enter a polarizing prism, and only vertically polarized light passes through. When the beams pass through the quarter(1/4)-wave plate, the beam's polarization plane is rotated by 45^0. The polarized light is then passed through the actuator (i.e., the tracking and focusing system) and focused on the CD surface. The laser beams are reflected by the CD/DVD lands and pits, although the lands produce a larger intensity than the pits. The reflected light is passed through ¼ wave plate. This second passing makes the laser beam 90^0 polarized, $i.e.$, the original vertically polarized laser beams are now horizontally polarized. The horizontally polarized light is reflected by the prism and hits the three photodetectors. The voltage generated at the photodetectors determines if a land or a pit has been read by the laser beam.

Table 14.4. Speed factors of CD and DVD drives

CD Factor	DVD Factor	Raw Data Rate
1x	0.17x	4.32 Mbit/s
4x	0.67x	17.3 Mbit/s
6x	1x	26.2 Mbit/s
12x	2x	52.4 Mbit/s
24x	4x	105 Mbit/s
30x	5x	131 Mbit/s
36x	6x	157 Mbit/s

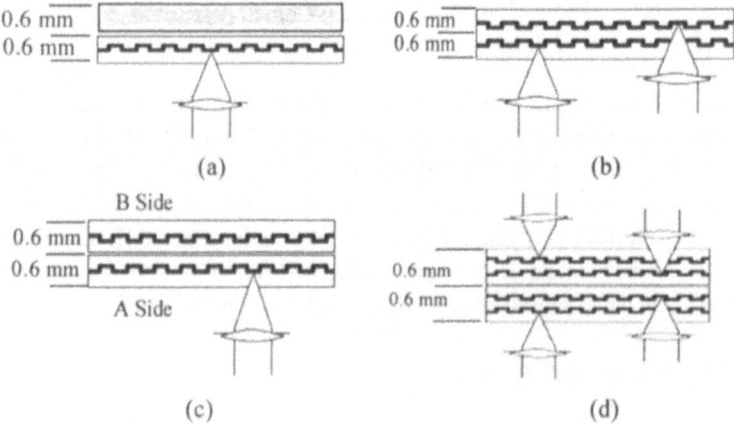

Figure 14.4. Layers in a DVD. a) single sided, single layer DVD (4.7 GB), b) single sided double layer (8.5 GB), c) double sided single layer (9.4 GB), and d) double sided, double layer (17 GB).

The three-beam principle employs three laser beams – the main laser beam along with two subbeams – to detect a spot. The two subbeams are on either side of the main beam. The reflection from the central spot (by the main beam) and the two side spots (by the side beams) are detected by three photodetectors. The central photodetector produces the digital signal corresponding to the central spot. The outputs of the two side photodetectors are compared, and an error signal is produced. This error signal is then fed to the tracking servomotor that moves the objective lens and keeps the main laser beam centered on the track. In addition, if the laser beam is not focused properly, the objective lens is moved towards or away from the CD (within $\pm 1\mu m$) to obtain a more accurate focus.

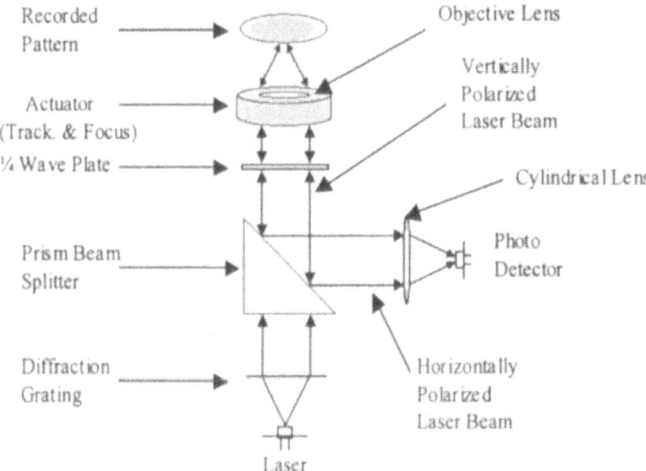

Figure 14.5. Playback of CD using the three-beam principle. Only one of the three laser beam and photo detector combinations is shown.

Reading CD/DVD by the Same Drive

Although the outer dimensions of a CD and DVD are identical, the inner dimensions are not the same. The layers in a DVD are thinner (0.6 mm) than a CD (1.2 mm). A DVD player is required to read both CD and DVD. This is generally achieved by controlling the focal length of the lens (see Fig. 14.6). A DVD drive generally distinguishes the separate layers of a multiple layer disc by selective focus. The separation between the two layers, about 20-70 μm, is sufficient for each layer to have a distinct focus for them to be distinguished.

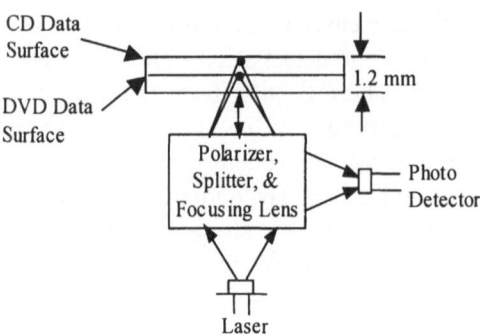

Figure 14.6. Reading CD and DVD by the same drive

14.1.5 CD-ROM, CD-R, and CD-RW

The CD, CD-R and CD-RW are employed in different applications, and hence have different requirements. The "read-only" media is used for mass distribution of data and software. This recording is fast and the CDs are durable. The CD-R media uses once-recordable material with a long life, whereas CD-RW media uses a medium that can be cycled between two states so that it can be erased and rewritten repeatedly.

CD-ROM

The CD-ROM recording is very similar to the gramophone (LP) recording process. A stamping master, which is usually made of nickel, is created such that it is the mirror image of the lands and pits that are to be created on a CD. To create the pits, the raised bumps on the stamper are pressed onto the liquid plastic (polycarbonate, polyolefine, acrylic or similar material) during the injection molding process. Once the pits are generated, the stamped surface is covered with a thin reflective layer of aluminum, which provides different reflectivity of lands and pits, and helps in the data reading process.

The DVD creation process is similar to that of a CD. However, more care is needed as the pits are smaller. In addition, there are more layers in a DVD. Hence, the recording is done layer-by-layer, and the substrates of different layers are glued.

CD-R Media

The CD-R recording process is distinct from the CD-ROM recording process. The CD-R recording is primarily done by the consumers, and therefore there is no mass production of the same content.

The CD-R disc has a special dye layer (see Fig. 14.7) that is not present in a CD-ROM. The dye is photoreactive and changes its reflectivity in response to the high power mode of the recorder's laser. The CD-R records information by burning the dye layer in the disc. There are three popular types of photoreactive dyes used by CD-R discs [7]. The green dye is based on cynanine compound, and has a useful life in the range of 75 years. The gold dye is based on phthalocynanine compound, and has a useful life of about 100 years and is better for high speed 2x-4x. The blue dye is based on cynanine compound with an alloyed silver substrate. These CDs are more resistant to UV radiation, and have a low block error rates.

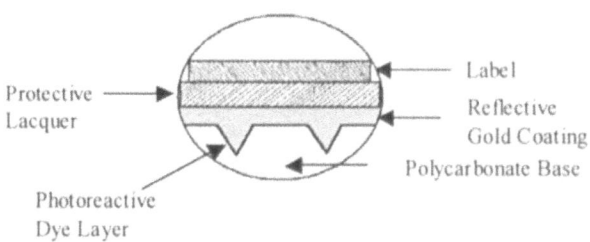

Figure 14.7. Different layers in a CD-R.

Rewritable CDs

CD-RW is based on the phase change principle. The reflective layer in the CD-RW is made from material that changes its reflectivity depending on whether it is in an amorphous or crystalline state. In its crystalline state, the medium has a reflectivity of about 15-25%. In its amorphous state, the reflectivity is reduced by a few %, which is enough for the detection of the state. The most common medium used in CD-RW is an alloy of antimony, indium, silver, and tellurium.

In a blank CD-RW, the entire reflective medium is in its crystalline state. When recording, the drive increases the laser power to between 8-15 mW, and heats the medium to above its 500-700^{0}C melting point. The operation is straightforward, and equivalent to the CD-R writing process except for laser power. To completely erase a disc and restore its original crystalline state, the disc is annealed. The reflective layer is heated to about 200^{0}C, and held at that temperature while the material recrystallizes.

Note that the recordable media is not as durable as the commercially stamped CDs. Recordable CDs are photosensitive, and should not be exposed to direct sunlight. The label side of a recordable CD is often protected only by a thin lacquer coating. This coating is susceptible to damage from solvents such as acetone and alcohol. Sharp-tipped pens should not be used to write on the labels as they may scratch the lacquer surface and damage the data.

14.1.6 Advantages of Optical Technology

Most multimedia storage devices used in our lives are either of magnetic (such as VHS video tapes, computer hard drives) or of optical type. It can be easily observed from our daily experience that the longevity of the magnetic storage media is not very good. For example, if we play a VHS cassette repeatedly, the video quality degrades. In this respect, optical technology has an important advantage over magnetic storage technology.

In magnetic storage, the magnetic property of the storage medium is exploited. The data is stored depending on the orientation of the tiny magnets present in the medium. The magnetic fields of hard disks are restricted to within a few millionths of an inch. Therefore, for recording or reading from a magnetic medium, the recording or the reading head must be very close (they virtually touch) to the magnetic medium. This decreases the longevity of the data stored in the magnetic medium. However, this is not the case with optical technology. The equipment that generates the beam of light that writes or reads optical storage need not be anywhere near the medium itself (see Fig. 14.8), thus providing significant flexibilities. Hence, the source and storage media generally have a small gap. In fact, a small gap helps in the playback process. If there is a small scratch on the CD/DVD surface (remember data is stored in the middle, and not in the surface), the laser light can still be focused on the data layer, and the data can be read. Since the optical head does not touch the substrate, the longevity of the optical media is very high. The CD or DVD has an added advantage. Since these disks are flat and work in the same way as a gramophone record, mass production of these disks are very convenient than the magnetic medium.

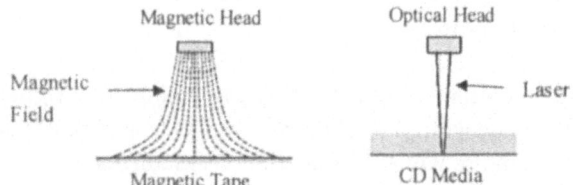

Figure 14.8. Comparison of playback in magnetic and optical storage.

14.2 CD AND DVD STANDARDS

There is a variety of CD and DVD standards targeted for various applications. The major CD and DVD formats are presented in Table 14.5. Note that for historical reasons the different standards are referred to as books with different (color) codes.

The CD-DA standard is primarily for recording audio data, and can be played back by home CD-players and computer CD-drives. The CD-ROM is targeted for mass distribution of data or software. CD-R and CD-RW are used for back up of large-size multimedia or computer data. The CD-V, also known as video-CD, is used for short low quality video. The CD-I (or CD-interactive) is the standard for the storage and distribution of interactive multimedia applications. SA-CD is the standard for recording very high quality audio signals.

There are primarily five DVD standards, documented in various Books, as shown in Table 14.5. Book A defines the physical specifications of the DVDs such as file system and data format. Book B defines the standard for video applications, whereas Book C defines the standard for audio applications. Books D and E defines the standard for DVD-R and DVD-RW storage specifications.

Table 14.5. Selected CD and DVD Standards

Book	Standards	Application
Red Book	CD-DA	Consumer Audio Playback
Yellow Book	CD-ROM	Data Distribution
Orange Book	CD-R, -RW	(Re) Writable Storage
White Book	CD-V	Consumer Video PB (with poor quality)
Green Book	CD-I (Interactive)	Interactive Multimedia Applications
Scarlet Book	SA-CD	Super Audio CD
Book A	DVD-ROM	Data Distribution, Computer Storage
Book B	DVD-Video	Consumer Video PB
Book C	DVD-Audio	Consumer Audio PB
Book D	DVD-R	Write Once Storage
Book E	DVD-RW	Rewritable Storage

In this section, we consider the three most popular standards: i) Audio CD, ii) CD-ROM, and iii) the DVD-video. The data format and their capacity in terms of length of audio and video data are briefly presented.

14.2.1 Audio CD

The CD was first introduced to the market with audio data represented using 16 bits/sample/channel precision and 44.1 KHz sampling rate. The data rate of a stereo audio CD can be calculated as follows:

$$\text{DataRate}_{CD-DA} = 16\frac{bits}{sample} \times 2\,channels \times 44100\frac{samples/s}{channel} = 1.4112 \times 10^6\,bits/s$$

$$= 176.4 \times 10^3\,bytes/s \approx 172.3KBytes/s \qquad (14.1)$$

The quantization of the analog audio signal to digital audio signal will introduce quantization noise in the audio signal and degrade the quality. It was mentioned in Chapter 4 that increasing the sample precision by 1 bit improves the SNR of the signal by approximately 6 dB. Hence, with 16 bits precision, the SNR of the digitized audio would be in the range of 96 dB. In practice, the CD-DA player manufacturers claim about a 90-95 dB dynamic range, which can be considered as excellent quality for most audio applications. Note that compared to CD, LP records provide a SNR of about 70 dB. There is another advantage of CD over LP. Each groove of an LP stereo record contains two signals, one each for the left and right channels. These signals are read and reproduced simultaneously by the turntable. In the case of CD, the left and right channel information is recorded separately. Therefore, the cross-talk between the two channels is minimized, improving the audio quality.

Eq. (14.1) shows that the CD-DA data rate is 1.4112×10^6 bits/s. However, the audio data needs to be encoded and formatted before storing it in the CD, resulting in an increase in the raw data rate. Fig. 14.9 shows the schematic of the CD-DA data encoding process. The encoder takes 24 bytes (*i.e.*, 192 bits) of raw audio data, and produces 588 bits of output, known as an audio *frame*. There are four steps in the encoding process, which are now briefly discussed.

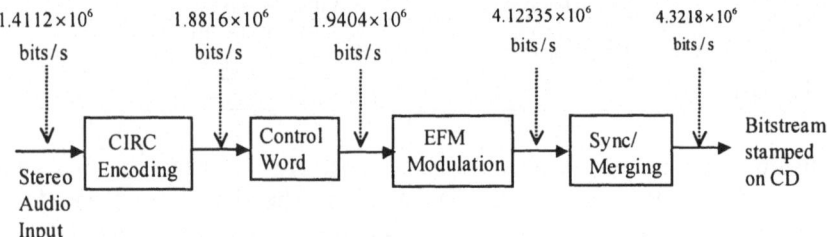

Figure 14.9. CD-DA data encoding process.

CIRC Encoding

In order to protect the data from bit errors, an error correction code is applied to the raw audio data. The bit errors may occur for various reasons, such as jitter, dust, scratches, and fingerprints. The CD-DA employs the cross-interleave Reed-Solomon code (CIRC) to correct the random and burst errors. The CIRC code converts 24 bytes of audio data into 32 bytes of CIRC encoded data, increasing the data rate by a ratio of 32:24. Hence, the data rate of the CIRC encoded audio data is 1.8816×10^6 bits/s. Note that although the CIRC code detects and corrects most bit-errors, there is a possibility that some bit errors may still go undetected.

Control Word

For each 32 bytes of CIRC encoded data, 8 bits control word is added, resulting in 33-bytes of output. The control word contains information such as the music track separator flag, track number and time.

EFM Modulation

The bits in a CD-DA are represented with lands and pits as shown in Fig. 14.10. Here, the ones are represented by a change from land-to-pit, or pit-to-land, and the zeros are represented by a continuation of the lands and pits. However, this scheme is not used exactly in practice. For example, consider the bit sequence 111111111. In this case, the pits and lands will occur alternately. It is difficult to design a laser system with sufficient resolution to read the sequence of lands and pits changing so frequently. When the CD standard was established, it was decided that at least two lands and two pits should always occur consecutively in a sequence. But, pits and lands should not be too long in order to correctly obtain synchronization signal. As a result, it was decided that at most ten zeros can follow one another.

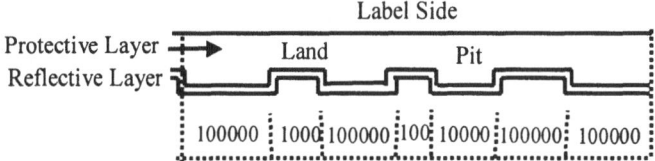

Figure 14.10. Bitstream representation with lands and pits.

In order to satisfy the above length constraints, the CD-DA system adopts a modulation scheme, known as *eight-to-fourteen modulation* (EFM). In this scheme, 8-bit words are coded as 14 bits. Note that with 14 bits, 16384 different symbols can be represented. Out of a possible 16384 symbols, only 256 symbols are chosen such that the minimum and maximum consecutive lengths of lands and pits are satisfied. An example of EFM is shown in Table 14.6. By limiting the maximum number of consecutive bits with same logic level, the overall DC content is reduced. In addition, by limiting the individual bit inversions (*i.e.*, $1 \rightarrow 0$ or $0 \rightarrow 1$) between consecutive bits, the overall bandwidth is lowered. Because of this modulation, the EFM produces 462-bit data from 264-bit (33-bytes) input.

The DVD also employs the EFM scheme, although with minor modifications to improve the robustness of the MPEG data recording. The scheme is known as EFM plus.

Sync/Merging

Even with the EFM, the conditions may not be satisfied at the beginning of the 14-bit symbol. Hence, in addition to the EFM, three extra bits are added to each 14-bit symbol. These three *merging* bits are selected such that the DC content of the signal is further reduced. The exact values of the merging bits depend on the adjacent bits of the 14-bit symbol. These extra merging bits produce 561-bit output for 462-bit input.

Table 14.6. Examples of eight-to-fourteen modulation

8 bit word	14-bit word		8 bit word	14-bit word
00000000 (0)	01001000100000		01100100 (100)	01000100100010
00000001 (1)	10000100000000		01110011 (115)	00100000100010
00000011 (2)	00100100000000		01111111 (127)	00100000000010
01001110 (80)	01000001001000		11110010 (242)	00000010001001

Finally, a 24-bit synchronization word along with 3 merging bits are added to the 462-bit input to complete a frame (consisting of a total of 588 bits) of audio data. The SYNC word has two functions. It indicates the start of each frame, as well as controlling the CD-player's motor speed.

Blocks and CD-DA Capacity

The smallest data unit in CD-DA is called a *block* and contains 2352 bytes of useful audio data. Since the stereo audio data rate is 1.4112×10^6 bits/s, the audio playback rate is 75 blocks/s. A typical CD-DA contains audio data for 74 minutes. Therefore, a CD-DA contains 74*60*75 or 333000 blocks. The storage capacity of the CD can then be calculated as

$$\text{Capacity}_{\text{CD-DA}} = 333,000 \text{blocks} \times 2352 \frac{\text{bytes}}{\text{block}} \approx 783.216 \times 10^6 \text{ bytes of audio data.}$$

The error correction and formatting bits increases the bit-rate significantly. Hence, if these bits are included, the raw capacity of the CD-DA is $783.216 \times 10^6 \times 4.3218/1.4112 \approx 2.4 \times 10^9$ bytes.

14.2.2 CD-ROM Standard

The CD-ROM format, used mostly for computer data storage, is similar to the CD-DA standard. The smallest data unit in both CD-DA and CD-ROM is a *block* that contains a total of 2352 bytes. The main difference between the two systems is the additional error detection and correction bits in the CD-ROM. The impact of a bit error in audio playback is not as crucial as in data storage application. A bit error in audio data will only degrade the audio quality marginally and may not even be detected by the listeners. However, a bit error in the computer data may have significant impact in many applications such as banking. The CD-DA employs EFM for error correction. The EFM has a low BER, but not small enough for computer storage application. Hence, 280 bytes (out of 2352) are used for additional error detection correction (in addition, 8 bytes are not used). In other words, 2048 bytes in each block are used for user data storage. The two formats are also known as Mode-1 (for computer storage) and Mode-2 (for audio recording and play back). With these extra error correction bytes, an average uncorrectable BER smaller than 10^{-12} can be achieved.

The user data storage capacity of the CD-ROM can be calculated as follows:

$$Capacity_{CD-ROM} = 333,000 blocks \times 2048 \frac{bytes}{block} \approx 681.984 \times 10^6 bytes \approx 650 \, MBytes$$

Although a CD has a raw capacity of more than 2 Gbytes, the effective data storage capacity is only 650 Mbytes. This is the storage capacity generally quoted on commercially available CD-Rs and CD-RWs.

14.2.3 DVD Data Storage Principles

The DVD data encoding process is very similar to that of the CD (see Fig. 14.9). However, as the DVD track pitch is thinner, and the pit lengths are smaller, more data can be packed in a DVD. There are some minor differences in the two data formats. First, CD has two main modes, one for the computer data storage and the other for the audio data storage. In audio CD, the data is stored without compression. If a bit error occurs, the neighboring samples are not affected significantly. Hence, a higher BER is tolerable. However, in the case of compressed video, bit-errors can significantly degrade the reconstructed video quality. Therefore, the DVD treats all data (*e.g.*, computer data, and compressed video bitstream) equally.

The DVD employs Reed-Solomon product code (RS-PC) instead of the CIRC code for error correction. The RS-PC code is more efficient for small blocks of data, which are more appropriate for computer data. In the case of DVD, the maximum correctable burst error is 2800 bytes, which corresponds to about 6 mm of track length (the maximum correctable burst error in a CD is about 500 bytes, which corresponds to about 2 mm of damage in a track).

The CD employs eight-to-fourteen modulation (EFM) where eight bits are converted to 14 bits to maintain the minimum and maximum number of zeros between two 1s. With the DVD, eight bits are converted to 16 bits, and the modulation scheme is known as EFMPlus (the *plus* indicates the addition of two extra bits).

In DVD, 2048 bytes of user data is stored in one sector. However, when header, EDC, ECC, and sync signals are added, the data size becomes 2418 bytes. The EFMplus (eight-to-sixteen) modulation again doubles the size, making the physical sector size 4836 bytes. The formatting and modulation comprise an effective overhead of 136%. Thus, although the DVD user data rate (1-speed) is 11.08 Mbits/s, the raw data rate is approximately 26.16 Mbits/s

14.2.4 Video-CD and DVD-Video Standards

The two most popular media for commercial video distribution are video-CD (or VCD in short) and DVD video. The two formats are compared in Table 14.7. It is observed that VCDs employ MPEG-1 algorithm, and have an overall (video and audio) data rate of 1.44 Mbits/sec. On the other hand, DVD-videos employ MPEG-2 compression algorithm, and have a variable data rate of 1-10 Mbits/sec, and an average data rate of 3.5 Mbits/sec. Note that DVD-video employ a higher bit-rate in addition to the more sophisticated (compared to MPEG-1) MPEG-2 encoder, and hence provides a superior video quality. As a result, the DVD has become very popular for movie distribution. In the following, we present more details about the DVD-Video standard.

Table 14.7. Comparison of video-CD and DVD video

	Video-CD	DVD-Video
Video Data Rate	1.44 Mbits/sec (video, audio)	1- 10 Mbits/s variable (video, audio, subtitles) Average = 3.5 Mbits/s
Video compression	MPEG-1	MPEG-2 (MPEG-1 is allowed)
Sound tracks	2 Channel-MPEG	NTSC: Dolby Digital PAL: MPEG-2
Subtitles	Open caption only	Up to 32 languages

There are primarily two types of DVD players available on the market. They may come as part of a personal computer, or they may come as a stand-alone DVD player. Both types of DVD players can generally play CD-Audio, and Video CD, in addition to DVD Video. With current technology, the DVD-ROM player in a PC works better with a separate hardware-based MPEG-2 decoder. The raw data rate of DVD video is 124 Mbits/sec, and the image resolution is approximately 480x720 pixels. This is about four times that of MPEG-1 or VHS (equivalent to 240 lines, compared to 480 lines quality of DVD) quality. The DVD Video supports both 4:3 and 16:9 aspect ratios. The associated audio can be in many forms. DVD video standard accommodates eight tracks of audio, with each track being a single data stream that may comprise one or more audio channels.

Each of these channels may use any one of the five specified encoding systems.

The average video data rate is approximately 3.5 Mbits/s. If audio and subtitles are added, the bit-rate is approximately 4.962 Mbits/s. A DVD with 4.7 GB net storage space can accommodate up to 133 minutes (=4.7*8*1024/(4.962*60)) of video. Hence, a single sided, single layer DVD can store up to 133 minutes of video. Although the "average" bit rate for digital video is often quoted as 3.5 Mbits/s, the actual bit-rate varies according to movie length, picture complexity and the number of audio channels required.

In order to prevent copyright violations, video DVDs may include copy protection. There are primarily three types of protection used in consumer video applications. In *analog copy protection*, the video output is altered so that the signals appears corrupt to VCRs when recording is attempted. The current analog protection system (APS) adds a rapid modulation to the colorburst signal and confuses the automatic gain control of the VCR. In *serial copy protection*, the copy generation management system adds information to line 21 of the NTSC video signal to tell the equipment if copying is permitted. In *digital encoding*, the digital media files are encrypted so that the video player cannot decode the signal without knowing the key.

Table 14.8. DVD regional codes

Code Number	Region
1	US, Canada
2	Europe, Japan, Middle East, South Africa
3	East and Southeast Asia
4	Central and South America, Caribbean, New Zealand, Australia
5	Africa, India, Korea, Soviet Union
6	China

Commercial DVDs are marked with a regional code (see Table 14.8). The DVD player checks to see if the region code in the DVD matches the code in its hardware. If it does not match, the DVD will not play.

A DVD drive does everything a CD drive does, and more. As DVD drives become inexpensive, it might be better to buy DVD drives. Although a DVD movie can be played using a computer drive and watched on the monitor, the computer DVD drives generally do not have a video jack (although, they typically provides audio output) as found in the home video players. Hence, these DVD drives cannot be directly used to watch a movie on television. In addition, computer DVD drives mostly use a software-based MPEG-2 decoder, which may be slow if the CPU is not powerful enough.

REFERENCES

1. J. D. Lenk, *Lenk's Laser Handbook*, McGraw Hill, 1992.
2. E. W. Williams, *The CD-ROM and Optical Disc Record System*, Oxford University Press, New York, 1994.
3. L. Boden, *Mastering CD-ROM technology*, John Wiley & Sons, New York, 1995.
4. J. Taylor, *DVD Demystified: The Guidebook for DVD-video and DVD-ROM*, McGraw-Hill, New York, 1998.
5. A. Khurshudov, *Essential Guide to Computer Data Storage: From Floppy to DVD*, Prentice Hall PTR, 2001.
6. L. Baert, L. Theunissen, and G. Vergult, *Digital Audio and Compact Disc Technology*, Second Edition, Newnes, Oxford, 1992.
7. W. L. Rosch, *Hardware Bible*, QUE, Fifth Edition, Indianapolis, 1999.

QUESTIONS

1. What are the advantages (and disadvantages) of optical storage media over magnetic storage media?
2. Compare and contrast the storage principle of the CD and gramophone record.
3. Compare and contrast the physical dimensions of CD and DVD.
4. How does a DVD drive read a CD?
5. Compare and contrast the writing principles of CD-ROM, CD-R, and CD-RW.
6. Why do we need CIRC encoding and EFM modulation? What is EFMPlus?
7. What is the storage capacity of a DVD? What is the raw storage capacity including the modulation bytes?
8. How many minutes of a digital movie can be stored in a DVD?
9. You have recorded your computer data in a CD-ROM (74 Minutes equivalent) with its full capacity. How long will it take if you read the CD at 12x speed? Assuming a BER of 10^{-12}, what is the probability that there will be a bit error while reading the entire CD?

Chapter 15

Electronic Displays

Electronic display is one of the most conspicuous electronic devices, and plays a significant role in visualizing multimedia information. The displays are used in a wide variety of consumer applications such as television, computer, calculators, scientific and military applications. Depending on the application, the display may also be called a monitor, a video display terminal, or a television receiver. Display devices have been available for several decades, and are continuously being improved. With the establishment of the high definition television standard, superior quality TV monitors are in high demand.

The cathode ray tube (CRT) displays have dominated the electronic display market for a long time. However, in the last few decades, several other types of display devices, such as vacuum-fluorescent display (VFD), field emissive display, plasma display panel (PDP), electroluminescent display, light emitting diode (LED) display, liquid crystal display (LCD), and digital micromirror display (DMD), have been developed [1, 2]. In this Chapter, a brief overview of a few select display technologies is presented.

15.1 IMPORTANT DISPLAY PARAMETERS

There is a wide range of applications for a display device. Hence, there is also a wide variety of display devices from which a suitable display can be chosen depending on the application requirements. Some important physical parameters for a display are its size, weight, resolution, gamma, color saturation, and refresh rate [1, 3]. On the other hand, the perceptual parameters such as brightness, flicker, contrast, and sharpness of a display are also important to obtain a better picture quality. A few select parameters are explained in the following.

Display size: This is one of the most important parameters of a display device. The display size is generally expressed in terms of the diagonal length of the viewing area. Some applications such as large TV monitors or projection TVs require a display size in the range 50-200". In the other extreme, calculators, digital watches and LCD viewfinders in video camera require a small display size of only a few inches. Typical computer monitors use a display size in the range 15-24", whereas typical home televisions use a display size in the range 13-48". Note that the weight and cost of display increases rapidly with the screen size. Hence, large displays (>70") are generally obtained by optically projecting the image from a smaller display. The cathode ray tube, liquid crystal display, and digital micro-mirror devices are typically used in projection systems.

Display Thickness and Weight: These parameters are also important to consumers, especially when there are constraints of physical space and weight. The current trend is to employ thinner [4, 5] and less bulky display devices.

Resolution: The resolution of a display device is expressed in terms of the number of pixels in the horizontal and vertical directions. The resolution is limited significantly by the display size. Typical computer monitors have resolutions from 640×480 to 1600×1200. On the other hand, typical digital TVs have resolutions in the range 720×480. While high definition TVs have resolutions of up to 1920×1080, the portable devices have resolutions as low as 50×50.

Color: This is another important aspect of a display device. Some devices such as digital watches, and calculators may not need a color display. However, entertainment applications generally require good color display. The number of colors supported in a typical display is in the range of 256-16 million colors. Note that 256, and 16 millions correspond to 8-bit and 24 bit/pixel resolutions, respectively.

Brightness and Contrast Ratio: Brightness is the result of all illumination falling on the surface, and the reflectance of the surface, whereas the contrast ratio is the ratio of maximum brightness and minimum brightness visible. Note that more ambient illumination decreases the contrast ratio.

Aspect Ratio: It is the ratio of the width and height of a display device. The aspect ratio of most computer monitors and analog TVs are 4:3. However, HDTV has an aspect ratio of 16:9.

Angle of View: Angle of view is the angle up to which the output of the display device can be read by a viewer. Computer monitors can have a narrow angle of view. However, TV and the projection displays must have a large angle of view since these are targeted for small or large groups of people.

Light Source: In all display devices, the light come from the display screen and enters our eyes, and we see the pictures. The display devices can be divided into two categories based on how the light come from the screen. In the emissive display, the light (or photons) is emitted by the display screen. This typically happens in most home TV receivers, which use the cathode ray tubes. In the non-emissive display, the display device does not emit any light. Rather, the display device transmits or reflects lights coming from another source. Note that LCD screens are of non-emissive type. The emissive devices generally require high power and hence are not efficient for portable application. On the other hand, non-emissive display devices depend on external light source for illumination and hence require lower power to operate.

15.2 CATHODE-RAY TUBE

The cathode ray tube (CRT) is the oldest and most widely used display device for a wide range of applications. The schematic of a CRT is shown in Fig. 15.1. The core of a CRT is an electron gun (which is basically a cathode) that emits electrons. A strong electric field is created by applying a high voltage (> 20000 V) between the anode and the cathode. The electric field is set up such that an electron beam is generated, which points towards the phosphor screen of the display. The electron beam impinges on the phosphor screen, which in turn emits light that is seen by a viewer. Hence the CRT is basically an emissive display device.

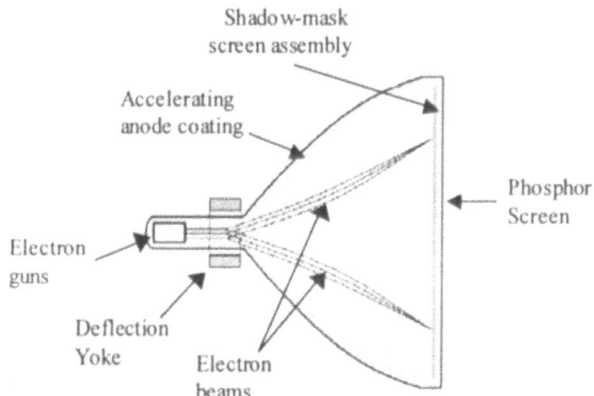

Figure 15.1. Different components of a cathode ray tube (CRT). Note that for color CRT, there will be three electron guns.

Note that the CRT is evacuated so that the electron beam emitted from the cathode passes through the tube easily. The raster scanning is accomplished by deflection of the electron beam using a set of magnetic coil around the neck of the CRT. The monochrome CRT has one electron gun, whereas a color CRT has three electron guns, one each for red, green and blue channels. The phosphor screen of a color display has three dots (red, green, and blue) for each pixel. These three phosphor dots are known as triplet or triad.

The electron guns point the beam at the individual phosphor dots while scanning. However, part of the beam can spill over and hit the other dots in the triplet, or even the neighboring pixel dots. This spilling over results in loss of color purity. CRT devices use a masking screen to stop these spilled-over electrons to improve the quality of the displayed image. There are primarily two types of masking screens – shadow mask and aperture grille (see Figs. 15.2 and 15.3).

The shadow mask (see Fig. 15.2) is a steel plate with a large number of holes (one hole for each color pixel). The three electron beams corresponding to each color pixel pass through these triplet holes and hit the phosphor screen. The spilled-over electrons are stopped by the shadow mask. The shadow mask is located very close to the phosphor screen, and hence the phosphor dots on the screen must be spaced at the same distance as the holes in the mask. Since the hole spacing is the same as the dot spacing, it is often called the *dot-pitch* of the CRT. Typical dot pitch is 0.28 mm or 0.40 mm.

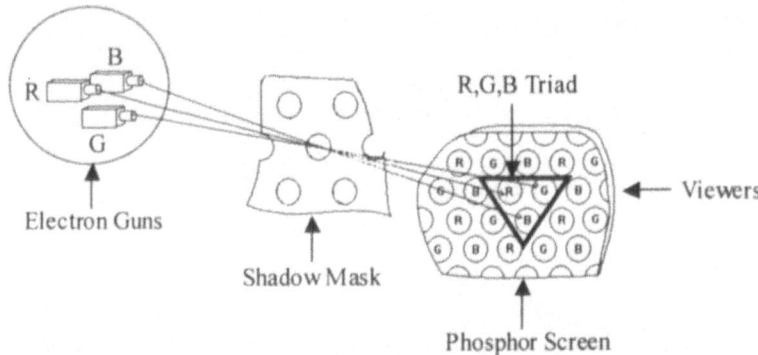

Figure 15.2. Three of phosphor dots and electron guns are arranged in a triangle.

Some CRTs, such as Sony Triniton CRTs, use an aperture grille instead of a shadow mask. Here, instead of circular holes, the electron beam passes through thin vertical windows (or slots) as shown in Fig. 15.3. The resolution of an aperture grille is specified in *slot-pitch*, which is the distance between the two slots.

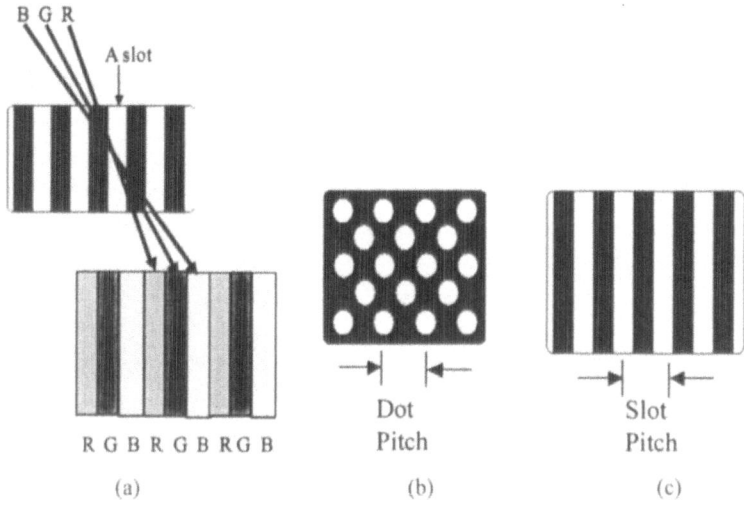

Figure 15.3. Masking of spilled-over electrons. a) aperture grille, b) dot-pitch for shadow mask, and c) slot-pitch for aperture grille. Note that in aperture grille, the red, green and blue phosphor dots are arranged in vertical direction.

A CRT can have different maximum deflection angles. A larger angle gives a smaller depth, but also makes the design more difficult to maintain focus and geometry over the full screen. As a result, larger deflection angles provide poorer picture performance. Computer displays generally have more demanding specifications and use a smaller angle CRT.

A major disadvantage of the CRT is that a large display typically has a large depth (*i.e.*, distance between the electron gun from the phosphor screen), resulting in a bulky display. The primary reason for the large depth is the single (or three for color CRTs) electron gun. A CRT of smaller depth can be designed if multiple electron guns are allowed. This is the basic principle of the field emission displays (FED), which is presented next.

15.3   FIELD EMISSION DISPLAY

The Field Emission Displays (FEDs) are high-resolution emissive flat panel displays [6, 7]. The FED works on the same principle as the CRT. However, instead of one (or three for color) electron gun, FED employs millions of micro cathodes. As a result, FEDs can produce large screen displays without increasing the depth of the display device.

Although, the concept of FED seems very attractive, there are several implementation issues that need attention. Some critical issues are microfabrication of cathodes, packaging of vacuum panel with thin glass substrates, assembly technology with micrometer level accuracy, and vacuum technology to keep the emission within the small space of flat panels.

A simplified schematic of the FED is shown in Fig. 15.4. The electrons are generated in high vacuum by the micro-tips of the cathode and are accelerated towards the phosphors by the positive voltage applied on the transparent anode. The electron beams hit the phosphor screen, and photons of different colors are emitted from the phosphors, which is observed by the viewers.

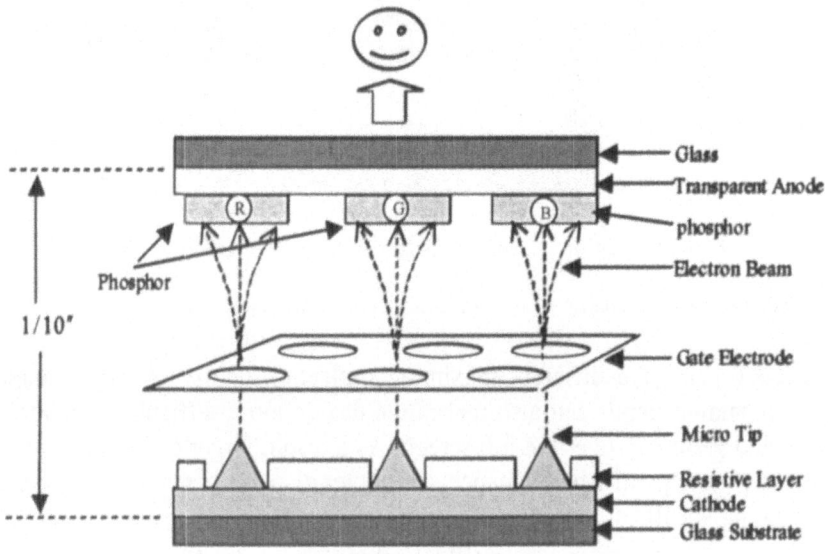

Figure 15.4. Simplified schematic of field emission display devices.

As mentioned earlier, FEDs employ multiple cathodes, one for each color channel of a pixel, unlike the single cathodes (or electron gun) in the case of the CRT. The distance between the cathode and the phosphor screens in FED can be made as small as $1/10''$ (as shown in Fig. 15.4) resulting in a flat panel display. Note that although the CRTs have high anode voltage, the anode voltage in typical FEDs can vary from one to several thousands volts.

Typical FEDs have several important features. They are thin, lightweight, and have low power consumption. High resolution, brightness and contrast can be easily achieved in FED. The FEDs can produce self-emissive distortion free images, and can provide a wide angle of view (as wide as 170^0). In addition, FEDS can withstand a wide range of temperature and humidity, and have good stable characteristics in severe environmental conditions.

Note that the FED technology is in its infancy, being only a few years old. Hence, only small screen displays (for automotive or small portable applications such as PDA) have been developed so far. However, it is expected to enter the consumer TV market in near future.

15.4 PLASMA DISPLAY

The plasma display technology is another leading contender for flat panel display. The plasma displays were used in the early laptop computers in the late 1980s. Although, it could not compete with the LCD technology for portable application, it has all essential advantages of large-size flat displays.

The plasma display panels (PDP) are based on radiation produced by an electronic discharge in a gas such as neon. When a large number of gas atoms loose one or more electrons, the gas is in an ionized state. With appropriate electric field, this process can be lead to discharge, and the gas is referred to as plasma, which emits radiation. This is the basic principle of all plasma displays.

A large number of monochrome PDPs use the discharge as the light source. However, in most color PDPs, the discharge is used to excite the (red, green and blue dots of a) phosphor screen. The schematic [3, 8] of a color PDP is shown in Fig. 15.5. The two glass plates (front and rear) contain a mixture of inert gases (such as neon and xenon). The inert gas is excited locally by electric fields set up by arrays of parallel and orthogonal electrodes. When sufficient voltage is applied to a pair of horizontal and vertical electrodes, a discharge occurs locally at the intersection of the electrodes. In the discharge, UV energy is radiated, which excites the color phosphor located nearby. The photons radiated by the phosphors are observed by the viewers.

The voltage-current relationship in PDP is shown in Fig. 15.6. When the voltage applied between the electrodes is small, there is little current. When the voltage reaches the breakdown point, the current in the cell increases rapidly. The voltage-current characteristic is highly nonlinear at the breakdown voltage V_f (see Fig. 15.6). Hence, a mechanism must be developed to control the current. In order to control the discharge, an dielectric layer is generally placed near the top electrode of the panel to form a capacitor [3]. When a cell is fired, the dielectric layer accumulates charge, and reduces the effective voltage applied to the electrodes. When enough charge accumulates at the dielectric layer, the voltage drops and the discharge is stopped. However, the charge on the dielectric, known as the wall voltage, remains in the dielectric layer. If this charge is not erased, it will fire the cell and emit a pulse of light for every successive cycle until the charge is erased by external means.

Note that the PDP is a truly digital display device where the cells are either ON or OFF, resulting in either dark or bright pixels. Hence, a mechanism must be devised to obtain a large number of brightness levels. This is done by a method similar to a digital to analog converter. For example, consider the decimal number 11 that can be represented as 1011 in binary format where the weights of 1, 0, 1, and 1 are 8, 4, 2, and 1, respectively (see Fig. 15.7(a)).

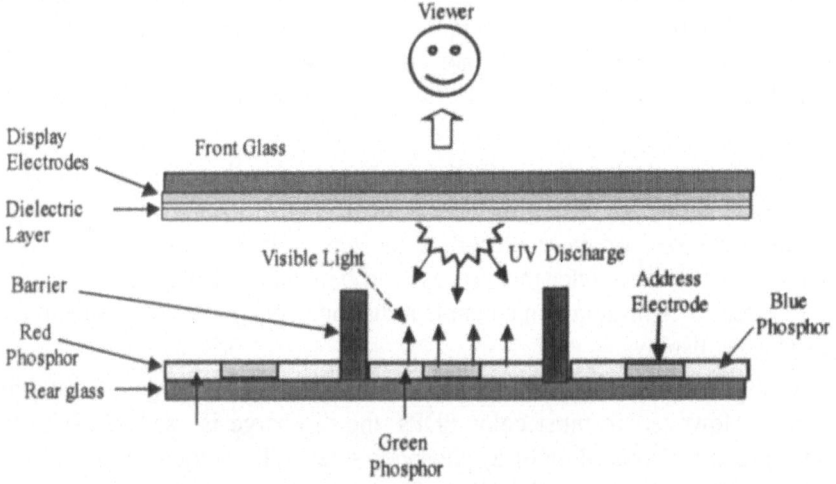

Figure 15.5. Simplified cross-section of a plasma display panel. Three dots (electrode combinations) are used for three color channels.

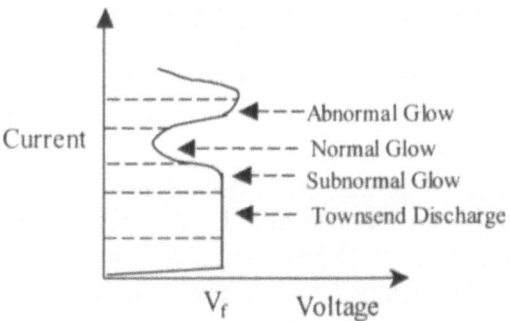

Figure 15.6. Voltage-current characteristics of a PDP Cell. V_f is the firing or breakdown voltage. In the Normal Glow region the cell is ON state.

The cell brightness in PDP can be controlled as follows. Assume that the PDP cell has to be made ON for τ time unit to obtain the brightness level of one (i.e., the level just brighter than a complete dark pixel). A brightness level corresponding to gray level 11 can be then obtained by first switching

ON the PDP for 8τ time units, switching it OFF for 4τ time units, and then switching it ON for 2τ, followed by another τ time units (see Fig. 15.7(b)). This switching ON and OFF is performed very fast, and due to the lowpass nature of the human visual system, it appears as a smooth brightness level. Even though at any time the pixel illumination is either on or off, the overall brightness would be corresponding to gray level 11.

$$1\ 0\ 1\ 1\ (\text{binary}) = 1 \times 2^3 + 0 \times 2^2 + 1 \times 2^1 + 1 \times 2^0 = 11\ (\text{decimal})$$

(a)

ON   OFF   ON

(b)

Figure 15.7. Controlling cell brightness in a PDP Cell

If the input image to be displayed has 8-bit resolution, the field interval is divided into eight periods (see Fig. 15.8). Each period contains a short time interval to erase the wall voltage in all cells. This is followed by a longer interval to scan all cells and specify whether they are to be ON or OFF (*i.e.*, write bit operation). This is followed by an interval called the *sustain interval*. In this period, an AC voltage is applied to the cells so that all cells that have wall voltages are fired repeatedly. The sustain interval of two consecutive bits has a ratio of 2:1 (as shown in Fig. 15.8). The light output from the eight sustain periods during each frame adds up with a power of two, similar to how the currents sum up in a conventional DAC.

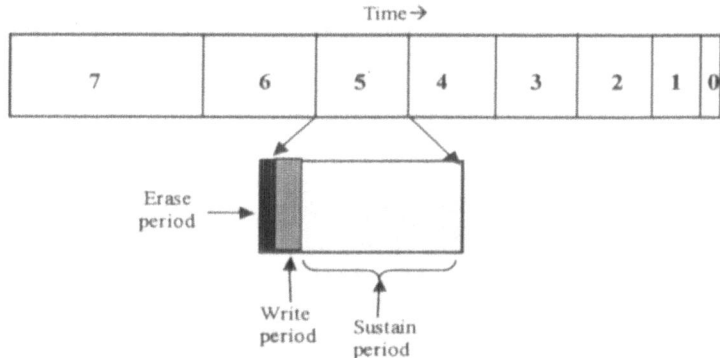

Figure 15.8. Timing diagram for the PDP display. The time intervals are not drawn to scale.

The angle of view in a PDP display is very wide and hence the display is suitable for any viewing situation. Although PDP principles are similar to the fluorescent lamp, unfortunately, the luminous efficiency is substantially poorer than a fluorescent lamp.

Note that the CRTs and the FEDs need high vacuum for their operations, and thus require use of thicker glass plates. The PDP, however, do not require thick glass plates as there is little difference between the inside and outside pressures. In addition, driving a large PDP display is not difficult as the capacitance between the PDP electrodes is small.

Unfortunately, even after 30+ years of research and development, PDPs are still expensive compared to the other flat panel displays. Hence, PDP technology is generally used in high-end display applications. For example, the Hitachi model CMP4121HDU is a 42″ plasma display TV with 1024×1024 display resolution, aspect ratio of 16:9, 16 million colors, 256 levels of gray, unit depth of 3.5″, and 68 lbs of weight. It costs about U.S $5000 (as of August 2002). However, when the PDP-based HDTV would be manufactured in large quantities, the unit prices are expected to drop.

15.5 LIQUID CRYSTAL DISPLAY

LCD displays are based on liquid crystal technology. Typical crystals are solid, and the molecular positions in a crystal are generally ordered, similar to the molecules in other solid materials. The molecules in a liquid crystal, however, do not exhibit any positional order, although they do possess a certain degree of orientational order.

The liquid crystals (LCs) can be classified according to the symmetry of their positional order. LCs can be nematic, chiral nematic, smectic or chiral smectic. In nematic liquid crystals, all positional order is lost, only the orientation order remains. Figure 15.9(a) shows the molecules in a nematic crystal [9, 10]. In a chiral nematic, the molecules are asymmetric and this causes an angular twist between the molecules (see Fig. 15.9(b)). In smectic, the molecules tend to exist in layers, and are oriented perpendicular to the plane of layers. Figure 15.9(c) shows molecules in a smectic material with two layers. The smectic material exhibits better viewing angle characteristics and contrast ratio. In chiral smectic, the molecules are in helical structure.

The nematic molecules of LCs have a special property – they can be aligned by grooves in the plastic to bend the polarity of the light that passes through them. The amount of bending depends on the electric field (or current) applied. Ordinary light has no particular orientation, so LCs do not visibly alter it. But in polarized light, the oscillations of the photons are aligned in a single direction. A polarizing filter creates polarized light by allowing light of a particular polarity to pass through. Polarization is the key to the function of LCDs.

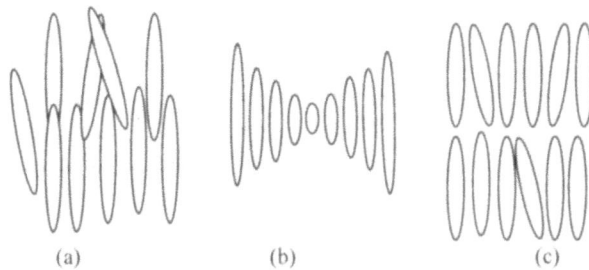

(a)   (b)   (c)

Figure 15.9. Different types of liquid crystals. a) nematic, b) chiral nematic, and c) smectic.

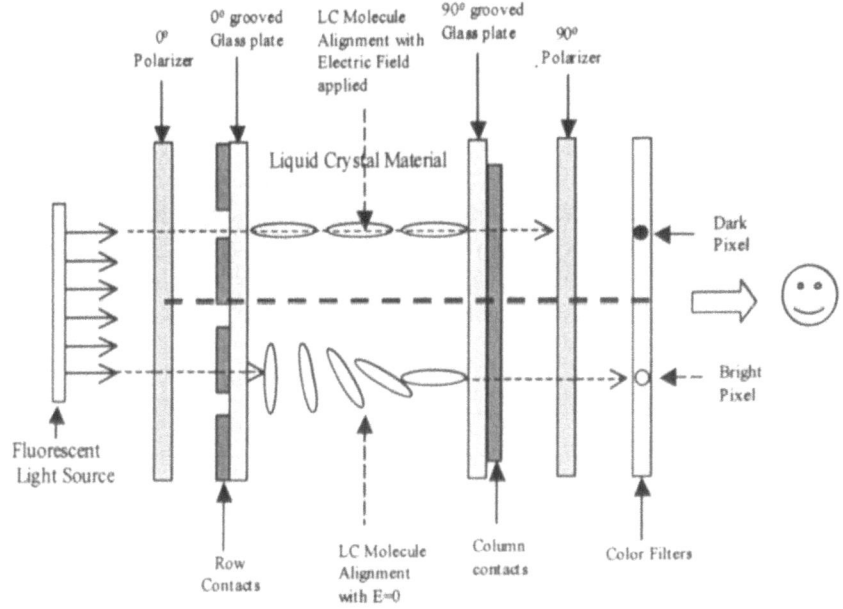

Figure 15.10. The LCD operation.

The operation of the LCD display is shown in Fig. 15.10. Fluorescent light source emits light that goes through a polarizer. The polarizer outputs the 0^0 polarized light that enters the LC. If there is no electric field (see the bottom-half of Fig. 15.10), the LC changes the polarization and makes the light 90^0 polarized. The second polarizer (90^0) passes this light, as the polarization matches. This light goes through the color filter, and we see a bright pixel. When an electric field is applied to the LC (see the top-half of Fig. 15.10), the molecular alignment changes. The LC passes the light without changing the polarization. The 0^0 polarized light is blocked by the second polarizer (90^0). Since no light comes out of the polarizer, we see a dark pixel.

The above operation is performed in parallel at each pixel (and at each color channel). A global control mechanism is employed to control the electric field through the local LC corresponding to each pixel. There are primarily two types of control matrices – passive and active. In a passive matrix LCD, each pixel is addressed in sequence by appropriately pulsing the row and column electrodes, and the pulse amplitude is modulated to control the density of each pixel in the display. Because the LC material decays slowly to the unexcited state, the pixel values can be set during the short pulse interval and they are held for some time to obtain a steady picture without flicker. However, because of this operation, the response of the passive-matrix LCD is slow and moving objects on the screen tend to smear.

In an active matrix LCD, also known as Thin Film Transistor (TFT) technology, a transistor switch is integrated on the bottom glass of the panel behind every pixel. The transistors set the voltage on each pixel of the panel. Since the transistors can be individually addressed, the voltage switching can be performed rapidly, resulting in significant improvement of the response speed (by a factor of 10) and the contrast ratio. However, adding the switching transistors has an overhead, and hence as a result active matrix costs more than the passive-matrix LCD of same display resolution. Table 15.1 shows the comparison between passive and active matrix LCDs.

Table 15.1. Comparison of passive and active matrix of LCDs.

	Passive	Active
Contrast	10-20	100+
Viewing Angle	Limited	Wide
Gray scale	16	256
Response Time	100-200ms	<50ms
Multiplex Ratio	480	>1000
Manufacturability	Simple	Complex
Cost	Moderate	High

Note that the LCD does not emit light; it merely modulates the ambient light, and hence the display power consumption is minimal. Because of its low power consumption and light weight, the LCD is very popular in portable applications. Displays may either be viewed directly or can be projected onto large screens.

LCDs, however, have two limitations. First, the angle of view of LCDs is limited ($<30^0$) because of the use of polarized light. Second, LCD panels are difficult to fabricate in larger sizes. Depending on the LCs, LCDs can be of different types such as nematic LCDs, twisted nematic LCDs, super twisted nematic LCDs, active addressing LCDs, and ferroelectric LCDs.

15.6 DIGITAL MICROMIRROR DISPLAY

It has been noted that the LCD is a nonemissive transmissive type of display where light is passed either through the LC or blocked. The digital micromirror display (DMD) device is another nonemissive type of display. However, it is of the reflective type where the light is reflected by a large number of mirrors.

The DMD is one of the latest display technologies [11, 12], developed by Texas Instruments in the mid-1990s. It employs sophisticated micro-electromechanical system technology to design a large number of micromirrors in a chip, such that the mirrors can individually be rotated by a small angle from their rest position. The micromirror array is attached and controlled by a static RAM array where each RAM cell and the associated mirror correspond to one pixel of the display. The aluminum mirrors can be made as small as 16×16 μm, and the number of mirrors in a DMD chip may be as high as a few millions. The mirrors in DMD are generally rotated through an electrostatic attraction. This electrostatic attraction is produced by the voltage difference developed across the air gap between a mirror and its RAM cell.

The operation of a single mirror in a DMD display is shown in Fig. 15.11. A light source (such as metal halide arc lamp) emits light that is focused on the mirror. The mirror position is controlled by the pixel intensity stored in RAM cells. In the ON state, the mirror is rotated 10^0 clockwise; the reflected light passes through the pupil of the projection lens, and a bright pixel is observed. In the OFF state, the mirror is rotated 10^0 anti-clockwise; the reflected light is not oriented towards the lens. As a result, the reflected light will not pass through, and a dark pixel is seen.

The DMD display is a true digital display, in the sense that the light output through the lens is either ON or OFF, resulting in bright and dark pixels, respectively. Gray scale capability in DMD can be obtained by using a digital pulse modulation technique similar to PDP. The mirror switching time can be less than 20 μs. Note that the DMD is basically a monochrome device that produces color identical to the source light. Hence, color is added in the DMD using (stationary or rotating) color filters (see Fig. 15.12).

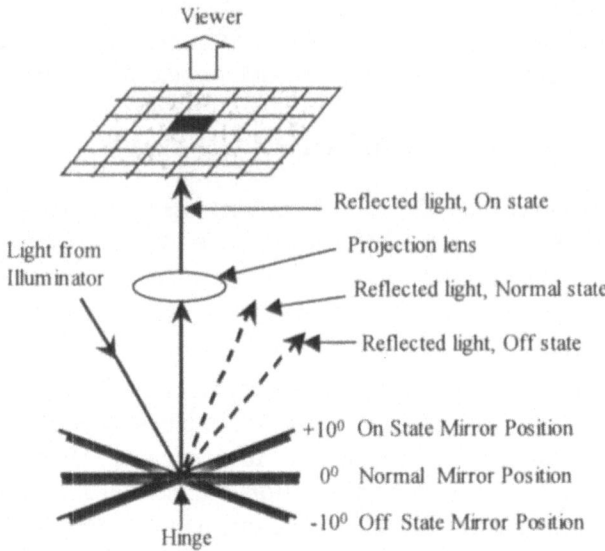

Figure 15.11. Mirror switching in a DMD display

Projection TV

Digital projection TV is gaining importance due to the growth of the home theaters and digital cinema [13]. There are several competitive technology for projection TV, such as CRT, active matrix LCD, and liquid crystal light valves (LCLV). Although these technologies are capable of producing good quality images, they also have their limitations. The CRT and LCD systems have limitations in producing high brightness, in addition to their stability and uniformity problems [3,14]. The LCLVs can produce high brightness, but they are expensive, and suffer from stability problem.

The DMD is a highly suitable display device for projection use, which is typically known *digital light processing* technology. Figure 15.12 shows the schematic of DMD-based projection TV. The light from the light source is passed through the color filter, which generates color light (red, green, and blue). The color light is reflected from the DMD mirrors onto the projection lens, which projects the image on to the screen.

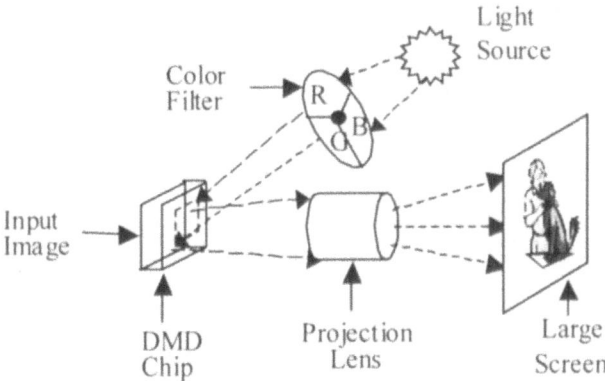

Figure 15.12. Projection display using DLP technology. The individual mirror positions in the DMD chip is controlled by the image pixel values.

An example of a DMD display is the Hitachi 55DMX01W model which is a 55″ rear projection digital TV, with 1280x720 display resolution, 350cd/m2 brightness and an aspect ratio of 16:9.

REFERENCES

1. J. S. Castellano, *Handbook of Display Technology*, Academic Press, 1992.
2. J. C. Whitaker, *Electronic Displays: Technology, Design, and Applications*, McGraw-Hill, New York, 1994.
3. A. C. Luther, *Principles of Digital Audio and Video*, Artech House, 1997.
4. M. Huang, M. Harrison, and P. Wilshaw, "Displays – the future is flat," *European Semiconductor*, Vol. 20, No. 2, p 15-16, Feb 1998.
5. S. Tominetti, and M. Amiotti, "Getters for flat-panel displays," *Proc. of the IEEE*, Vol. 90, No. 4, pp. 540-558, April 2002.
6. C. Ajluni, "FED technology takes display industry by storm," *Electronic- Design*, Vol. 42, No. 22, pp. 56-66, Oct 25 1994.
7. S. Itoh and M. Tanaka, "Current status of field-emission displays," *Proc. of the IEEE*, Vol. 90, No. 4, pp. 514-520, April 2002.
8. H. Uchike and T. Hirakawa, "Color plasma displays," *Proc. of the IEEE*, Vol. 90, No. 4, pp. 533-539, April 2002.
9. V. G. Chigrinov, *Liquid Crystal Devices: Physics and Applications*, Artech House, 1999.
10. H. Kawamoto, "The history of liquid-crystal displays," *Proc. of the IEEE*, Vol. 90, No. 4, pp. 460-500, April 2002.
11. L. J. Hornbeck, "Digital light processing for high brightness, high-resolution applications," *Proc. of SPIE: Projection Displays III*, Vol. 3013, pp. 27-40, 1997.
12. P. F. V. KesselL. J. Hornbeck, R. E. Meier, and M. R. Douglass, "A MEMS-based projection display," *Proc. of the IEEE*, Vol. 86, No. 8, pp. 1687-1704, August 1998.

13. E. H. Stupp, M. S. Brennesholtz, and M. Brenner, *Projection Displays*, John Wiley & Son Ltd, 1998.

QUESTIONS

1. Describe different parameters that are important for a display device.

2. Explain the working principle of CRT.

3. What is a shadow mask, and an aperture grille? What is meant by the statement "The dot pitch of a TV is 0.30 mm"?

4. What is the working principle of the Field Emission displays?

5. Compare and contrast the CRT display and FED.

6. Explain briefly the working principle of a plasma display panel (PDP).

7. How do we achieve gray scale in a PDP? How do we achieve color?

8. How does LCD work? Is it an emissive display? How do we achieve gray scale in LCD? How is color reproduced in LCD?

9. Compare passive and active Matrix LCDs.

10. Draw the schematic of the optical switching in a DMD. What is a typical mirror size in DMD? Which technology is used to fabricate a DMD?

11. Compare and contrast LCD and DMD.

12. Explain the principle of DMD based projection display.

Appendix

A CD-ROM has been included in this book to provide readers with supplementary reading material, computer programs, and selected digital audio and image data. The MATLAB programs used to demonstrate several examples have been included so that readers can run the programs themselves to process multimedia signals. The MATLAB code has been chosen since it is versatile tool for signal processing and graphical plots. Two HTML codes (corresponding to Chapter 13) have also been included in the CD. In addition, two theses written by the author have been included for those interested to learn more about visual compression and retrieval, especially in the wavelet transform framework. The materials included in the CD are listed below:

A.1 MATLAB Functions

These are functions called by main MATLAB programs.

```
CD:\MATLAB\dwtorthogonal       % Functions for orthog. DWT calculation
CD:\MATLAB\dwtbiorthogonal     % Functions for biorth. DWT calculation
CD:\MATLAB\motion estimation   % Functions for motion estimation
CD:\MATLAB\misc                % Miscellaneous MATLAB Functions
```

A.2 MATLAB Examples (CD:\programs)

These are the original MATLAB programs used in various examples.

```
Example2_1.m    % audio noise masking
Example2_2.m    % MIDI file
%
Example4_6.m    % SNR of quantized audio signal
%
Example5_2.m    % DFT of 1-D signal
Example5_4.m    % DCT of 1-D signal
Example5_10.m    % 2-D DFT Spectrum
%
Example7_3.m    % Companding of audio signal
Example7_4.m    % Calculation of the LPC coefficients
```

Example7_6.m	% DPCM coding of audio signal
%	
Example8_2.m	% DPCM coding of image
Example8_3.m	% DPCM coding of image
Example8_6.m	% Energy compaction using DCT and wavelet
Example8_7.m	% Performance of block-DCT coding
%	
Example9_3.m	% Full search motion vector calculation
Example9_4.m	% Fast motion vector calculation
Example9_5.m	% Motion vector calculation for Claire sequence
%	
Example10_1.m	% Audio filtering
Example10_2.m	% Audio equalization
Example10_3.m	% Noise suppression by digital filtering
Example10_4.m	% Spectral subtraction method
Example10_5.m	% MIDI file
Example10_6.m	% MIDI file
%	
Example11_1.m	% Image interpolation
Example11_2.m	% Image cropping
Example11_3.m	% Image contrast stretching
Example11_4.m	% Histogram Equalization
Example11_5.m	% Image Sharpening
Example11_6.m	% Wipe operation
Example11_7.m	% Dissolve operation
Example11_8.m	% Fade in and out operations
%	
Example12_1.m	% Energy compaction in the YIQ/YUV color space

A.3 Hypertext Examples (CD:\programs)

These are the HTML/XML programs used in Chapter 13.

Example13_1.htm	% HTML code
Example13_2.htm	% XML code embedded in HTML

A.4 Supplementary Chapters (CD:\supplementary chapters)

Some color figures were originally intended to be included in the book. However, they were ultimately not included in order to reduce printing costs. These figures (Chapters 3 and 8) have been included in the CD.

A.5 Theses (CD:\documents)

The following theses were written by the author, and included for interested readers who want to learn more about visual compression and retrieval, especially in the wavelet transform framework.

1. M. K. Mandal, *Wavelets for Image Compression*, M.A.Sc Thesis, University of Ottawa, 1995.
2. M. K. Mandal, *Wavelet Based Coding and Indexing of Images and Video*, Ph.D. Thesis, University of Ottawa, Fall 1998.

A.6 Input Data Files

Audio (CD:\data\audio)

bell.wav	% An 8-bit, 22.05 KHz audio signal
test44k.wav	% An 8-bit, 44.1 KHz audio signal
noisy_audio1.wav	% Audio signal with narrowband noise
noisy_audio2.wav	% Audio signal with wideband noise

Images (CD:\data\images)

{airplane, baboon, Lena}.tif	% standard 512x512 gray level images
{banff1, banff2, lakelouise,niagra, geeta}.tif	% Miscellaneous images
lenablur.tif	% blurred Lena image
airplane256.tif	% 256x256 airplane image

Video (CD:\data\video)

{claire1,claire2}.tif	% two frames from Claire sequence
{football000,football002}.tif	% two frames from football sequence
{shot1,shot3}.tif	% frames from two video shots

A.7 Output Data Files (CD:\data\)

Chapter 2

test{1,2,3,4,5}.wav	% Output of Example 2.1
Examp2_2.mid	% Output of Example 2.2

Chapter 10

bell1_lpf.wav	% LPF output of Example 10.1
bell1_hpf.wav	% HPF output of Example 10.1
bell1_bpf.wav	% BPF output of Example 10.1
Examp10_2.wav	% Output of Example 10.2

Examp10_3_128tap.wav   % Output of Example 10.3 for 128 tap filter
Examp10_3_200tap.wav   % Output of Example 10.3 for 200 tap filter
Examp10_4.wav   % Output of Example 10.4
Examp10_5.mid   % Output MIDI file for Example 10.5
Examp10_6.mid   % Output MIDI file for Example 10.6

Video

CD:\data\chap11\wipe1   % Transition frames in Example 11.6
CD:\data\chap11\wipe2   % Transition frames in Example 11.6
CD:\data\chap11\dissolve % Transition frames in Example 11.7
CD:\data\chap11\fade   % Transition frames in Example 11.8

MATLAB® is a registered trademark of the MathWorks, Inc.

Index